ALLE · ZEIT · WACH · 1842

Rudolf Klute
Hermann H. Hahn (Eds.)

CHEMICAL WATER AND WASTEWATER TREATMENT III

Proceedings of the
6th Gothenburg Symposium 1994
June 20 - 22, 1994
Gothenburg, Sweden

In memoriam Rudolf Klute

Springer-Verlag
Berlin Heidelberg NewYork London Paris
Tokyo HongKong Barcelona Budapest

Prof. Dr. Rudolf Klute †
Technische Hochschule Darmstadt
Institut WAR
Wasserversorgung · Abwassertechnik
Abfalltechnik · Umwelt- und Raumplanung
Petersenstraße 13
D - 64287 Darmstadt

Prof. Dr. Hermann H. Hahn
Universität Karlsruhe
Institut für Siedlungswasserwirtschaft
Am Fasanengarten
D - 76128 Karlsruhe

ISBN-13:978-3-642-79112-3 e-ISBN-13:978-3-642-79110-9
DOI: 10.1007/978-3-642-79110-9

Cip data applied for

Softcover reprint of the hardcover 1st edition 1994

Typesetting: Camera-ready by authors
SPIN:10468682 51/3020-5 4 3 2 1 0 - Printed on acid -free paper

Preface

Exactly ten years ago an experiment was started that proved to be extremely successful: the First Gothenburg Symposium. Its intent was to further the understanding of all processes pertaining to Chemical Water and Wastewater Treatment, and to bring together specialists working in basic research as well as in development and administration. Now, the Proceedings of the Sixth Symposium are about to be published, clearly proving that there is a need for this forum. They dramatically illustrate the significance and the dynamic development of the topics of these symposia.

It is fascinating to witness that in this time of reduced economic growth or even standstill, the environmental drive has not come to a halt, as many anticipated or feared. It is accepted more and more that the protection of the environment, a constant theme in all the Gothenburg Symposia, is not only a topic to be dealt with in times of affluence; it is now also seen as an instrument for cutting expenditure, saving energy, and husbanding resources. The ever growing interest in these Gothenburg Symposia, documented by the large number of contributions the scientific panel received and the large demand for the books of this series that always exceeds the supply, testify to this commitment.

These proceedings of the Sixth Gothenburg Symposium show new developments and new directions. Besides the ever present topics such as chemical selection and dosing or the treatment of drinking water or wastewater, new technological developments are reported for the control of non-point sources such as stormwater runoff. Furthermore, many of the contributions contained in this volume appear to be of a more basic nature while at the same time the papers reporting solutions for very practical problems seem to have become more numerous: this signalizes that theory and practice have interacted even more effectively than previously. This is one of the main objectives of these symposia. The scientific committee also states with some pride that the contributions of this volume reflect work done in many of the six continents of our globe: the symposia have become a truly international meeting place.

The editors express their deep gratitude to all who made this book possible: Karin Knisely with her careful editing of the manuscript and Adam Leinz and the Springer Publishing Company with their high standards for a perfect book. The cooperation of Kemira Kemi of Helsingborg with the members of the scientific committee proved again to be a fruitful one, giving a voice to some thirty authors of outstanding contributions.

April 1994

Karlsruhe, Germany H. H. Hahn
Darmstadt, Germany R. Klute

Members of the Scientific Committee

Prof. T. Asano, USA

Dr. D. Averill, Canada

Prof. P. Balmér, Sweden

Dr. M. Boller, Switzerland

Dr. P. Dolejs, Czech Republic

Dr. N. Graham, England

Prof. P. Harremoës, Denmark

Prof. R. Klute, Germany

Prof. L. Lijklema, The Netherlands

Dr. M. H. Marecos do Monte, Portugal

Prof. R. Mujeriego, Spain

Prof. H. Ødegaard, Norway

Mr. J. M. Rovel, France

Mr. J. Sibony, France

Mr. K. Stendahl, Sweden

Prof. N. Tambo, Japan

Dr. G. Tiravanti, Italy

Prof. M. Viitasaari, Finland

Contents

Floc Formation and Separation

Chemicals and Dosing Control

Drinking Water Treatment

Wastewater Treatment

Stormwater Treatment

Sludge Handling

Floc Formation and Separation

Dissolved Air Flotation: Pretreatment and Comparisons to Sedimentation

J. K. Edzwald, D. Q. Bunker, Jr., J. Dahlquist, L. Gillberg, and T. Hedberg

Abstract

Coagulants, flocculation time and mixing intensity were examined in laboratory studies using a synthetic fulvic acid water and three Swedish water supplies. Alum and polyaluminium chlorides (PACs) of various chemistries were effective when used under favorable pretreatment conditions. Alum and PACs had approximately the same optimum dosages for the fulvic acid water when expressed as mg Al added per mg TOC. An exception was the Ca PAC had a lower optimum dose at higher pH. The PACs were effective at flocculation times as low as 5 min while 10 min was needed with alum, especially in cold waters. Dissolved air flotation (DAF) performance improved slightly with increasing flocculation mixing intensity. DAF performed better than settling for aquatic humic or algae type supplies especially with lower flocculation times, and performed better for a river water with turbidity under cold water conditions. Excellent DAF treatment was achieved with Ca, Si, and sulfate PACs.

Introduction

The research described here had three purposes: first, to evaluate various coagulants in the dissolved air flotation (DAF) treatment of three very different water types; second, to examine the effects of flocculation time and mixing intensity on DAF performance; and finally, to compare DAF performance to sedimentation for three Swedish water supplies.

DAF is a proven water treatment technology in Europe, South Africa, and Australia, and an emerging technology in North America. It is receiving considerable attention because it is more efficient in separating low density particles than sedimentation processes. Pretreatment of coagulant addition and flocculation prepares particles (particle bubble stability and size) for flotation. Effective flotation requires destabilized particles (low or no electrical charge and hydrophobic particles). Thus particle destabilization requirements for DAF are similar to other downstream particle separation processes such as sedimentation or direct filtration. Laboratory [1–3] pilot and full scale [4–6] studies have documented that good coagulation conditions are necessary for DAF. Large floc particles are not needed in flotation, however, so the flocculation requirements are more analogous to direct filtration.

R. Klute / H.H. Hahn (Eds.)
Chemical Water and Wastewater Treatment III

European DAF practice evolved with flocculation tank detention times of 20 to 45 min [5,7] similar to sedimentation plants. Edzwald et al. [1] modeled particle bubble collisions in the reaction zone of the flotation tank and concluded that large floc size particles are not needed. Their calculations suggested that floc particles of 10 to 30 μm (*pinpoint floc*) should be prepared for flotation. Laboratory and pilot studies have confirmed that small floc particles are effectively removed by flotation and flocculation tanks with detention times as low as 5 minutes may be feasible [8,9]. A full-scale DAF plant in Antwerp (Belgium) has been built with a detention time less than 10 min. Less research has been done on flocculation mixing intensity for DAF systems. Janssens [9,10] found in pilot plant studies that short flocculation times of 5 to 6 min combined with G values (G is the mean velocity gradient) of 30 to 80 sec^{-1} provided good DAF treatment in treating drinking water supplies containing algae. Ødegaard [11] found in a wastewater treatment application of DAF that the optimum mixing intensity was at G values of 70 to 100 sec^{-1}, higher than for values used in sedimentation.

Experiments

Watertypes and Coagulants. Three different water types were examined: 1) waters containing dissolved organic carbon (DOC) composed largely of aquatic humic matter – a synthetic water and Ljungskile Reservoir (Village of Ljungskile, Uddevalla Municipality), 2) a low to moderate turbidity river water of low color, the Göta River (Göteborg), and 3) a lake of low turbidity and color that experiences seasonal algae problems, Lake Mälaren (Stockholm). Raw water characteristics are summarized in Table 1. The aquatic humic matter for the synthetic water was extracted from the Great Dismal Swamp (Virginia, USA), and was composed almost entirely of aquatic fulvic acid. Specific UV absorbance (SUVA) data were used to characterize the nature of the DOC of the three water supplies [12]. A SUVA value of about 4 or higher indicates the DOC is composed largely of aquatic humic matter (fulvic and humic acids). The data in Table 1 indicate that Ljungskile Reservoir contains natural color and the DOC is high in aquatic humic matter. On the other hand, the DOC of the Göta River and Lake Mälaren has a low fraction of aquatic humics. The SUVA measurements also indicate that the DOC of these two supplies is relatively low in molecular weight, less hydrophobic than humics and less reactive with metal coagulants. The DOC concentration measurement for Lake Mälaren is based on one sample and is higher than expected.

The polyaluminium (PAC) coagulants studied are listed in Table 2. They were prepared by Kemira Kemi AB (Helsingborg, Sweden). Alum was used as a reference coagulant in the research. Ferric chloride was used in some experiments.

DAF Experiments with Great Dismal Swamp Fulvic Acid. These experiments were carried out at the University of Massachusetts (Amherst, MA, USA) using a bench scale DAF system (Aztec Env. Controls Ltd., UK.) consisting of four parallel batch units. The rapid mixing conditions were 10 seconds at a G (mean velocity gradient) of 380 sec^{-1} followed by flocculation mixing at various G

Table 1. Raw water characteristics

	Great Dismal Swamp FA	Ljungskile reservoir	Göta River	Lake Mälaren
pH	6.3	7.1	7.1	7.7
Turbidity [NTU]	0.15	0.6	5.7	1.5
Color [Pt-Co]	NM	30	15	17
UV @ 254 nm [cm^{-1}]	0.225	0.21	0.13	0.16
DOC [mg/L]	5.0	5.8	4.8	6.7
SUVA [m^{-1}/mg/L]	4.4	3.6	2.7	2.4
Alkalinity [mg/L $CaCO_3$]	20	20	15	70
Hardness [mg/L $CaCO_3$]	NM	30	30	90

Table 2. List of polyaluminium coagulants

General Name	Product	Al Wt. [%]	Basicity [%]	Other
	Used with Great Dismal Swamp fulvic acid			
Low basicity PAC	Kemwater PAX-14	7.7	33	
High basicity PAC	UPAX-20	9.8	81	
PAC with Ca	UPAX-5	1.3	33	11.2 % Ca
PAC with Si	Kemwater PAX-XL 60	6.7	Not appl.	1.2 % Si
PAC with SO_4	Kemwater PAX-11	5.2–5.4	24–36	1.7–2 % SO_4
	Used with Swedish water supplies			
Low basicity PAC	Kemwater PAX-14	8.0	37	
High basicity PAC	UPAX-10	11.5	73	
PAC with Ca	Kemwater PAX-XL 1	4.3	68	3.2 % Ca
PAC with Ca	UPAX-5	1.3	33	11.2 % Ca
PAC with Si	Kemwater PAX-XL 60	8.7	Not appl.	1.8 % Si
PAC with SO_4	Kemwater PAX-11	4.9	37	1.8 % SO_4

conditions of 20, 55 and 140 sec^{-1}. The flocculation detention time was also varied. Flotation time, saturator pressure, and recycle ratio were held constant at 10 min, 480 kPa (70 psig) and at 8 %, respectively. Treatment performance was monitored by measuring UV absorbance at 254 nm (UV_{254}) a surrogate measure for DOC, residual turbidity, and residual coagulant (soluble metal, samples filtered with a 0.45 μm membrane filter).

DAF and Sedimentation Experiments with the Swedish Water Supplies. The experimental research was done at Chalmers University and at the Görvaln Water Treatment Plant (near Stockholm). The flotation equipment was provided by Purac AB, and experiments were done at the bench scale level. The rapid mixing conditions were 10 seconds at 400 rpm followed by flocculation mixing conditions of 10, 30, or 50 rpm. Flocculation detention time was varied in some experiments. Flotation time, saturator pressure, and recycle ratio were held constant at 10 min, 480 kPa (70 psig), and 10%, respectively. For some experiments with Lake Mälaren, algae were spiked into the raw water to increase the algae concentration. The algae had been previously extracted from Lake Mälaren by passing water through a 100 μm sieve.

DAF was compared to sedimentation in side by side paired experiments in which pretreatment conditions of coagulant, pH, coagulant dosage, and flocculation mixing intensity and time were identical (in some cases, flocculation times prior to settling were longer). What differed were the clarification times or equivalent overflow rates, expressed as the ratio of the DAF rate to the sedimentation rate. These were examined at two ratios of 5.5 and 11 to 1. Comparisons between DAF and sedimentation are presented graphically using bivariate plots in which each data point is a paired experiment. The 45 degree line on each figure is the *"line of equal value"*; if data falls on the line then DAF and sedimentation performed equally. Data below the *"equal value line"* means DAF had better performance while data above the *"equal value line"* indicate that settling was better. Treatment performance for both DAF and sedimentation was evaluated using turbidity, UV absorbance at 254 nm, and residual coagulant.

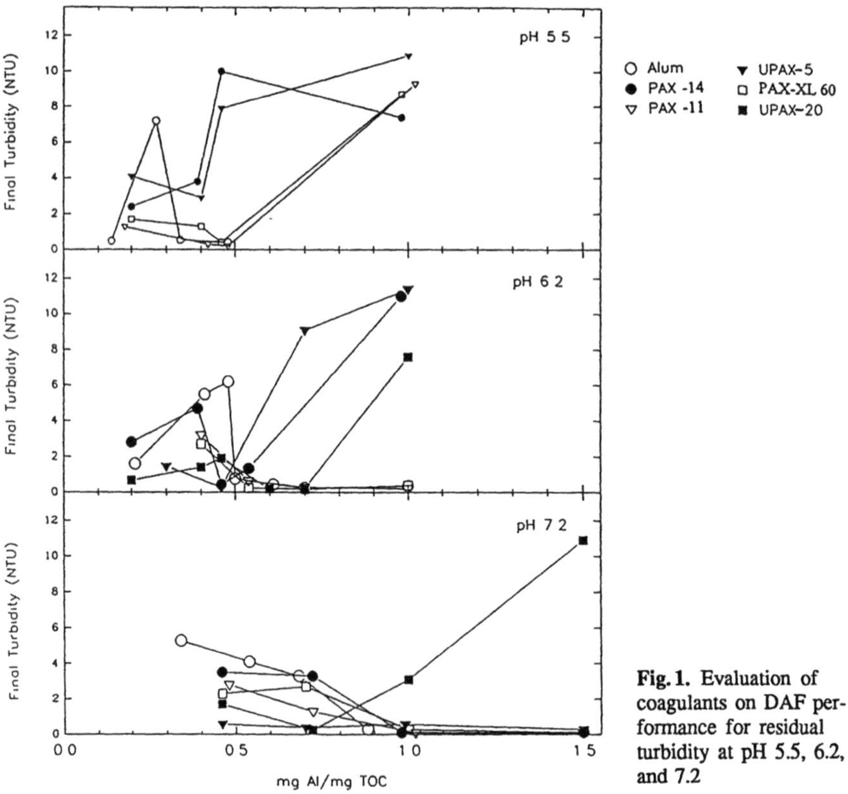

Fig. 1. Evaluation of coagulants on DAF performance for residual turbidity at pH 5.5, 6.2, and 7.2

Results and Discussion

Great Dismal Swamp Fulvic Acid

Coagulation. Experiments were performed at three pH conditions (5.5, 6.2, and 7.2) at fixed flocculation conditions (20 min flocculation time and a mixing intensity of 20 sec^{-1}). The water temperature was 20°C. Flotation conditions are described above. Results are presented in Figures 1–3. The coagulant dosages have been expressed on a common basis as mg Al added per mg TOC of the raw water. This allows direct comparison of the coagulants and evaluation of dosage requirements on a common basis. Optimum coagulant doses are defined as those that produced low turbidity (1 NTU or less), UV removal of 80% or greater, and residual soluble Al of 100 μg/L or less, while higher doses produced only marginal improvements in turbidity and UV removal. Specific highlights of the results follow.

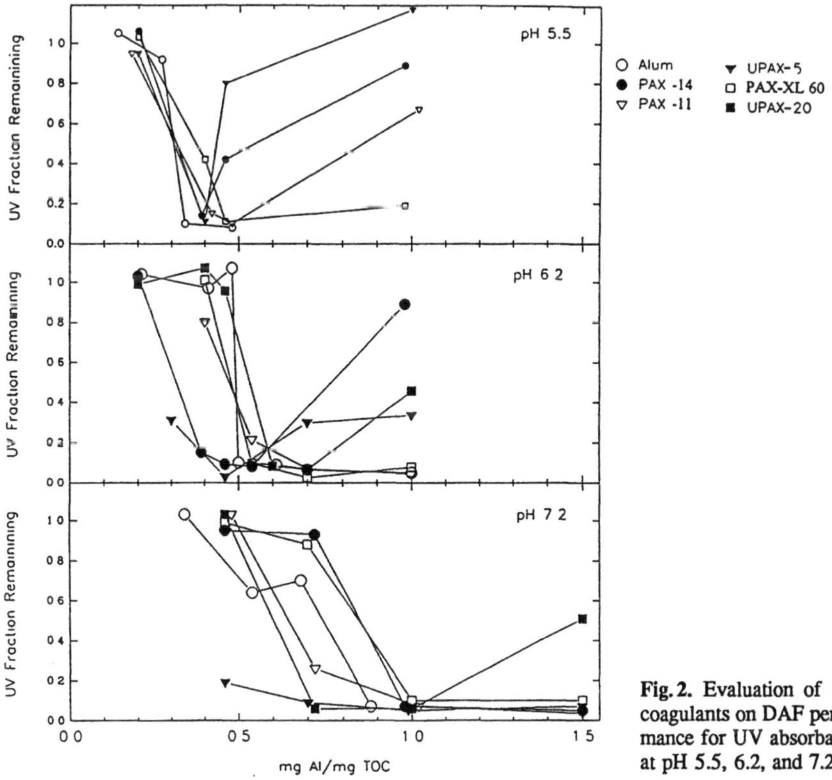

Fig. 2. Evaluation of coagulants on DAF performance for UV absorbance at pH 5.5, 6.2, and 7.2

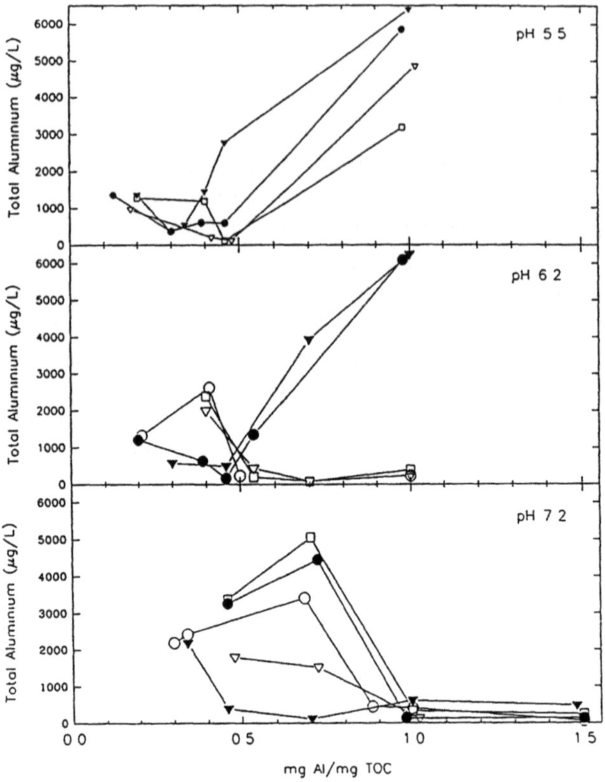

Fig. 3. Evaluation of coagulants on DAF performance for residual total Al at pH 5.5, 6.2, and 7.2

pH 5.5. Optimum doses are about 0.4 to 0.5 mg Al per mg TOC for the coagulants tested: alum, low basicity PAC (PAX-14), sulfate PAC (PAX-11), calcium PAC (UPAX-5), and the Si PAC (PAX-XL60). At the optimum dose, low turbidities were obtained for alum, the sulfate and silica PACs. Overdosing occurred for all coagulants.

pH 6.2. Optimum doses are about 0.6 to 0.7 mg Al per mg TOC for alum, high basicity PAC (UPAX-20), sulfate PAC, and the silica PAC. The calcium and low basicity PACs had lower optimum doses of 0.4 to 0.5. No overdosing occurred for the sulfate and silica PACs. The low basicity and the calcium PACs had a narrow optimum range prior to overdosing.

pH 7.2. Optimum doses are about 0.9 to 1 mg Al per mg TOC for alum, low basicity PAC, sulfate PAC, and the silica PAC. The high basicity and calcium PACs had lower optimum doses of about 0.7. Overdosing occurred only with the high basicity PAC.

Some general comments follow. The optimum coagulant dosage depends strongly on pH. When coagulant dosages are expressed on a common basis (mg Al/mg TOC), most of the coagulants (alum and various PACs) have approximately the same optimum dose. An exception is the calcium based PAC has lower optimum dosage requirements at pH 6.2 and 7.2.

The conventional wisdom is that optimum pH conditions with alum is at pH 5.5. Note that this pH does have the lowest optimum Al dosages, but that there is a narrow optimum range followed by overdosing. Also, these experiments were done at a water temperature of 20 °C. Colder water temperature,which would be common in practice shifts the optimum pH higher for alum by a few tenths to a half pH unit [1, 13]. UV removal is good at about 90 % at optimum coagulant dosages for all coagulants and pH conditions. The sulfate and silica PACs gave good overall results especially at pH 6.2 and 7.2. Considering Al dosage requirements, optimum dosing ranges, residual Al, and water temperature, we recommend operation at pH 6 to 6.5 (toward the higher value in colder waters). An exception would be with the use of a calcium based PAC where good results were obtained above pH 7. Overdosing occurred readily at pH 5.5 and 6.2, but no overdosing was observed at pH 7.2 and the optimum dosage was low at 0.7 mg Al/mg TOC. At 5 mg/L TOC, the calcium PAC dose is 3.5 mg/L as Al. The Ca added is 30 mg/L which not only lowered the Al based coagulant dose for removal of natural organic matter, but it would also decrease the corrosion potential of the water.

Great Dismal Swamp Fulvic Acid and Ljungskile Reservoir

Flocculation. Figure 4 shows the effect of flocculation time for cold water conditions (4–6 °C) in the DAF treatment of the fulvic acid water. The sulfate PAC (PAX-11) produced low turbidities (near 1 NTU or less) at flocculation times as low as 2.5 to 5 minutes. The alum and the silica PAC (PAX-XL60) needed a flocculation time greater than 5 min; however, the Si PAC produced low turbidities

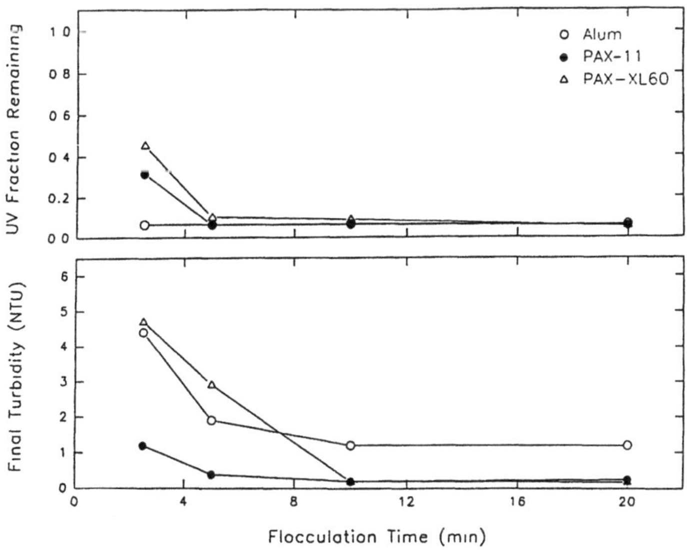

Fig. 4. Effect of flocculation time on DAF performance for Great Dismal Swamp Fulvic Acid (Coagulant dosages fixed at 3.5 mg/L as Al^{+3}, pH 6.2, floc. mixing intensity at 55 sec^{-1}, water temperature at 4 °C)

at flocculation times of 2.5 to 5 min when used at other flocculation mixing intensities (to be shown later). Alum yielded high turbidities compared to the other coagulants, especially for flocculation times greater than 5 minutes. These data support the premise that alum floc particles may not attach effectively to bubbles at the zero point of charge due to hydrophilic effects and that this effect increases in cold water [1]. UV removal was independent of flocculation time, if at least 2.5 minutes of flocculation was provided. The data indicate that removal of dissolved organic species is primarily dependent on a phase change (from solution to particles which are separated by filter paper in the analytical analysis), is controlled by coagulation, and occurs quickly even in cold waters [2]. DOC is removed from solution by coagulation in which precipitated particles are formed (aluminium-fulvate precipitate or DOC adsorbed on aluminium hydroxide particles). Because the laboratory analysis for dissolved UV involves a filtration step, dissolved UV removal is not dependent upon flotation.

The results for Ljungskile Reservoir, a supply high in aquatic humics and thus a test case for the fulvic acid synthetic water work, are presented in Figure 5 and confirm the fulvic acid data. A flocculation time of 10 min was needed with alum while excellent results were achieved at flocculation times of 5 min for the sulfate PAC, calcium PAC (PAX-XL1), and the low basicity PAC (PAX-14). The UV data also support the results obtained with the fulvic acid water.

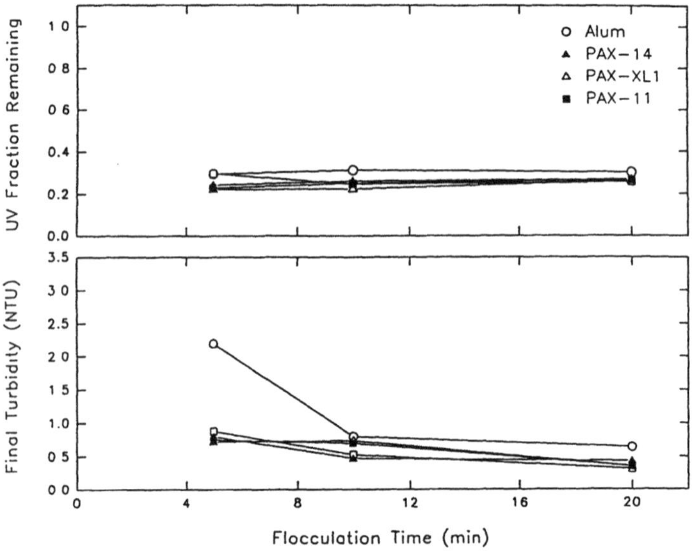

Fig. 5. Effect of flocculation time on DAF performance for Ljungskile Reservoir (Coagulant dosages fixed at 3.5 mg/L as Al^{+3}, pH 6.3, floc. mixing intensity at 30 rpm, water temperature at 20 °C)

The effect of flocculation mixing intensity under cold water conditions when treating Great Dismal Swamp FA are presented in Figure 6a. In general, alum gave the poorest performance and the sulfate PAC (PAX-11) the best performance.

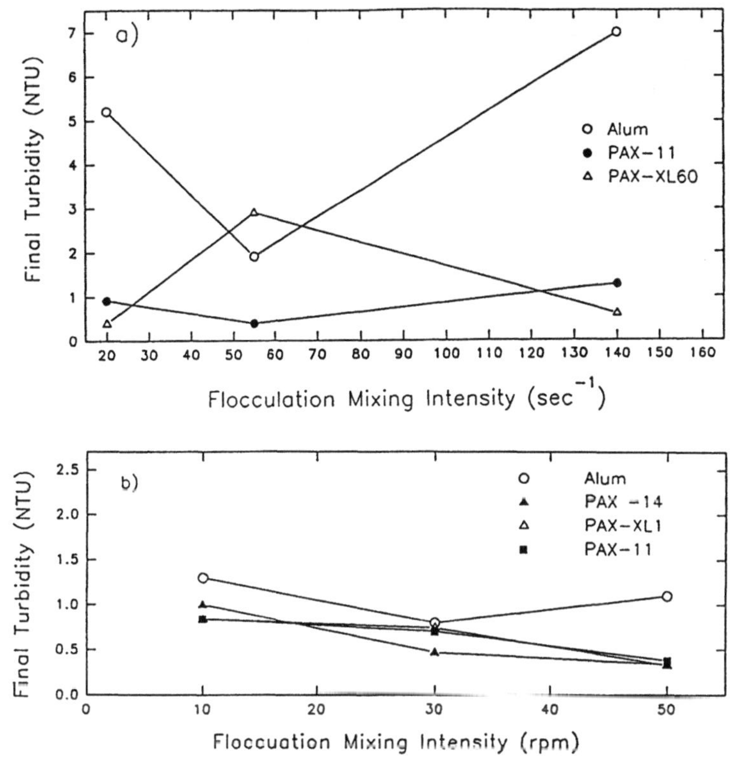

Fig. 6a, b. Effect of flocculation mixing intensity on DAF performance (a: Great Dismal Swamp Fulvic Acid, coagulant dosages fixed at 3.5 mg/L as Al^{+3}, pH 6.2, floc. time at 5 min, water temperature at 4 °C; **b**: Ljungskile Reservoir, coagulant dosages fixed at 3.5 mg/L as Al^{+3}, pH 6.3, floc. time at 10 min, water temperature at 20 °C)

Note that the silica PAC (PAX-XL60) gave good results at low and high mixing intensities for the 5 min flocculation time. Alum performed poorly at both high and low mixing intensities while the sulfate PAC worked well over the entire range of mixing intensities. Results for Ljungskile Reservoir are presented in Figure 6b. The alum data confirm the prior results with best results at an intermediate mixing intensity. The sulfate PAC, the low basicity PAC (PAX-14), and the calcium PAC (PAX-XL1) gave good results for all mixing intensities with slight improvement with increasing intensity.

DAF Comparison to Sedimentation for the Ljungskile Reservoir. DAF was compared to sedimentation for two DAF to settling overflow rate ratios (5.5 and 11 to 1) with a flocculation time held constant at 10 min. The data are presented in Figure 7. DAF performed better than sedimentation with turbidities of 0.8 NTU or less for all the coagulants and at both overflow rate ratios. The low basicity PAC (PAX-14) produced the lowest turbidity waters (0.4 NTU) for DAF. As expected, sedimentation improved at the higher DAF overflow rate ratio of 11 to 1, but

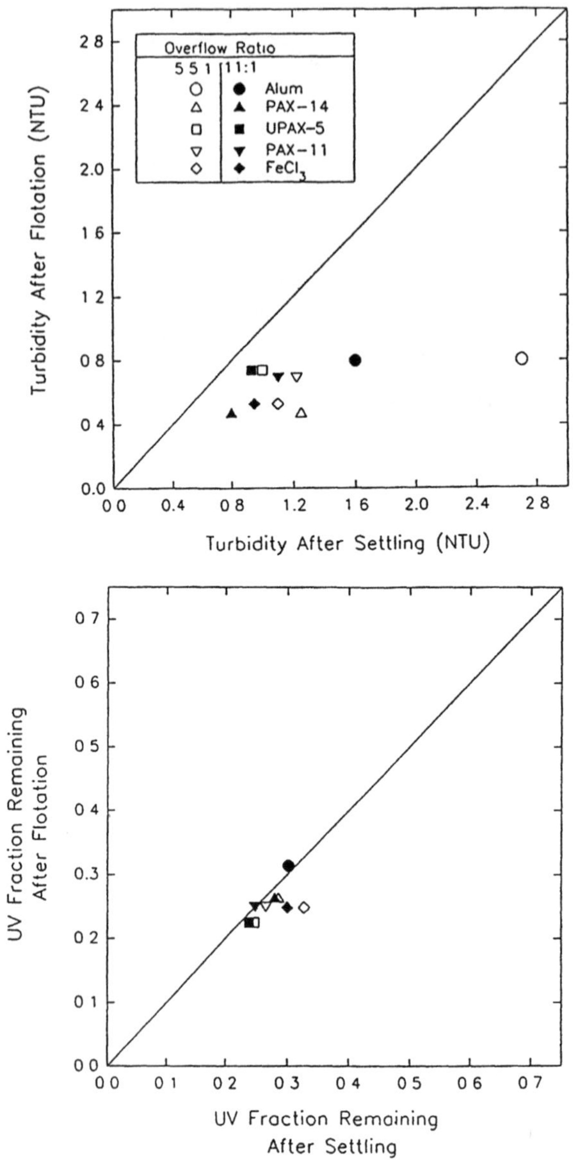

Fig. 7. Comparison of DAF to sedimentation for Ljungskile Reservoir (Al based coagulant dosages fixed at 3.5 mg/L as Al^{+3} and pH 6.3, $FeCl_3$ at 6.5 mg/L as Fe^{+3} and pH 5.5, floc. time at 10 min, water temperature at 20 °C)

still did not perform as well as DAF – all turbidity data below the *"line of equal value"*.

UV data generally lie on the *"line of equal value"* indicating no difference between DAF and settling. This is because dissolved UV or DOC removal depends on coagulation as explained above. The Ca PAC (UPAX-5) produced the greatest UV removal (about 80 %).

Lake Mälaren and Göta River Waters

Flocculation. Figures 8 and 9 illustrate the results obtained for flocculation time for the two water supplies. Flocculation time does not affect UV removal (exception might be the one data point in Figure 9 for UPAX-10 (high basicity PAC), but this is most likely an error). UV removal, as explained above, involves a phase change and should be rapid (minutes). In these experiments, samples were taken after DAF so the total time was an additional 10 minutes to the indicated flocculation times. For Lake Mälaren (Fig. 8), all the coagulants produced 60 to 70 percent UV removal. There are some differences in the ability of the coagulants to remove UV from the Göta River (Fig. 9). Alum performed slightly poorer compared to the other coagulants.

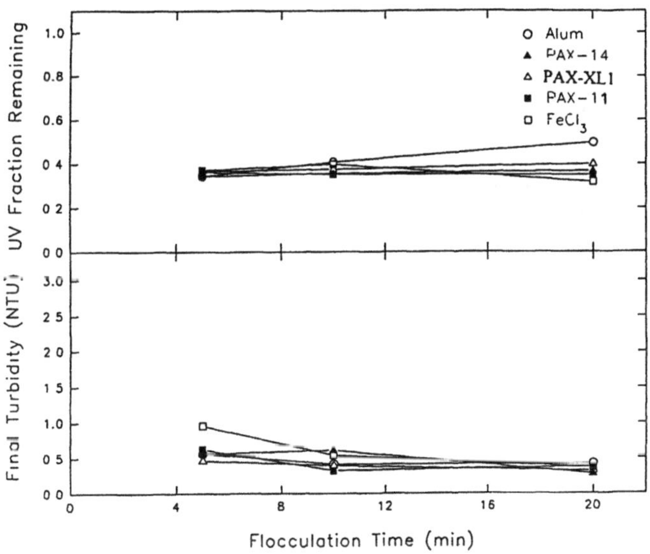

Fig. 8. Effect of flocculation time on DAF performance for Lake Mälaren (Al based coagulant dosages fixed at 3.5 mg/L as Al^{+3} and $FeCl_3$ at 6.5 mg/L as Fe^{+3}, pH 6.3, floc. mixing intensity at 30 rpm, water temperature at 20°C)

DAF performed well at flocculation times as low as 5 min (turbidity of 1 NTU or less) for all coagulants on both supplies as shown in Figures 8 and 9. The data in Figure 8 for Lake Mälaren show that DAF improved with flocculation time when $FeCl_3$ was used. Alum and the PACs produced turbidities of 0.5 NTU or less for all flocculation times. The data in Figure 9 for the Göta River at a flocculation time of 10 min show differences in DAF performance for the coagulants. The high basicity PAC (UPAX-10), the sulfate PAC (PAX-11), and $FeCl_3$ produced lower turbidities than alum and the low basicity PAC (PAX-14).

Figure 10 shows that good DAF performance was obtained for all flocculation mixing intensities tested with the exception of $FeCl_3$ at the lowest intensity. DAF performance generally improved (lower turbidity) with increasing intensity for all

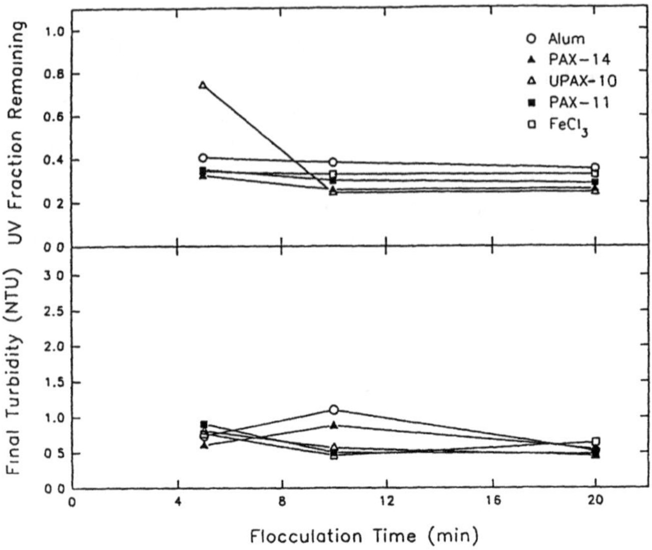

Fig. 9. Effect of flocculation time on DAF performance for the Göta River (Al based coagulant dosages fixed at 3.5 mg/L as Al^{+3} and $FeCl_3$ at 6.5 mg/L as Fe^{+3}, pH 6.8, floc. mixing intensity at 30 rpm, water temperature at 20 °C)

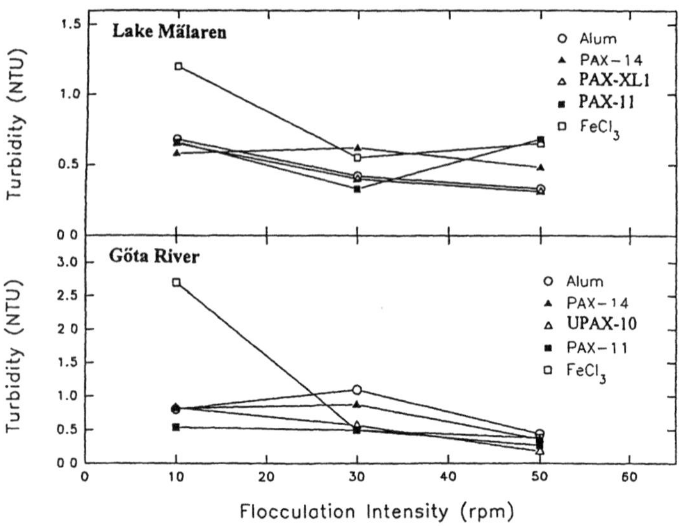

Fig. 10. Effect of flocculation mixing intensity on DAF performance (Al based coagulant dosages fixed at 3.5 mg/L as Al^{+3} and $FeCl_3$ at 6.5 mg/L as Fe^{+3}, floc. time 10 min water temperature at 20 °C, Lake Mälaren at pH 6.3, Göta River at pH 6.8)

coagulants for both water supplies. Generally, the lowest turbidities were obtained with the sulfate PAC (PAX-11) and Ca PAC (PAX-XL1) for Lake Mälaren and with the sulfate PAC and high basicity PAC (UPAX-10) on Göta River water.

DAF Comparison to Sedimentation. The results are shown in Figures 11 and 12. The Lake Mälaren data in Figure 11 were obtained for experiments where algal sized particles were spiked into the test waters to simulate conditions of an algae bloom. The data show that with a 20 min flocculation time, DAF and sedimentation performed about the same (data on the "equal value line") except DAF performed better with alum. DAF performed better than sedimentation for all coagulants when a pretreatment flocculation time of 10 min was used. All coagulants produced low turbidities following DAF with only 10 min of flocculation, particularly the Ca PAC (PAX-XL1), the low basicity PAC (PAX-14), and $FeCl_3$.

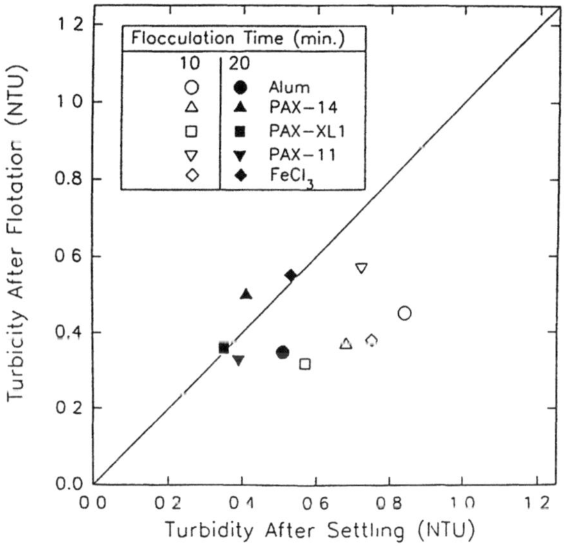

Fig. 11. Comparison of DAF to sedimentation for Lake Mälaren for two flocculation times (Al based coagulant dosages fixed at 3.5 mg/L as Al^{+3} and $FeCl_3$ at 6.5 mg/L as Fe^{+3}, pH 6.3, floc. mixing intensity at 30 rpm, water temperature at 20 °C)

Figure 12 shows data for the Göta River for cold water temperatures (4 °C) and at a DAF to settling overflow rate ratio of 11 to 1. The UV data show again that removal of soluble organic matter depends on the coagulant, not the solid/liquid separation step, with the PACs performing better than alum. The turbidity data show for cold water conditions that DAF performed much better than sedimentation for all coagulants. The turbidities following DAF were about 1 NTU or less for all coagulants. For sedimentation all residual turbidities were above 2 NTU with alum the least effective coagulant. The most effective coagulant for sedimentation was the low basicity Ca PAC (PAX-XL1).

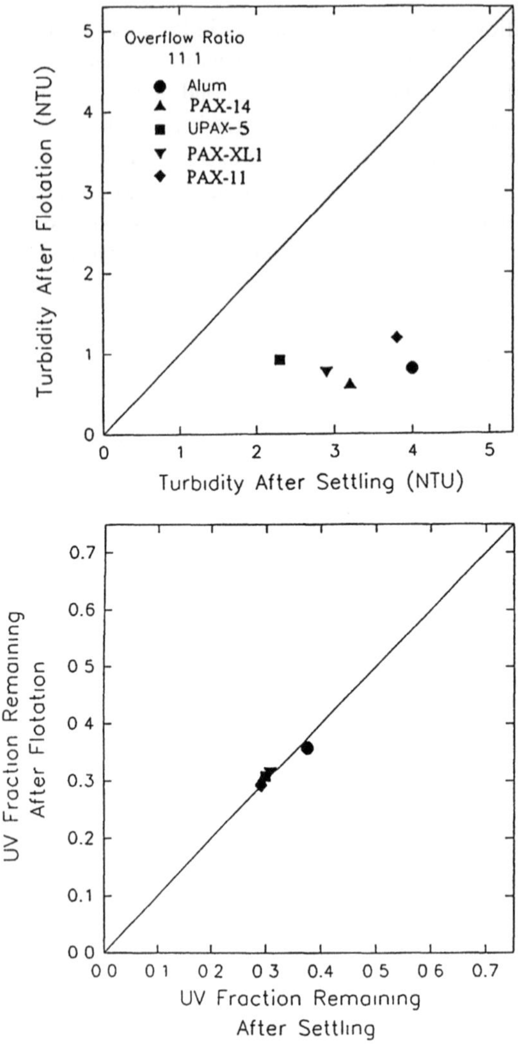

Fig. 12. Comparison of DAF to sedimentation for the Göta River (Al based coagulant dosages fixed at 3.5 mg/L as Al^{+3}, pH 6.8, floc. mixing intensity at 30 rpm, water temperature at 4 °C)

Conclusions

- Alum and the PACs with differnt chemical properties (basicity, % Al, and with or without Si, Ca, or sulfate) were all effective in dissolved air flotation (DAF) when used under favorable pretreatment conditions of coagulant dosage, pH, flocculation time, and flocculation mixing intensity. Overall, excellant DAF treatment was achieved with Ca, Si, and sulfate based PACs.
- In treating an aquatic humic type water, the optimum coagulant dosages for alum and the PACs were approximately equal when expressed as mg Al added per TOC of the raw water, and increased with pH. An exception was the Ca based PAC, which had a lower optimum dosage at higher pH compared to the

other coagulants. Considering Al dosage requirements, optimum dosing ranges, residual Al, and water temperature, operation at pH 6 to 6.5 is recommended for coagulation-flocculation.

- Alum is slower to react than the PACs, especially in cold waters, and requires longer flocculation times. The PACs were effective at flocculation times of 5 minutes.
- DAF performed better than sedimentation in treating humic or algae type waters especially at lower flocculation times. DAF was more effective than sedimentation under cold water conditions in treating a river water containing turbidity.

Acknowledgements. The authors are pleased to acknowledge the financial support of Purac AB and Kemira Kemi AB, and from the Swedish National Board for Industrial and Technical Development (NUTEK). We thank William R. Knocke of Virginia Polytechnic Institute and State University for supplying the aquatic humic matter for the synthetic water experiments. This research was carried out at the University of Massachusetts and at Chalmers Tekniska Hogskola. We thank Evy Axen, Gabriella Kaffehr, Mona Zanders, and Inger Hessel for their help at Chalmers. We also appreciate the cooperation and assistance of the following Swedish water utilities: Norrvatten (Gorvaln Water Treatment Plant) the City of Göteborg's Water and Sewage Works, and the Uddevalla Municipality.

Literature

[1] Edzwald, J.K., Malley, J.P. Jr., Yu, C.: A Conceptual Model for Dissolved Air Flotation in Water Treatment. Water Supply *8* (1990) 141

[2] Malley, J.P. Jr., Edzwald, J.K.: Laboratory Comparison of DAF with Conventional Treatment. J. Amer. Wat. Works Assoc. *83* (9) (1991) 56

[3] Edzwald, J.K., Wingler, B.J.: Chemical and Physical Aspects of Dissolved Air Flotation for the Removal of Algae. J. Wat. Supply Research Techn. (Aqua) *39* (1990) 24

[4] Hyde, R.A., et al.: Clarification by Flotation. J. Amer. Wat. Works Assoc. *69* (7) (1977) 69

[5] Zabel, T.: Flotation in Water Treatment. In: The Scientific Basis of Flotation, K.J. Ives (Ed). Martinas Nijhoff, Boston 1984, pp. 349–377

[6] Zabel, T.: The Advantages of Dissolved Air Flotation for Water Treatment. J. Amer. Wat. Works Assoc. *77* (5) (1985) 42

[7] Heinänen, J.: Use of Dissolved Air Flotation in Potable Water Treatment in Finland. Aqua Fennica *18* (2) (1988) 113

[8] Edzwald, J.K., Walsh, J.P., Kaminski, G.S., Dunn, H.J.: Flocculation and Air Requirements for Dissolved Air Flotation. J. Amer. Wat. Works Assoc. *84* (3) (1992) 92

[9] Janssens, J.G.: Developments in Coagulation, Flocculation and Dissolved Air Flotation. Wat. Engineer. Management *139* (1) (1992) 26

[10] Janssens, J.G.: The Application of Dissolved Air Flotation in Drinking Water Production. In: Particular for Removing Algae. Proceedings of DVGW-Schriftenreihe Wasser Nr. 67, Eschborn 1991

[11] Ødegaard, H.: Engineering Aspects of Flocculation. In: Chemical Water and Wastewater Treatment, A. Grohmann, H.H. Hahn and R. Klute (Eds.). Fischer, New York 1985, pp. 81–102

[12] Edzwald, J.K., Van Benschoten, J.B.: Aluminium Coagulation of Natural Organic Matter. In: Chemical Water and Wastewater Treatment, H.H. Hahn and R. Klute (Eds). Springer, Berlin Heidelberg New York 1990, pp. 341–359

[13] Van Benschoten, J.E., Edzwald, J.K.: Chemical Aspects of Coagulation Using Aluminium Salts: 2. Coagulation of Fulvic Acid Using Alum and Polyaluminium Chloride. Wat. Res. *24* (1990) 1527

James K. Edzwald
Department of Civil and
Environmental Engineering
University of Massachusetts
Amherst, MA 01003
USA

Donald Q. Bunker
Tighe and Bond Inc.
53 Southampton Road
Westfield, MA 01085
USA

Jan Dahlquist
Purac AB
Box 1146
S-221 05 Lund
Sweden

Lars Gillberg
Kemira Kemi AB
Box 902
S-251 09 Helsingborg
Sweden

Torsten Hedberg
Department of Sanitary Engineering
Chalmers Tekniska Hogskola
S-412 96 Göteborg
Sweden

The Use of Factorial Designs to Identify Factors Affecting Particulate Removal in a Full Scale Tertiary Precipitation Plant with Flotation

H. Aspegren, B. Andersson, and E. Arvin

Abstract

The Sjölunda wastewater treatment plant comprises mechanical, biological and chemical treatment. The chemical treatment step includes flocculation and flotation. The concentration of suspended solids has been higher than expected when the plant was built. In this study factorial designs were used in combination with laboratory tests in order to identify factors of significance for the removal of suspended solids. The primary aim of the project was to reduce the suspended solids to about 10 mg/l.

The investigation showed that the type of water, secondary effluent from trickling filters or an activated sludge process, and the flocculation energy were important for the process performance. The dosage of alum and polymer and the dosage point of the polymer were also significant. A high flocculation energy, i.e. a case with relatively small flocs, improved the overall removal. It also seems possible to decrease the concentration in the tertiary effluent to below 10 mg/l.

Introduction

The Sjölunda wastewater treatment plant in Malmö, Sweden comprises mechanical, biological and chemical treatment. The chemical treatment step includes flocculation and flotation. A concentration of suspended solids of less than 10 mg/l in the tertiary effluent was expected from past experiences when the tertiary plant was commissioned in 1979. However, the concentration has been substantially higher. The plant has been subjected to numerous comprehensive investigations but it has not been possible to fully explain or clarify the reason for the unexpectedly poor performance.

One of the problems during these investigations has been that several factors may influence the full scale process performance simultaneously and it is consequently very difficult to identify the relevant factors. The use of factorial designs has in many investigations been a successful methodology for solving problems by identifying and quantifying significant factors and possible interactions. This technique has also been applied at treatment plants [1]. Factorial designs were consequently used in combination with laboratory batch tests as part of an overall

R. Klute / H.H. Hahn (Eds.)
Chemical Water and Wastewater Treatment III

investigation with the primary aim of reducing the concentration of suspended solids in the tertiary effluent from the flotation plant to about 10 mg/l.

The Sjölunda Wastewater Tratment Plant

In order to understand the present role of the chemical precipitation step and its significance to the overall process performance, an introductory description of the plant will be presented. A schematic flow sheet of the plant is shown in Figure 1. The plant is designed for an average flow capacity of 1650 l/s and effluent standards of 15 mg/l for BOD_7 and 0.5 mg/l for total phosphorus. The maximum flow capacity for complete treatment is 4400 l/s.

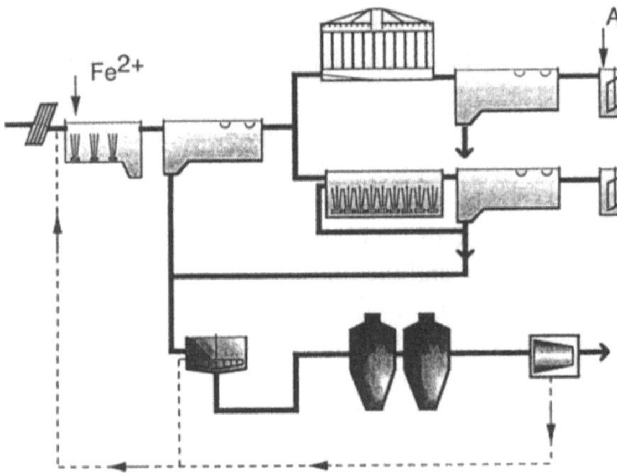

Fig. 1. Process configuration at the Sjölunda wastewater treatment plant

After primary treatment the flow is divided into two parallel treatment lines, one high loaded activated sludge line and one plastic packed trickling filter line. The two lines are designed and operated at almost the same hydraulic load.

Phosphorus removal was originally accomplished by adding alum to both lines in the tertiary treatment step. However, soon after the plant was taken into operation it was realized that it was not possible to comply with the effluent standards for phosphorus with this configuration due to relatively high concentrations of suspended solids in the tertiary effluent. Numerous studies were undertaken in order to explain and solve this problem. No direct explanation could, however, be found and instead two-point precipitation was introduced in order to decrease the phosphorus load on the precipitation plant. Ferrous sulphate was used as coagulant in the pre-precipitation step and alum and a cationic polymer were added to the secondary effluent from the trickling filter line in order to flocculate and remove non-settleable suspended material (pinpoint floc). In the activated sludge line the effluent requirements could be fullfilled without coagulant addition. The flotation process is still in operation, though, and is used as a security and polishing step for removal of suspended solids.

A schematic outline of the tertiary treatment step is shown in Figure 2. The plant is divided into two parallel lines with secondary effluent from the trickling filter line and the activated sludge line. Each line is further divided into 8 parallel units. Each unit comprises two flocculation reactors in series and one flotation basin. About 10–15 % of the flow from the tertiary effluent is recycled, pressurized and saturated with air and then used for the flotation process. The detention time in the flocculation reactor is about 50 minutes based on the average flow.

In the line with secondary effluent from the trickling filters, around 125 g/m^3 of alum is added at the very beginning of the distribution channel. The polymer (0.5 mg/l) is added at the inlet to each of the 8 units. The fact that alum is added at one point means that the coagulation time will not be the same for the different units. Based on an average flow the coagulation time varies between 20 to 140 seconds. It should also be mentioned that the water quality at the Sjölunda plant is not typical for a Swedish wastewater. The hardness is substantially higher, the concentration of calcium is on the average around 100 mg/l, and the concentration of magnesium is around 16 mg/l. The alkalinity of the water is also relatively high, around 5–6 meqv/l, and the pH of the water is stable around 7–8.

Despite the fact that it is possible to meet the effluent standard for phosphorus with the present process configuration, the problem with rather high concentrations of suspended solids in the tertiary effluent remains. These problems are more pronounced in the trickling filter line. As the demands for treatment can be expected to become more stringent in the future, the incentive for improving the performance and thus preparing the plant for these future demands is strong.

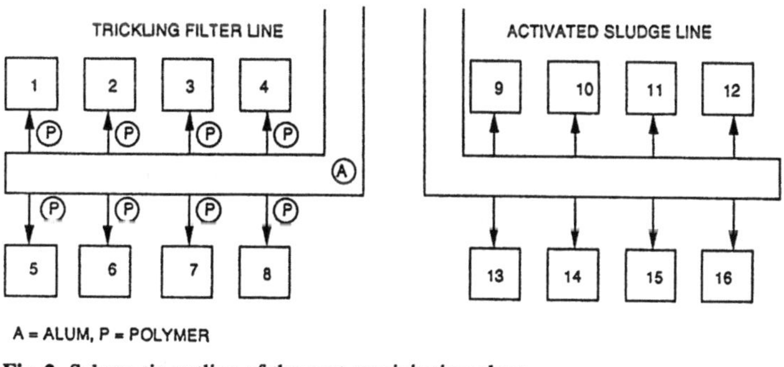

Fig. 2. Schematic outline of the post-precipitation plant

Methodology

Based on the description of the overall plant structure and the post-precipitation plant it is quite obvious that several factors may be of significance for the overall removal of suspended solids. There is little quantitative information available on how these factors will affect the process and what interactions exist between the different factors. As a consequence this means that the evaluation procedure and the experimental procedure must be chosen and combined in such a way that the significant factors can actually be identified. In this case a laboratory procedure was combined with two level factorial designs and central composite designs.

The laboratory test procedure was mainly designed to study the effect of the flocculation step. Factors related to the air-saturated water, e.g. the ratio of air-saturated water to the water sample, the pressure etc, were kept constant. As a basis for the experiments the Kemira CAF coagulation and flocculation equipment [6] was used. This laboratory equipment was developed for flocculation-settling experiments and it was necessary to adapt it to flocculation-flotation experiments, mainly by replacing the original 1 l reactor beaker with a 2 l beaker in order to accommodate the air-saturated water for flotation. Furthermore the 2 l beaker was equipped with a valve along the side towards the bottom of the reactor. This valve was used for introducing the air-saturated water and to extract a water sample for further analysis. The original sample volume was 1 l and 0.2 l of air-saturated water was used for flotation. Air-saturated water was prepared from tertiary effluent from the trickling filter line which was filtered (GF/A) prior to its use. The water was saturated with air in a conventional laboratory pressure chamber at a pressure of 0.5 MPa. The coagulation and flocculation energy was measured indirectly by measuring the rotational speed of the mixing paddle of the laboratory equipment.

The experiments were evaluated with respect to suspended solids, turbidity and floc size. Water samples extracted from the experiments were analysed for suspended solids and turbidity by the Sjölunda laboratory (accredited as testing laboratory no. 1033) within 24 hours of the time they were sampled. The floc size was assessed by visual inspection during the experiment.

The experimental program was divided into two phases. During the first phase the primary aim was to decide what factors were of principal importance to the process by using factorial designs at two levels. During the second phase the identified factors were investigated more thoroughly at five levels by using central composite designs.

Factorial designs at two levels are especially useful when a great number of factors and their effect on a certain variable is to be explored and directions for future experiments are to be determined. The name of this design implies that each factor is investigated only at two levels, at a low and a high level. These levels are also referred to as the -1 and $+1$ level. By performing factorial designs the number of experiments that is required in order to identify main effects is minimized and, furthermore, interaction effects can be detected [2, 3] and [5]. A main effect directly affects the result, whereas an interaction effect implies that the effect of a variable can change as other variables are changed. As interaction effects tend to be smaller with higher order, they can in some cases be disregarded. This will lead to a further reduction of the necessary number of experiments. This possibility is exploited in fractional factorial designs.

By using central composite designs the relationship between one or more measured responses and a number of input variables is studied at five levels. Typically these designs answer questions concerning how a particular response is affected by a given set of variables over a specified region of interest.

The factorial designs and central composite designs were produced and evaluated by means of two software products [8] and [9].

The test series during this investigation was initiated with two screening tests. The first screening test was a fractional factorial design, a resolution IV design [2]. Eight factors were studied at two levels in 32 experiments. The experiments were divided into two blocks. All main effects were clear of two factor interactions. Based on the results, a more refined second screening test was planned. This test was based on a resolution V design [2]. Five factors were studied at two levels in 16 experiments. The design was repeated twice on two consecutive days. All main effects were clear of two factor interactions and all two factor interactions were clear of one another.

Based on the screening tests the effect of two selected factors was elucidated and studied by using a central composite design. The factors were studied in 13 experiments with each factor at five levels. This design was repeated five times.

The tests and the results will be more thoroughly discussed below.

Results

The first screening test was divided into two blocks and the experiments were performed on two days, a Monday and a Wednesday during the same week. Monday water may be said to represent the weekend quality and the Wednesday water may be said to represent the working day quality. The investigated factors and their levels are presented in Table 1.

Table 1. Investigated factors and selected levels in first screening test

Factor	Low level -1	High level 1
Secondary effluent	Activated sludge	Trickling filter
Coagulation time [s]	30	120
Coagulation energy [rpm]	250	400
Flocculation time [min]	15	60
Flocculation energy [rpm]	20	40
Polymer [mg/l]	0.0	0.5
Detergent [ml/l]	0.0	0.05
Coagulant*	Al(III)	Fe(III)

* Same dosage as Me(III)

The eight factors were selected with previous investigations in mind. For example, it has long been discussed to what extent the water quality, i.e. day of the week and biological treatment method, determines the resulting concentration of suspended solids. Factors related to the coagulation and flocculation process were obviously of interest. One of the main issues was, for example, the relatively long detention time in the full scale reactor. The effect of using iron instead of alum and the role of the polymer also had to be determined. Finally, the use of a commercial detergent was to be assessed. The levels of the different factors were selected with respect to the prevailing conditions in the full scale plant.

The test showed that the type of water was of great significance for the concentration of suspended solids. Secondary effluent from the trickling filter line was more difficult to treat than secondary effluent from the activated sludge line. Furthermore the day of the week affected the concentration of suspended solids. The lowest concentration was obtained with Monday water.

The test also indicated that improvements of the coagulation and flocculation system could be accomplished mainly by manipulating the coagulation time and flocculation energy. In absolute numbers an improvement of around 2 mg/l of suspended solids could be anticipated. This number might seem rather moderate. However, it actually corresponds to an improvement of the removal of suspended solids of up to 40 % in this test. The results also indicated that the lowest concentration that could be expected in the full scale plant was around 10 mg/l. The type of coagulant did not affect the result. The use of a polymer did not have a significant effect on the concentration of suspended solids. However, the polymer seemed to stabilize the flotated sludge layer, making it less vulnerable to mechanical abrasion. This is, of course, of great importance in a full scale application. Finally, the detergent addition had a negative effect on the result.

The second screening design was repeated on two consecutive days, one Wednesday and the following Thursday. The factors were selected based on the results of the first screening test. Only secondary effluent from the trickling filters was used and the primary aim was to study the flocculation step. The factors and the selected levels are presented in Table 2 and the resulting significant factors, at a 5 % level or below, are shown in Table 3.

Table 2. Investigated factors and selected levels in second screening test. (Coagulation energy = 400 rpm. Flocculation time = 60 minutes and secondary effluent from the trickling filters)

Factor	Low level −1	High level 1
A Coagulation time [s]	30	120
B Flocculation energy [rpm]	20	40
C Dosage point of polymer* [min]	2	58
D Polymer dosage [mg/l]	0.5	1.5
E Alum dosage [mg/l]	125	175

* Time for dosage of polymer after coagulation

The test indicated, with respect to suspended solids that the flocculation time, the polymer and the alum dosage should be kept at a high level and that the polymer should be added at a relatively early stage during flocculation. The polymer seemed to stabilize the sludge layer in the previous test. In this test, however, it directly affected the concentration of suspended solids. Furthermore, the interaction between the coagulation time and the flocculation energy was again detected.

Table 3. Results of second screening test

Factor		Evaluation criteria (level)	
		Suspended solids	Turbidity
	Block	–	–
A	Coagulation time	–	–
B	Flocculation energy	+1	–
C	Dosage point of polymer	−1	−1
D	Polymer dosage	+1	–
E	Alum dosage	+1	–
	Coagulation time* Flocculation energy	+1 * +1	+1 * +1

– : No significant effect
+1 : Effluent improved / lower concentration of suspended solids or turbidity by increasing level of factor
−1 : Effluent improved / lower concentration of suspended solids or turbidity by decreasing level of factor

It has to be realized, though, that the concentration of suspended solids was low, ranging from 3 to 11 mg/l. This fact probably explains the rather low correlation coefficient (R^2 = 0.6), i.e. the significant factors and interactions only explain about 60 % of the observed variation.

By using factorial designs it is also possible to describe quantitatively the relation between the independent variables and the concentration of suspended solids by using a polynomial. This is shown below for the concentration of suspended solids.

$$\text{SS (mg/l)} = 5.4 - 0.5B + 1.1C - 0.5D - 0.6E - 0.4AB\,.$$

The variables A–E are explained in Table 3. The variables can attain values of −1 (low level) or +1 (high level).

The polynomial suggests that it is possible to affect the result with about 2–3 mg/l suspended solids by manipulating the flocculation process. Quantitatively the most important factor seems to be the dosage point for the polymer (factor C). However, it can also be seen that the flocculation energy (factor B) may be an important factor when the two factor interaction is considered.

The influence of different factors on the concentration of suspended solids and turbidity was apparent in this experiment. It has to be realized, though, that turbidity is not a measurement of suspended solids, which is the primary variable. Furthermore the turbidity was very low, ranging between 0.7 and 3.3 FTU.

In order to elucidate the effect of the flocculation energy the next step in the investigation was to perform central composite designs with the flocculation energy and flocculation time as the primary factors. The actual design of the experiment is shown in Table 4. The flocculation energy was varied at five levels within a substantially broader range in comparison to previous tests. The range was varied from 35 to 200 rpm.

Table 4. Design of laboratory experiment

Factor	
Flocculation time [min]	6–35
Flocculation energy [rpm]	35–200
Coagulation time [s]	120
Flocculation energy [rpm]	400
Dosage point of polymer* [min]	2
Polymer dosage [mg/l]	1.0
Alum dosage [mg/l]	150

* Time for dosage of polymer after coagulation

The result from this first test showed that a quadratic model was significant at a 5 % level or below with respect to suspended solids ($R^2 = 0.75$). The lowest concentration was obtained when the flocculation energy was kept high (140 rpm) and when the detention time was kept moderate (25 minutes). The concentration of suspended solids was low, ranging from 3 to 7 mg/l.

A quadratic model was significant for the turbidity as well ($R^2 = 0.84$). However, the overall picture was not as clear as for the concentration of suspended solids. The optimum for turbidity and suspended solids did not coincide. Two optima could actually be seen for turbidity. Either the flocculation energy should be kept very high and the flocculation time short or the flocculation energy should be kept low and the flocculation time long.

Apart from these results it was also observed that the size of the floc varied with the flocculation energy. The floc size might, of course, affect the overall result. Furthermore there was also a growing concern that the reason why the turbidity might differ from the concentration of suspended solids is the presence of air bubbles in the measuring flask for turbidity. The sample bottle was shaken before the sample was poured into the measuring flask.

In the next set of experiments it was decided that two new dependent variables were to be included, floc size and turbidity after "rest". The latter way of measuring turbidity means that the sample has been allowed to rest for some time (more than 5 minutes) before it is measured.

The floc size was assessed manually during the flocculation experiments by using a 10 grade scale. Grade 3 represents small flocs, approximate size 0.5–1 mm, and grade 8 represents large flocs, approximate size 1–1.5 mm. It must be stressed that this scale must be regarded as a qualitative one and that a comparison between different days is not recommended.

The first two replicates were not very conclusive and no significant factors could be detected. The reason for these rather disappointing results could not be determined. However, the result was possibly affected by the fact that the experimenters were new persons, probably leading to higher variability.

The next two replicates which were made soon afterwards were generally consistent with the first test. In both replicates a linear model was found to be

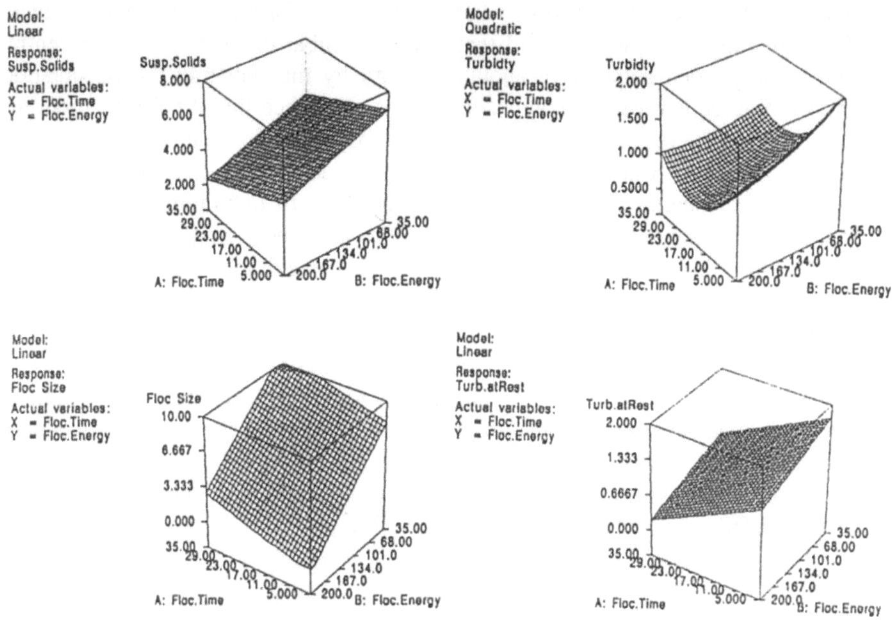

Fig. 3. Dependent variables as a function of flocculation energy and flocculation time

significant for the concentration of suspended solids with R^2 values of 0.6 and 0.7. In the first test a quadratic model was significant. The model indicated again that the flocculation energy should be kept high. The result is illustrated with one of the data sets in Figure 3 for the four dependent variables. The result was similar for the other data set.

It appears from the figure that a consistent pattern evolves and that the lowest concentration of suspended solids is obtained when the flocs are comparatively small, i.e. at high flocculation energies.

This is rather intriguing as it is usually assumed that a large floc is preferred in the flotation process [4], [7] and [10]. The range for suspended solids is about 4 mg/l between the highest and lowest concentration. It is also interesting to notice that turbidity after rest conforms to the general pattern found for suspended solids with a linear model being significant, whereas for "normal" turbidity a quadratic model is significant.

Conclusion

The methodology which was used in this investigation where laboratory tests were combined with factorial designs has clearly been successful in identifying significant factors. The results with respect to the concentration of suspended solids has been consistent throughout the study. The first screening test indicated that the type of water, secondary effluent from the trickling filters or the activated sludge units, and the flocculation energy were of importance for the process performance. The screening test also indicated that it was possible to decrease the concentration of suspended solids with about 2–3 mg/l by manipulating the flocculation energy. Subsequent central composite designs confirmed that the flocculation energy affected the result. The concentration of suspended solids could be decreased with about 3 mg/l when the flocculation energy was increased. This number may seem rather moderate. However, it actually corresponds to an improvement of the removal of suspended solids by about 40 to 50 % in these experiments.

The tests also indicated, with respect to suspended solids, that an increased polymer and alum dosage would improve the removal. Furthermore the polymer should be added at an early stage during flocculation.

The central composite designs also revealed that the lowest concentration of suspended solids was obtained when the flocs were relatively small. The smallest flocs and lowest concentrations were in turn obtained when the flocculation energy was high.

Finally the laboratory tests showed that it was possible to attain a concentration of suspended solids which was well below 10 mg/l.

References

[1] Arvin, E., Brøndum, L., Monrad, J.D.: Afvandelighed af mekanisk-biologisk råslam ved Lynetten Renseanlaeg. (Dewaterability of Mixed Primary/Secondary Raw Sludge at the Lynetten Sewage Treatment Plant). Proceedings VAR-dagene at NTH, August 27-29, NTH, Trondheim, Norway, 1990 (in Danish)

[2] Box, E.F., Hunter, W.G., Hunter, J.S.: Statistics for Experimenters. An Introduction to Design, Data Analysis and Model Building. Wiley, New York 1978

[3] Brøndum, L., Monrad, J.D.: Statistisk forsøgsplanlaegning I–III. Den Private Ingeniørfond, København, Denmark, 1989 (in Danish)

[4] Heinänen, J., Jokela, P., Peltokangas, J.: Experimental Studies on the Kinetics of Flotation. In: Chemical Water and Wastewater Treatment II, R. Klute and H.H. Hahn (Eds.). Springer, Berlin Heidelberg New York 1992, pp. 247–262

[5] Hicks, C.R.: Fundamental Concepts in the Design of Experiments. 2nd Ed. Holt, Rinehart and Winston, New York 1973

[6] Kemwater: Handbok i Vattenvård. Kemwater, Helsingborg, Sweden, 1989 (in Swedish)

[7] Schers, G.J. and Dijk J.C. van: Dissolved-Air Flotation: Theory and Practice. In Chemical Water and Wastewater Treatment II, R. Klute H.H. Hahn (Eds.). Springer, Berlin, Heidelberg, 1992, pp. 223- 246.

[8] Stat-Ease Inc.: Design-Ease Software, Version 2.0, Reference Manual. Minneapolis, MN, USA, 1991

[9] Stat-Ease Inc.: Design-Expert Software, Version 3.0, User's Guide. Minneapolis, MN, USA, 1992

[10] Tambo, N.: Optimization of Flocculation in Connection with Various Solid-Liquid Separation Processes. In: Chemical Water and Wastewater Treatment, H.H. Hahn and R. Klute (Eds.). Springer, Berlin Heidelberg New York 1990, pp. 17–32

Bengt Andersson and Henrik Aspegren
Malmö Water & Sewage Works
S-205 80 Malmö
Sweden

Erik Arvin
Department of Environmental Engineering
Technical University of Denmark
Building 115
DK-2800 Lyngby
Denmark

Membrane Microfiltration of Secondary Wastewater Effluent

R. Patel[1], A. C. Penisson[1], R. D. Hill[2], and M. R. Wiesner

Abstract

This pilot study was designed to evaluate the efficacy of membrane microfiltration in treating secondary wastewater effluent. An 8 liter per minute pilot was operated at the City of Houston's Southwest Wastewater Treatment Plant. Membrane permeates were of similar or better quality than filtrates from the plant's conventional sand filters. Good removal of particulate contaminants, including coliform bacteria, was observed. In comparison with the plant's filtrate, permeate quality was also much less variable. Operating the pilot in a crossflow configuration did not improve the permeate quality and did not retard the increase in transmembrane pressure in comparison with dead-end operation. Chemical cleaning of the membrane with caustic solution was required approximately once every 5 days. Preliminary results indicate that coagulation pretreatment in conjunction with membrane microfiltration can also be used to remove phosphorus from wastewater.

Introduction

Only a few studies have evaluated membrane microfiltration of secondary wastewater effluent. Microfiltration membranes might be used to achieve very low turbidity effluents with very little variance in treated water quality. Because bacteria and many other microorganisms are also removed, such membrane disinfection might avoid the need for chlorine and subsequent dechlorination. Metal salts of iron or aluminium may also be added to enhance membrane performance. For example, iron or aluminium coagulants may be added to precipitate otherwise soluble species such as phosphorus and arsenic as well as improving the removal of viral particles. Coagulation of colloidal materials may also increase the effective size of particles applied to membranes and increase permeate flux by 1) reducing foulant penetration into membrane pores, 2) forming a more porous cake on the membrane surface, 3) decreasing the accumulation of materials on the membrane due to particle size effects of particle transport, and 4) improving the backflushing characteristics of the membrane.

[1] Graduate Student, Department of Environmental Science and Engineering, Rice University, Houston, Texas
[2] Chief Engineer, Wastewater Operations Division, Houston Department of Public Works

R. Klute / H.H. Hahn (Eds.)
Chemical Water and Wastewater Treatment III

In this paper we report on pilot experiments in which secondary wastewater effluent was filtered using membrane microfiltration. Some preliminary results using a commercial polyaluminium coagulant (Kemira Water Treatment, Inc.)[3] to pretreat the feed water are also reported.

Materials and Methods

A skid-mounted microfiltration membrane pilot[4] with a nominal capacity of 8 liters/minute was used in this study. This pilot uses 4 membrane modules, each with 1 m^2 of outside-in hollow fiber membrane area with an effective pore size of 0.2 μm. Feed water to the pilot was drawn from a secondary clarifier at the Houston Southwest Wastewater Treatment Plant (WWTP). Clarified water at the WWTP is subsequently chlorinated and filtered using conventional sand filters before being dechlorinated. The dechlorinated effluent is discharged into Brays Bayou, a river which runs through residential portions of the City of Houston and is bordered with parks, jogging paths and bike trails.

The membrane unit was operated in a constant permeate flux mode in which the transmembrane pressure (TMP) increases over time as materials deposit on and in the membrane. Permeate flux was set at 340 L/hr and readjusted as needed to maintain this flow. The unit was operated in both dead-end and crossflow filtration modes. During dead-end filtration, all water entering the membrane module exits as permeate. In the crossflow mode, a portion of the water entering the module flows across the membrane fibers and exits the module as concentrate. In both modes of operation, the pilot was automatically backflushed every 20 minutes, resulting in a recovery (permeate flow divided by feed flow) for the unit of 82 %. One run was also performed with backflushing once every 5 hours corresponding to a recovery of 98 %.

Samples from the feed and permeate flows were analyzed for turbidity, total organic carbon (TOC), phosphorus concentration, and particle size distribution. TOC was measured using a Shimadzu TOC 500 total organic carbon analyzer. Phosphorus concentrations were measured colorimetrically. The particle size distributions of the feed and permeates were measured using an electronic particle counter (Coulter Multisizer) which operates on the electrical sensing zone principle. Samples were also analyzed for total and fecal coliforms. Similar analyses were performed on samples of influent and effluent water from the packed bed filters at the treatment plant during the same period of operation.

---

3 PAX-XL2, PAX-105, and PAX-18

4 Memtec America Corporation, Timonium, MD 21093

Results and Discussion

Permeate Quality

Dead-End Filtration. The membrane unit showed excellent turbidity removal when operated in the dead-end filtration mode. Permeate turbidity was approximately 0.1 NTU throughout the entirety of both pilot runs in the dead-end mode. This represents a reduction of approximately 1 log unit in turbidity below the levels in the feed water (Fig. 1). In contrast, filtrate turbidity from the wastewater treatment plant was never lower than 0.4 NTU during the sampling periods. The turbidity of the plant filtrate was highly variable, ranging from 0.4 to as high as 17.4 NTU. Typically, the plant produced a filtrate with a turbidity near 1 NTU.

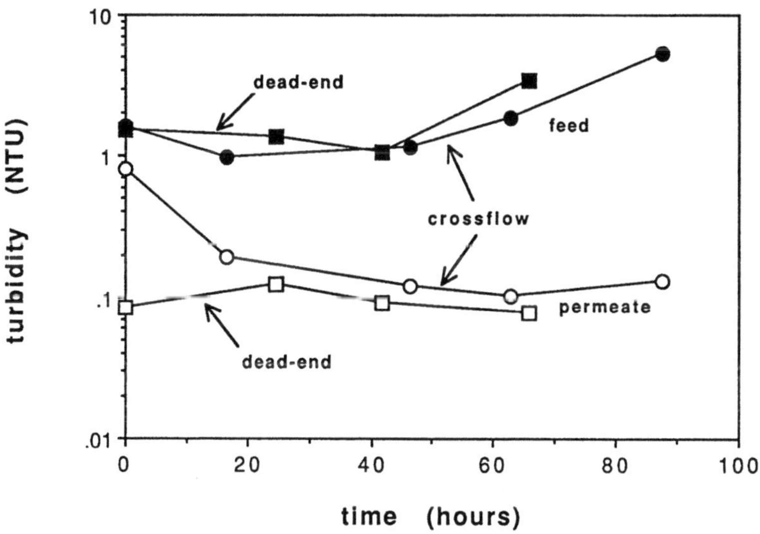

Fig. 1. A comparison of turbidity in the feed and membrane permeates when the pilot was operated in dead-end and crossflow modes

Particle counts also indicated good removal of colloidal materials by the membrane. Particles larger than 0.4 μm were reduced by approximately 2 log units in number concentration across the membrane (Fig. 2). Backflush frequency (or equivalently recovery) did not appear to impact permeate quality significantly. During the same periods particle counts were often observed to increase across the plant's conventional filters. These data are consistent with data on filtrate turbidity for the plant that indicated occasional increases in turbidity across the filters.

Good reductions in total coliforms were also observed across the membrane. However, the number in the permeate tended to increase over time. This is attributed in part to contamination and growth of bacteria in the sample line. No

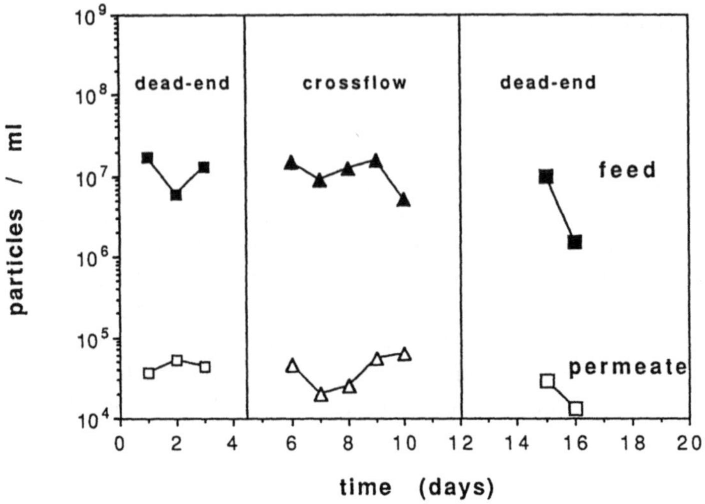

Fig. 2. Particle removal across the microfiltration membrane during periods of dead-end and crossflow filtration without coagulant addition

fecal coliforms were detected in the corresponding samples of either plant filtrate or membrane permeate. Comparable removals were observed in dead-end and crossflow filtration (Fig. 3).

A reduction in the TOC concentration across the membrane was observed when the pilot was operated in dead-end filtration. Removal of TOC is due to retention of particulate materials by the membrane. None of the dissolved organic carbon should be removed. TOC removals were variable, perhaps due to changes in the composition of the feed water. The filter initially showed TOC removal of close to 50 % for the first twenty-four hours of operation. After this time, the TOC removal dropped and permeate concentrations were very close to concentrations in the feed water (Fig. 4).

Crossflow Filtration. When operated in a crossflow configuration, the membrane produced permeates of low turbidity. However, turbidities tended to be slightly higher than those obtained when the pilot was operated in a dead-end filtration mode. Initial permeate turbidities were just under 1 NTU. After 16 hours of operation, permeate turbidity dropped to approximately 0.1 NTU and remained at this low level throughout the duration of the pilot experiment (Fig. 1). Particle counts indicated that the membrane was effective in removing particles. The number concentration of particles larger than 0.4 μm was reduced by approximately 2 logs (99 %) relative to the feed flow (Fig. 2). When the concentration of the feed water due to recycling of the crossflow is taken into account, the reduction of particle counts across the membrane increases to nearly 3 logs.

TOC measurements of the feed and permeate waters indicated considerable variation in permeate quality during the filter run (Fig. 4). The TOC in the membrane permeate was sometimes higher than that in the feed. For example, while

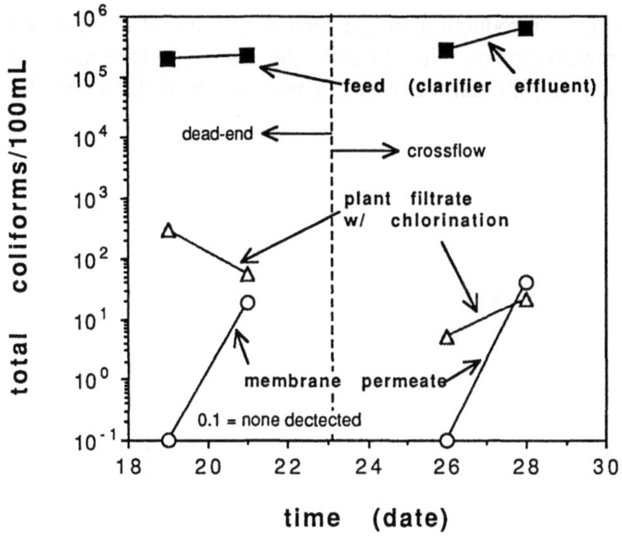

Fig. 3. Removal of total coliforms by microfiltration and by conventional chlorination followed by filtration during two periods in January 1994 when the membrane pilot was operated in dead-end and crossflow filtration

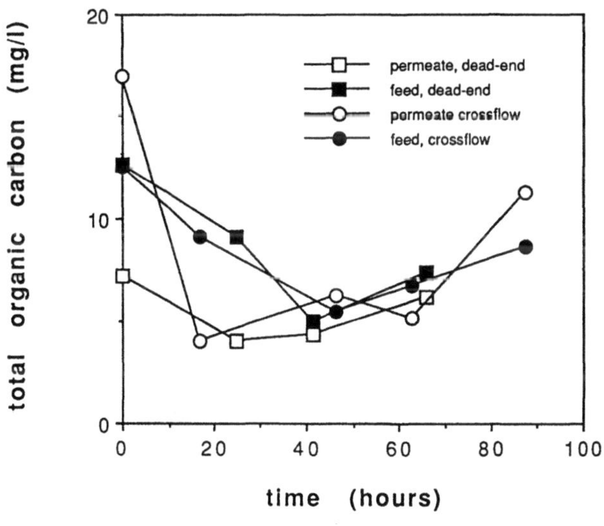

Fig. 4. Evolution of TOC in the membrane feed and permeates during dead-end and crossflow filtration

removals of 25 to 50% typified membrane performance, after 46 hours of crossflow filtration the TOC in the permeate was slightly higher than that measured in the feed water. However, these values are within the range of analytical error indicating that essentially no TOC was removed by the membrane. In addition to changes in the feed water quality, fluctuations in TOC removal may be attributable to the effect of concentrating organic materials in the break tank during the filter run. Occasional "breakthrough" of fine colloidal material may occur as loadings on the membrane increase. Furthermore, there was significant bacterial growth on the interior of lines supplying clarified water to the pilot unit. Sloughing of bacteria and their subsequent introduction to the feed may have also increased the variability of TOC measurements. Nonetheless, the overall performance of the membrane in removing organic carbon was not remarkably different in the dead-end and crossflow modes.

Permeate Flow

The permeate flow rate, normalized for TMP (specific permeate flow), did not differ substantially between crossflow and dead-end filtration mode (Fig. 5). In both cases, specific permeate flow tended to stabilize at a value of approximately 200 L/hr/kPa. This permeate flow corresponds to a permeate flux of 50 L/m^2/hr/kPa. The initial TMP was typically near 27 kPa, increasing to 68 to 100 kPa during a pilot run. The crossflow filtration mode may have initially impeded the decline in permeate flow in comparison with dead-end filtration. However, after 2 days of operation specific permeate flows were similar in both operating modes.

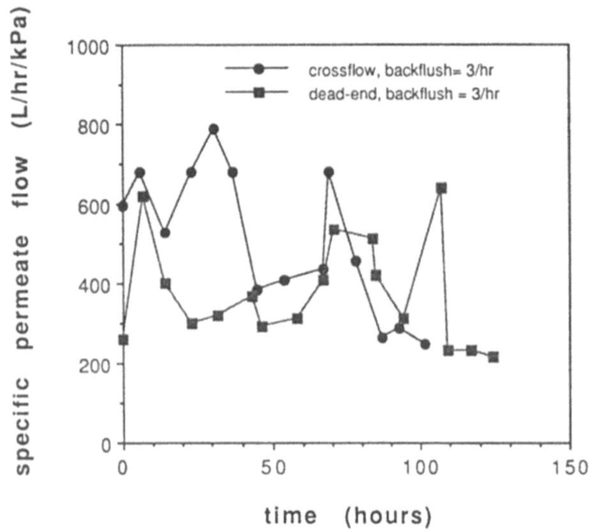

Fig. 5. A comparison of specific permeate flow when the microfiltration unit was operated in dead-end and crossflow modes

For the same mode of operation (dead-end) but a different backflush frequency and corresponding recovery for the membrane unit, there was a small but significant effect on permeate flow (Fig. 6). The initial decline in permeate flow at the two recoveries appeared to be similar. However, over a longer period, permeate flux at a recovery of 98 % was somewhat less than that obtained at a recovery of 82 %.

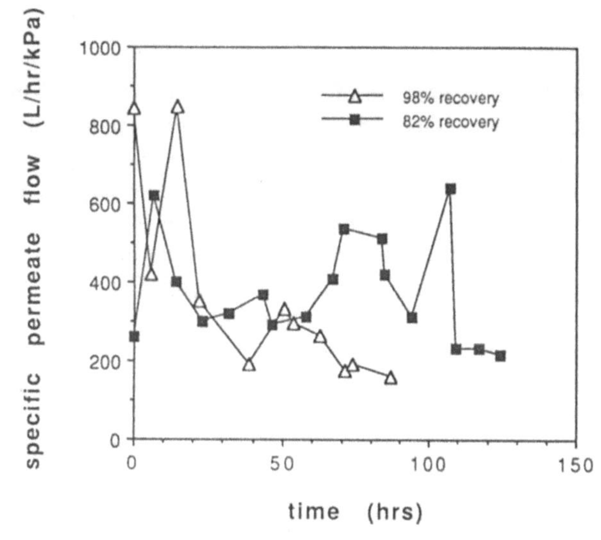

Fig. 6. Specific permeate flow observed during operation of the pilot at two recoveries

Coagulation Pretreatment and Permeate Quality

A preliminary run using a polymeric aluminium coagulant (PAX 18, Kemira Water Treatment Inc., Whippany, NJ) yielded some improvement in permeate quality. During this run, the pilot was operated in the crossflow filtration mode. As anticipated, coagulation pretreatment removed some of the phosphorus from the wastewater (Fig. 7). No attempt was made to optimize coagulant addition for either phosphorus removal or permeate flow. Rather, a dose of 10 mg/l as aluminium was added based on settleability observed in simple jar tests. More extensive runs, designed to evaluate the effect of chemical pretreatment of permeate flux, are in progress.

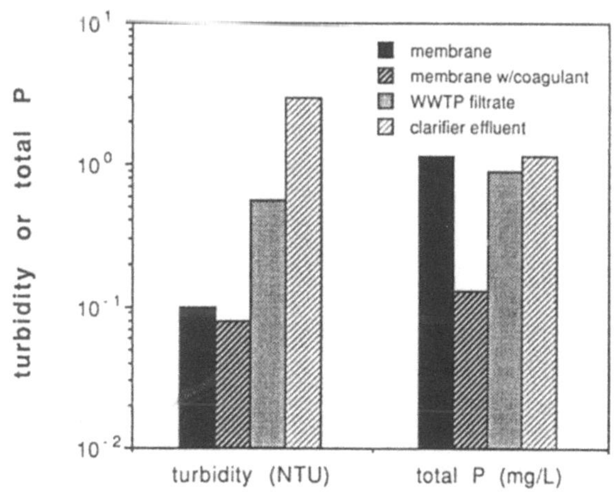

Fig. 7. Comparison of turbidity and total P in the membrane permeates with and without coagulation pretreatment with the wastewater treatment plant (WWTP) filtrate. Effluent from the secondary clarifiers served as feed water for these filters

Conclusions

Membrane microfiltration produced a permeate of similar or better quality than that produced by conventional filtration. Good removal of particulate contaminants, including coliform bacteria, was observed. In this regard, the process appears to be as effective as chlorination for the removal of coliforms from secondary waste effluent. A key advantage is the ability to filter and disinfect in a single step without the need for subsequent dechlorination. Preliminary results indicate that coagulation pretreatment in conjunction with membrane microfiltration can be used to reduce phosphorus concentrations as well. In contrast with the investigator's expectations, there did not appear to be any advantage in running the microfiltration unit in a crossflow mode and there may even be some disadvantages. The permeate quality and evolution of pressure drop obtained from the membrane operated in the dead-end mode was similar or superior to that obtained under crossflow conditions.

Acknowledgments. The authors gratefully acknowledge the materials and technical assistance provided by Memtec America Corporation and Kemira Water Treatment, Inc.; most notably, the efforts of Frank Nabozny and Anthony Stashick of Memtec and Jeff Roberts of Kemira. We also extend our appreciation to James Vickers, formerly of Memtec America Inc.

Mark R. Wiesner
Department of Environmental Science and Engineering
Rice University
P.O. Box 1892
Houston, Texas 77251
USA

Chemicals and Dosing Control

Influence of the Basicity of Polyaluminium Chlorides When Cleaning Municipal Wastewater

L. Gillberg

Abstract

The influence of different doses of aluminium sulphate and polyaluminium chlorides when cleaning a thin and a thick sewage on the residual turbidity, concentrations of TOC, ortho-P, non-ortho-P, SS, and alkalinity and sludge volume produced was studied.

Both the alkalinity consumed and the amount of ortho-phosphate precipitated decreased with increasing basicity of the coagulant. At low coagulant doses no or little alkalinity was consumed. When the coagulants were added above a threshold concentration the alkalinity consumed as well as the amount of sludge produced increased strongly with the dose used.

When the thick water was cleaned to a constant phosphorus concentration the type of coagulant giving lowest sludge volume depended on the amount of phosphorus removed. At high removal rates the lowest basicity coagulants gave the lowest sludge volume. When the residual P concentration was 1.5 mg/l the coagulant with OH/Al of 1.1 gave the lowest sludge volume.

When waters were cleaned to a constant turbidity the sludge volume decreased somewhat with increasing basicity of the coagulant.

Introduction

Different types of aluminium-based coagulants are used for the precipitation of particles and ortho-phosphate from municipal wastewater. Aluminium sulphate has been used for this purpose since around the beginning of this century and polyaluminium chlorides with different basicities have been gaining increased popularity during the last few decades.

The influence of the basicity of polyaluminium chlorides when cleaning a synthetic wastewater was studied by Ødegaard et al (1990). Fiessinger and Bersillon (1977) and Bottero et al. (1980) reported the turbidity reducing effects of polyaluminium chlorides of different basicities on bentonite suspensions.

At each municipal wastewater plant the coagulant is added in such an amount that the requirements of the authorities are fulfilled. These requirements may stipulate that the cleaned water shall have a maximum phosphorus or COD content.

R. Klute / H.H. Hahn (Eds.)
Chemical Water and Wastewater Treatment III

The aim of this paper is to examine more fully the influence of aluminium sulphate and polyaluminium chlorides of different basicities when cleaning municipal wastewaters of two thicknesses to a constant turbidity or phosphorus concentration.

Materials and Methods

Wastewaters Used. This study was made on two waters, one collected at VEAS in Oslo, Norway and one collected in Helsingborg, Sweden. In this paper the Norwegian water is called thin and the Swedish water is called thick. Data of the waters are shown in the Table 1 below.

Table 1. Composition of the wastewaters

Date of collection (year–month–day) Place of collection	1992–03–05 Oslo	1993–10–08 Helsingborg
Time collected	Day mean sample	9.30 am
Turbidity [FTU]		
Non-sedimented sample	49	94
3 cm from the surface of sample sedimented for 10 minutes	–	35
Suspended solids [mg/l]		
Non-sedimented sample	–	100
Sedimented sample	–	34.5
pH	7.4	8.34
Alkalinity [mekv/l]	2.32	5.55
TOC [mg C/l]	30.8	–
P concentration [mg P/l]		
Non-sedimented sample	2.45	9.03
Sedimented sample		3.62
Ortho-P concentration [mg P/l]	1.00	1.80
Per cent of the total phosphorus as ortho-P		
Non-sedimented sample	40.8	20
Sedimented sample	–	49.7

Analysis of the wastewater from Norway showed that its Ca^{2+} and Mg^{2+} hardness was 5.38 °dH of which 21 % was caused by magnesium. The tap water in Helsingborg has a hardness of 5.5 °dH of which 10 % is caused by magnesium and the rest by calcium. Both the waters thus had about the same hardness.

Jar Test Procedure. All experiments were carried out in 1 liter glass beakers with Kemira's Flocculator 90. This flocculator can handle up to 6 glass beakers. Each glass beaker has a stirrer with an individual motor equipped with a tachometer.

The rotational speeds of the propeller and the time of rotation can be individually controlled for each beaker. The stirring devices were set so that each sample was treated identically.

The coagulants were mixed into the wastewaters in 10 seconds with a propeller speed of 350 rpm. Coagulation was carried out in 10 minutes with a propeller speed of 40 rpm. The samples were allowed to sediment for 10 minutes in Norway and for 20 minutes in Sweden. Samples for analysis were withdrawn 3 cm below the water surface with a plastic syringe.

Turbidity. Turbidity was measured with a Hach ratio turbidimeter. Before measuring the turbidity the samples were sieved through a brass sieve with openings of 60 micrometers. The reason for the sieving procedure is that an unpublished study has shown that the accuracy of the turbidity measurements increased substantially (the standard deviation of 15 repeated turbidity measurements decreased by 70 %) if the samples were sieved before the analysis. The reason for this effect is that it is mainly the small particles, with diameters around the wavelength of the light (0.4–0.7 micrometers) which cause the turbidity. Larger particles, say with diameters exceeding 50 micrometers, contribute little to the turbidity but, when sedimenting, block the passage of light in an irregular way and thereby to a large extent contribute to the imprecision of the analysis. Sieving the sample before measuring its turbidity thus eliminates this effect.

Suspended Solids. Suspended solids were analysed in duplicate by sieving 100 ml samples of cleaned water and 50 ml samples of non-coagulant treated water through dried and weighed Whatman GF/C filters. Only suspended solids in the Swedish water were measured.

TOC (Total Amount of Organic Carbon). Each sample was analyzed in duplicate with an ASTRO 2001 system 2 instrument. Only TOC analyses of the the Norwegian water were carried out.

Total Phosphorus. In Norway the colour developed after adding ascorbic molybdate reagent to samples oxidized with potassium peroxide sulphate was analysed at 880 nm. In Sweden Dr. Langes LCK603 method was used.

Ortho-phosphate. Ortho-phosphate was analysed in samples filtered through Whatman GF/C filters. In Sweden the samples were analysed by Dr. Langes method and in Norway non-oxidized samples were analysed using the ascorbic molybdate reagent.

Alkalinity. The alkalinity was analysed on samples filtered through a Whatman GF/C filter by titrating the samples with 0.05 M (Sweden) or 0.1 M (Norway) HCl until pH 4.5 was reached.

In Norway the titration was done automaticlly with a Radiometer titrator and in Sweden it was done manually. In Sweden only non-treated water and the samples with the lowest and the two highest coagulant doses were analysed for alkalinity.

Sludge Volume. The sludge volume was measured after 1 hour sedimentation in Imhoff cones.

Coagulants. Aluminium sulphate (ALS), produced by Kemira Chemicals, was used. Two samples of Kemwater PAX-14 with OH/Al ratios of 0.9–1.1 were used as polyaluminium chlorides. The other polyaluminium chlorides were prepared in the laboratory. The coagulants used and their OH/Al ratios are listed below:

Place of experiment	Molar OH/Al ratio of the coagulants
Norway	0.9; 1.9; 2.0; 2.1; 2.2; 2.34; 2.47
Sweden	0 (ALS); 1.1; 1.5; 1.8; 2.05; 2.15; 2.30; 2.44.

All polyaluminium chlorides were in the form of liquids with aluminium concentrations 8–12 weight %. They were added undiluted to the sewage waters with micropipettes with disposable plastic tips.

Results

Turbidity. Figure 1 illustrates the influence of the aluminium dose on the residual turbidity of the treated waters. It can be seen that the polyaluminium chlorides cleaned the water more efficiently than aluminium sulphate in terms of turbidity. The turbidity reducing effect of the polyaluminium chlorides increased with increasing basicity of the salts. Of the polyaluminium chlorides the ones with lowest basicity were the least efficient in decreasing turbidity.

Suspended Solids. Figure 2 illustrates the influence of the aluminium dose on the residual content of suspended solids in the Helsingborg water. For the sake of clarity, the results from only three coagulants are shown. Aluminium sulphate gave the highest, and polyaluminium chloride with an OH/Al ratio of 1.1 the second highest, content of residual suspended solids. The rest of the coagulants produced water with a residual concentration of suspended solids of about the same magnitude.

TOC (Total Amount of Organic Carbon). Figure 3 illustrates the influence of the polyaluminium chloride dose on the residual TOC content of the Norwegian water. Here also, for the sake of clarity, some of the curves are omitted. The best TOC removal was obtained with the coagulant with an OH/Al ratio of 2.34.

Phosphorus. Sewage contains phosphorus both in the form of ortho-phosphate and in the form of particles. Both the total phosphorus concentration and the concentration of ortho-phosphate was determined. The difference between these two figures constitutes the non-orthophosphate phosphorus, which is present mainly in the form of particles.

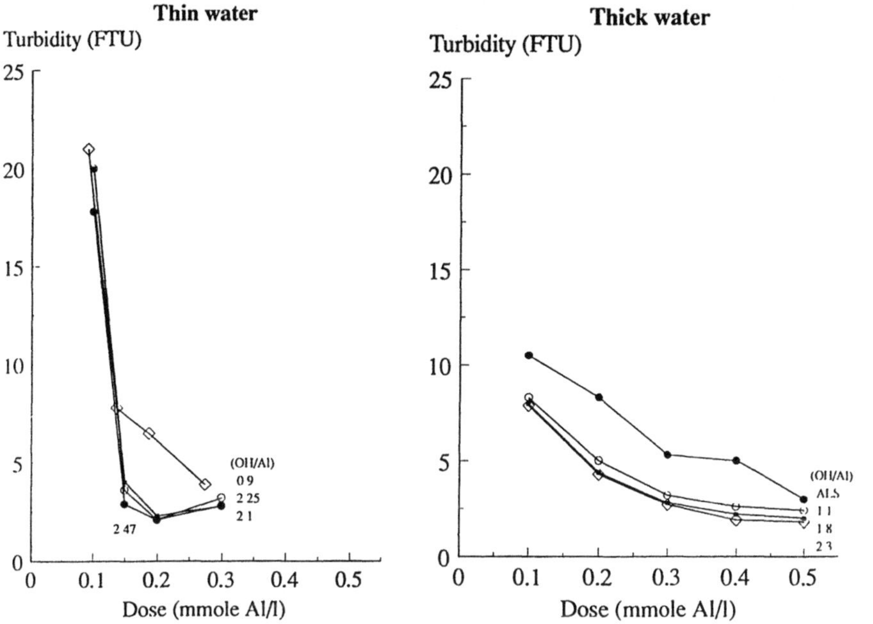

Fig. 1. The influence of the aluminium dose, given as different coagulants, on the residual turbidity of the waters

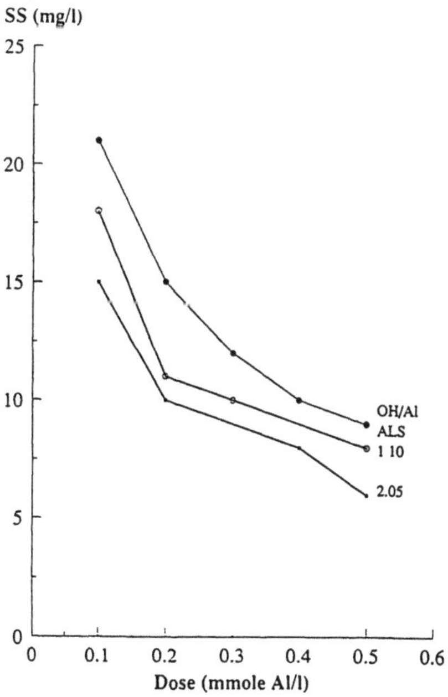

Fig. 2. The influence of the aluminium dose, given as aluminium sulphate and polyaluminium chlorides with molar OH/Al ratios of 1.10 and 2.05 on, the residual concentration of suspended solids in the Helsingborg wastewater

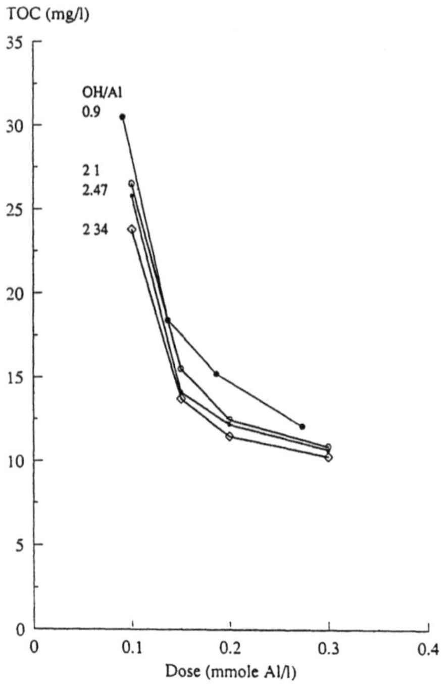

Fig. 3. The influence of the aluminium dose, given as polyaluminium chlorides with different molar OH/Al ratios, on the residual TOC concentration in the Norwegian wastewater

In order to understand the particle and ortho-phosphate reducing effect of the different coagulants it is useful to have an understanding of the chemistry of polyaluminium ions. A variety of aluminium and polyaluminium ions exist. The table below illustrates the composition and properties of four well described forms.

Table 2. The molar OH/Al ratio, basicity and average charge per aluminium atom and the ion charge for aluminium, dimeric and tridecameric aluminium ions

Ion formula	OH/Al	Basicity [%]	Average charge per aluminium atom	Ion charge
$Al(H_2O)_6^{3+}$	0	0	+3.00	+3
$Al(OH)(H_2O)_5^{2+}$	1	33.3	+2.00	+2
$Al_2(OH)_2(H_2O)_8^{4+}$	1	33.3	+2.00	+4
$Al_{13}O_4(OH)_{24}(H_2O)_{12}^{7+}$	2.46	82.1	+0.53	+7

The ions described above are only a fraction of the polyaluminium ions that exist. Aluminium ions may add to dimers and form polymeric structures (Brosset et al., 1954). The tridecameric ion (Al_{13}) can also polymerise into different structures (Bottero et al., 1988).

The influence of the coagulant dose on the per cent ortho-phosphate and non-orthophosphate removed from the Helsingborg water is illustrated in Figure 4. A number of interesting conclusions can be drawn from this figure.

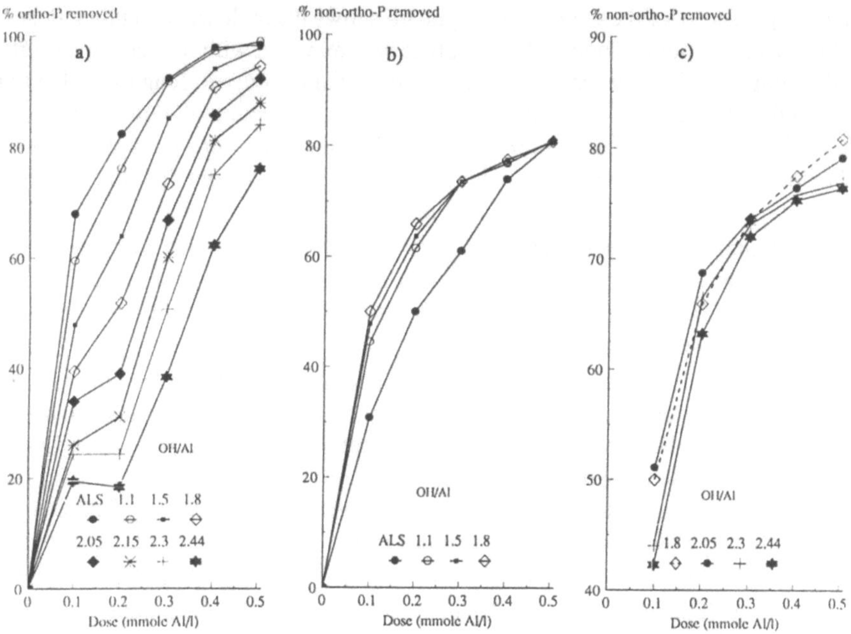

Fig. 4. The influence of the dose of coagulants with differentOH/Al ratios on the per cent removal of ortho- and non-ortho-phosphorus from the Helsingborg wastewater. The per cent removal of the non-ortho-phosphorus was calculated from raw wastewater which was sedimented for 10 minutes. The samplings were done 3 cm below the surface of the waters

In order to precipitate ortho-phosphate the negative charge of the ortho-phosphate has to be neutralised. From Table 2 it can be seen that the average positive charge of the aluminium atoms decreases with increasing basicity of the salts. This explains why ortho-phosphate was precipitated to a decreasing degree with increasing basicity of the salts, as can be seen from Figure 4a.

However, the difference in ortho-phosphate reducing capacity of aluminium sulphate and the polyaluminium chloride with an OH/Al ratio of 1.1 was not large. At the highest dose tested this polyaluminium salt even precipitated ortho-phosphate to a slightly higher extent than did aluminium sulphate. Unpublished studies made at our laboratory have showed that dialuminiumhydroxy phosphate is more insoluble than aluminiumhydroxy phosphate in the pH range of the experiment (almost neutral). The differences in phosphate solubility and in pH explain why aluminium sulphate and the polyaluminium salt with lowest basicity precipitated ortho-phosphate to about the same degree at the highest doses.

Figure 4b shows how the coagulant dose with OH/Al ratios from 0 to 1.8 removed the non-orthophosphate-P. (Without any coagulant added 25.2% of the total amount of non-ortho-P sedimented.) The largest proportion of the non-ortho-P is in the form of particles. According to Schultze-Hardy's rule, the removal of lyophobic particles is immensely increased (z^6) with increasing charge (z) of the coagulant. Table 2 shows that the charge of polyaluminium ions increases with increasing basicity of the coagulant. Figure 4b shows that at lower coagulant doses, an increasing amount of non-orthophosphate was removed with increasing basicity of the coagulants. However, at the highest coagulant dose, all coagulants having molar OH/Al ratios of 0–1.8 precipitated the same amount of non-ortho-P. At this high coagulant concentration most particles probably were precipitated. The behaviour described above agrees with the effects expected from Schultze-Hardy's rule.

A comparison of the slopes in Figures 4a and 4b shows that at some doses the aluminium sulphate and polyaluminium chlorides with lowest basicities had similar slopes. However, with increasing basicity of the polyaluminium chlorides, the slopes at lower doses differed markedly.

This means that aluminium sulphate and polyaluminium chlorides with OH/Al ratios in the range of about 0–1.5 precipitate ortho-phosphate and particles to about the same degree. However, when the OH/Al ratio of the polyaluminium salt is 2.05 and higher, the coagulant at low doses preferentially precipitates particles. At higher doses, ortho-phosphate is precipitated to an increasing degree. The OH/Al range with this effect corresponds to the range where the polyaluminium ions contain an increasing amount of polymerized Al_{13} (Bottero and Bersillon, 1989).

Figure 4c illustrates the removal of non-ortho-P by the coagulants with the highest basicities. The result of the coagulant with an OH/Al ratio of 1.8 is illustrated both in Figure 4b and as a broken line in Figure 4c and it is clear that this coagulant was the most effective in precipitating non-ortho-phosphate. Coagulants with higher basicity precipitated the non-ortho-P material to a decreasing degree. This implies that the charge of the coagulant decreases with OH/Al ratios above 1.8. This agrees with the findings of Bottero and Bersillon (1989) that the charge of polymerized Al_{13} ions decreases with increasing OH/Al ratios.

A similar analysis was done on the Norwegian water, see Figures 5. Orthophosphate was precipitated more efficiently with decreasing OH/Al ratio of the coagulant in this water as well. The major proportion of the non-ortho-phosphorus containing material was precipitated with the lowest coagulant doses when the OH/Al ratios were 2.34–2.47. The amount of non-orthophosphate precipitated went through a maximum. In this thin water, the highest coagulant dose used was thus so large that some material was restabilised.

In most wastewater treatment plants the total phosphorus concentration is a more interesting parameter than the two ones mentioned above. Figure 6 shows the influence of the coagulant dose on the per cent of the total phosphorus removed in the two waters. There seems to be a range of OH/Al ratios inside which coagulants precipitate phosphorus most efficiently. This range seems to differ from water to water. The OH/Al ratio that precipitated the total amount of phosphorus to the largest extent and the coagulant OH/Al ranges producing good removal of phosphorus is presented in Table 3.

It seems that the thickness of the waters treated had an influence on the type of coagulant that most efficiently reduced the total phosphorus concentration of the waters. The total phosphorus in the water with the highest original phosphorus

Table 3. The OH/Al ratio of the coagulant and of the three coagulants that most efficiently precipitated the total amount of phosphorus in the waters and the original phosphorus concentration of the two waters

Water	mg P/L originally present in water	OH/Al range giving good removal of phosphorus	Best OH/Al
Norwegian	2.45	1.9–2.1	1.9
Swedish	9.03	0–1.5	1.1

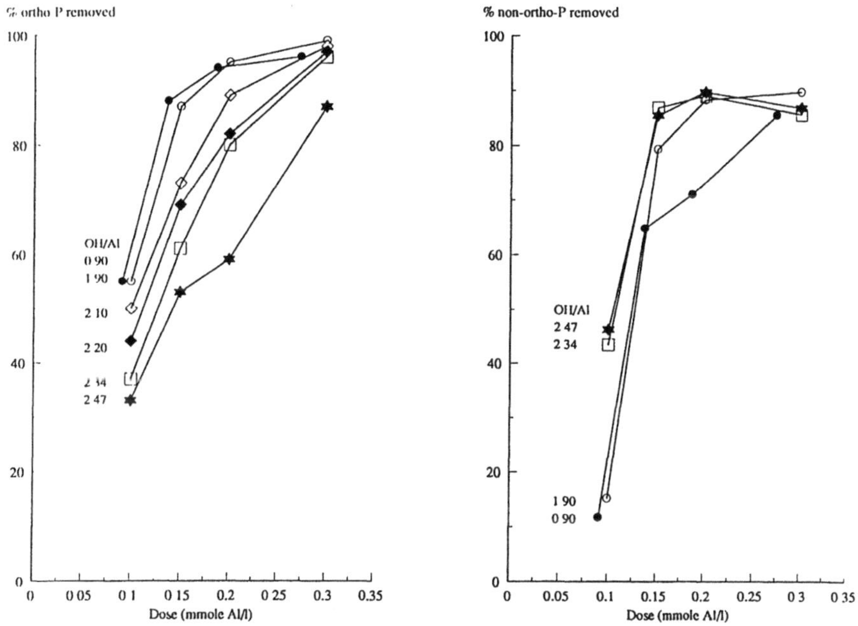

Fig. 5. The influence of the coagulant dose on the per cent removal of ortho-and non-ortho-phosphate in the thin wastewater. The per cent removal of the non-ortho-phosphate was calculated based on the concentration of non-ortho-phosphate in non-sedimented raw wastewater

concentration was most efficiently removed by the coagulant with an OH/Al ratio of 1.1 while the phosphorus in the water with the lowest original phosphorus concentration was most efficiently removed with the coagulant with an OH/Al ratio of 1.9.

Alkalinity. Due mainly to an increased use of fossil fuels, our planet is subjected to acidification. In some places wastewater must be cleaned with a low alkalinity consumption.

The alkalinity consumed is due to the formation of hydrogen ions when the coagulant reacts with water. The number of hydrogen ions which can be formed when a mole of an (poly)aluminium salt is added to water is proportional to the

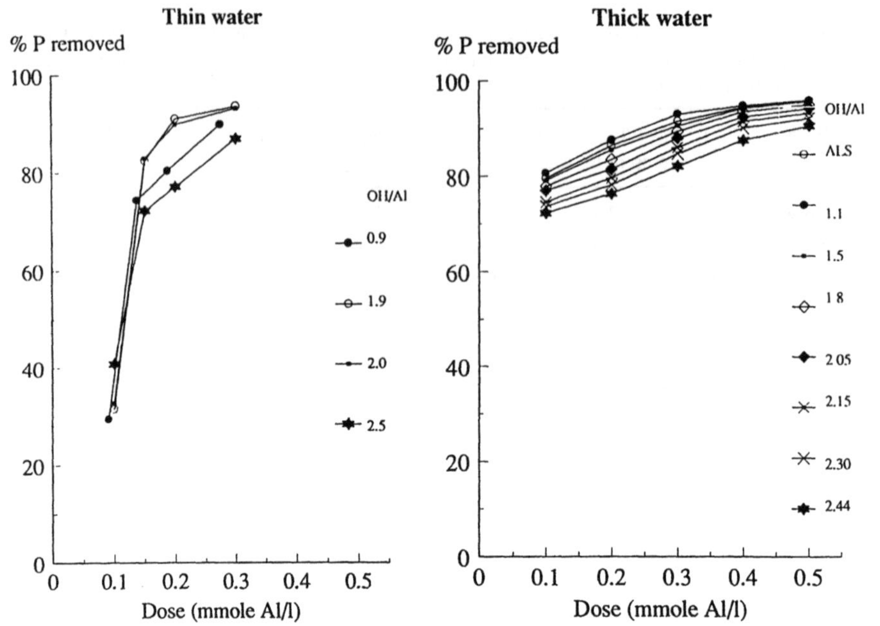

Fig. 6. The influence of the aluminium dose, given as aluminium sulphate or polyaluminium chlorides with different OH/Al ratios, on the per cent removal of phosphorus in the two wastewaters. The per cent removed was calculated based on the phosphorus concentration in non-sedimented raw wastewater

average charge of the (poly)aluminium atoms. With increasing basicity of the coagulant, the lower the average positive charge of the atoms of the coagulant and

- the fewer the hydrogen ions that can be formed when the coagulant reacts with water, i.e. the lower the alkalinity consumed and
- the less ortho-phosphate that can be neutralised and precipitated.

Therefore, as can be seen from Figure 7, the alkalinity consumption decreases with increasing basicity of the salt. Decreasing ortho-phosphate precipitation and alkalinity consumption are related to increasing basicity of the coagulant, as can be seen by comparing Figures 4, 5 and 7.

The alkalinity consumption is caused by the reactions between water and coagulant molecules. No alkalinity is consumed when the coagulant reacts with particle surfaces or forms precipitates with ortho-phosphates or other soluble substances. The thicker a sewage water, the less alkalinity is consumed by a coagulant. From Figure 7 it can be seen that the thickest water consumed the lowest amount of alkalinity per dose of coagulant added.

At low coagulant doses the likelihood of coagulant molecules reacting with ortho-phosphate or particle surfaces is relatively large. At increasing coagulant doses there are fewer and fewer particle surfaces or ortho-phosphate molecules left to be neutralised. The coagulant molecules then react with water to an increasing degree and the alkalinity consumption increases. Figures 8 and 9 show

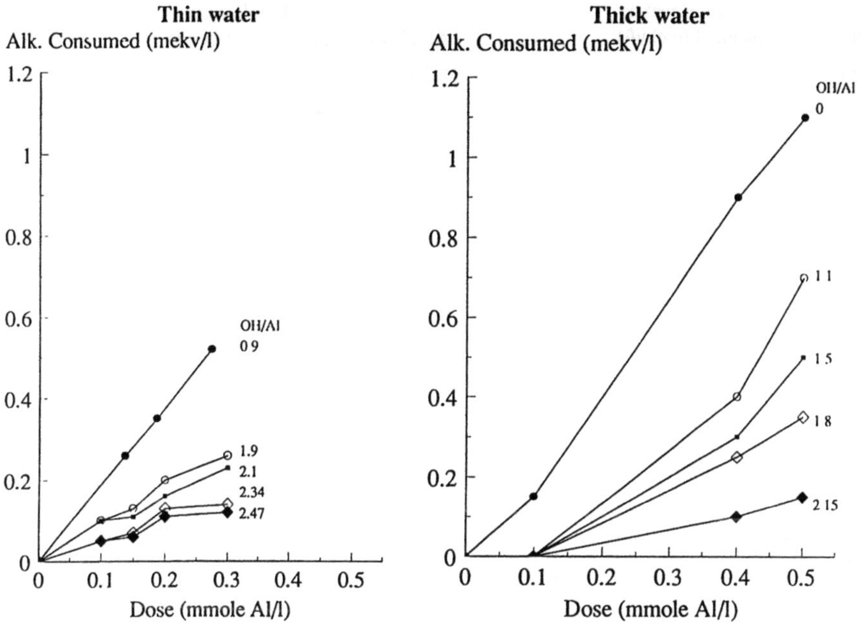

Fig. 7. The influence of the aluminium dose, given as aluminium sulphate or polyaluminium chlorides with different OH/Al ratios, on the alkalinity consumed in the two waters studied

the influence of the residual turbidity and TOC content of the waters on the alkalinity consumed. From the figures it can be seen that at low coagulant additions the turbidity or TOC is decreased to a substantial degree without or with low alkalinity consumption. The same phenomenon was observed with the effect of residual phosphorus concentration on the alkalinity consumption. It can be seen from the figures that there seems to be a threshold concentration below which little alkalinity is consumed. Above the threshold concentration, the alkalinity consumption increases with increasing coagulant dose.

In conclusion, therefore, if the alkalinity consumption should be kept to a minimum then:

1) a coagulant with high basicity should be used and
2) the coagulant should not be overdosed.

In each wastewater plant the coagulant is added in such an amount that legal requirements are fullfilled. These requirements can be limits on COD, TOC or total-phosphorus concentration, for example.

In Sweden and Norway the coagulant is often dosed so that the treated water has a phosphorus concentration of 0.5 mg P/l. In central Europe higher phosphorus concentrations are permitted.

The type of diagrams illustrated in Figures 8 and 9 can be used as a basis to illustrate the influence of the OH/Al ratio of the coagulants on the alkalinity consumed when the waters are cleaned to a constant TOC concentration or turbidity.

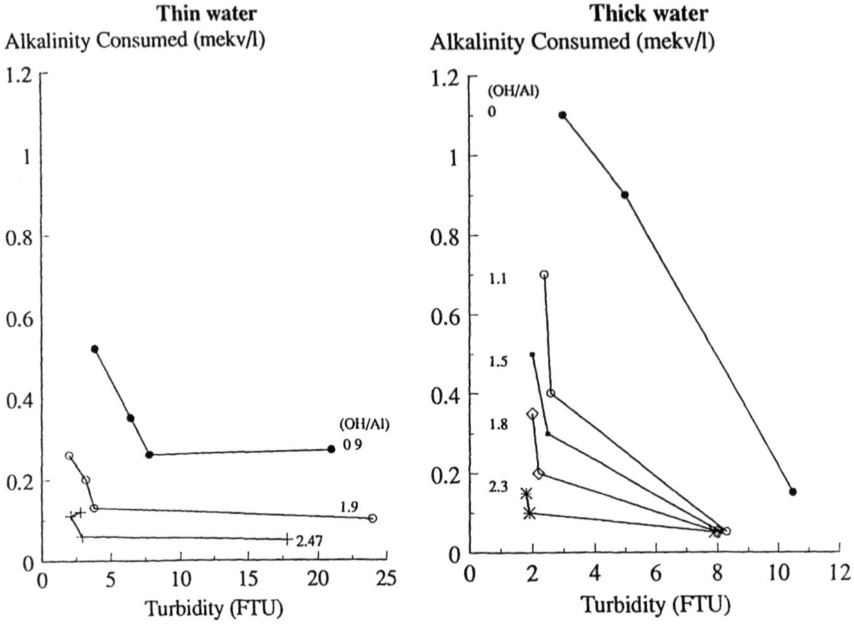

Fig. 8. The influence of the residual turbidity of the two waters treated with aluminium sulphate and polyaluminium chlorides with different OH/Al ratios on the amount of alkalinity consumed by the coagulants

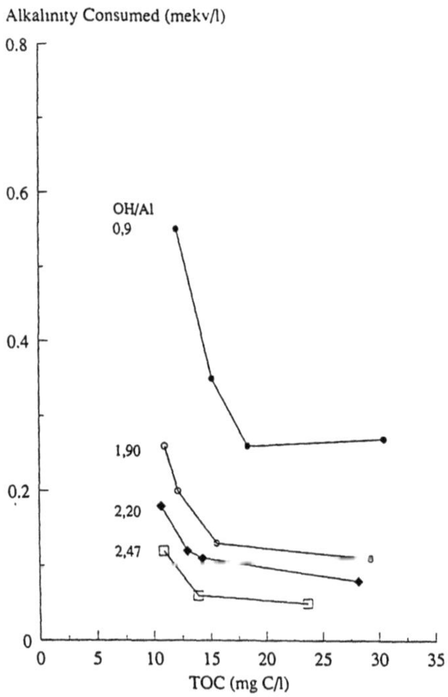

Fig. 9. The influence of the residual TOC concentration on the alkalinity consumed when the thin wastewater was treated with different doses of polyaluminium chlorides with different OH/Al ratios

This type of analysis was done from hand drawn curves. Figure 10 shows the influence of the OH/Al ratio of the coagulant on the alkalinity consumed when the wastewaters were cleaned to phosphorus concentrations of 0.5 and 1.0 mg P/l respectively. It can be seen that the lowest alkalinity was consumed by the coagulants with the highest basicity. Figure 11 shows that the lowest alkalinity was also consumed by the highest basicity coagulant when the water was cleaned to a constant residual TOC content. The same was the case when water was cleaned to a constant turbidity. All these results indicate that a substantial amount of alkalinity is saved when a coagulant with low basicity is replaced by a coagulant with high basicity.

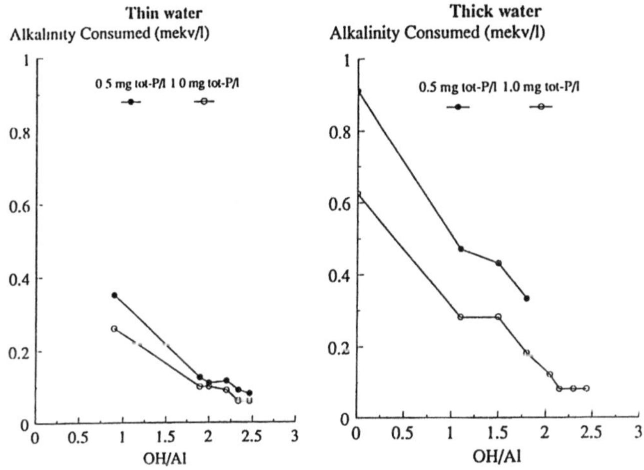

Fig. 10. The influence of the OH/Al ratio of the coagulant used on the alkalinity consumed when the two wastewaters were treated to a residual phosphorus concentration of 0.5 (solid points) and 1.0 mg P/l (open circles), respectively

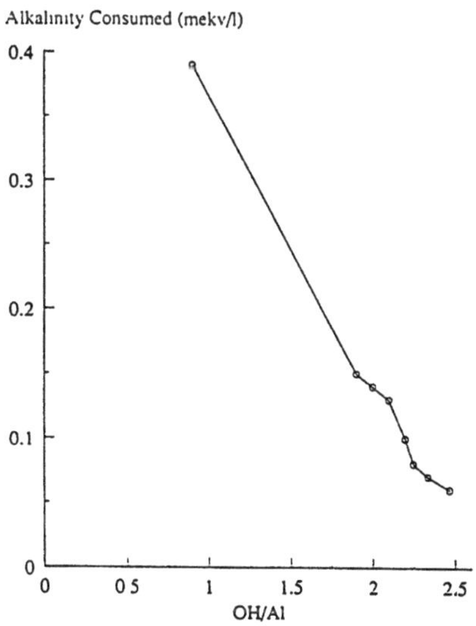

Fig. 11. The influence of the OH/Al ratio of the polyaluminium chlorides used on the alkalinity consumed when treating the Norwegian wastewater to a residual TOC concentration of 14 mg C/l

Sludge. Sludge handling consumes a considerable part of the total costs of a wastewater treatment plant. It is therefore of great interest to understand the influence of different parameters on the volume of sludge produced.

Figure 12 illustrates the influence of the residual turbidity of the treated waters on the sludge volume produced. The lines in the figure connect the effects of increasing coagulant doses. It can be seen that the sludge volume strongly depends on the thickness of the water. With increasing thickness of the water, the larger the volume of sludge produced. With decreasing residual turbidity of the cleaned water the sludge volume generally increased. There seems to be a coagulant dose above which the coagulant cannot clean the water to lower turbidity. If the coagulant dose exceeds that dose, the sludge volume increases drastically.

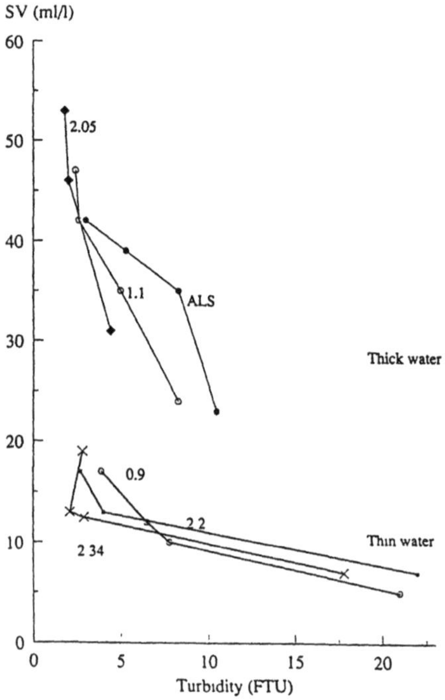

Fig. 12. The influence of the residual turbidity of the two wastewaters on the sludge volume produced when they were cleaned by aluminium sulphate and polyaluminium chlorides with different OH/Al ratios

Figure 13 illustrates the influence of the residual phosphorus concentration of the treated waters on the volume of sludge produced. It can be seen that both the thickness of the water and the basicity of the coagulant had a large influence on the sludge volume produced. These phenomenoma were especially pronounced with the thickest water.

Figure 14 shows the influence of the OH/Al ratios of the coagulants when they were added in such an amount that a constant phosphorus concentration of 0.5 mg P/l was reached. It can be seen that in the thin water the sludge volume was only slightly influenced by the basicity of the coagulant used. When the OH/Al ratio of the coagulant was larger than 2.2, a somewhat increased sludge volume was obtained.

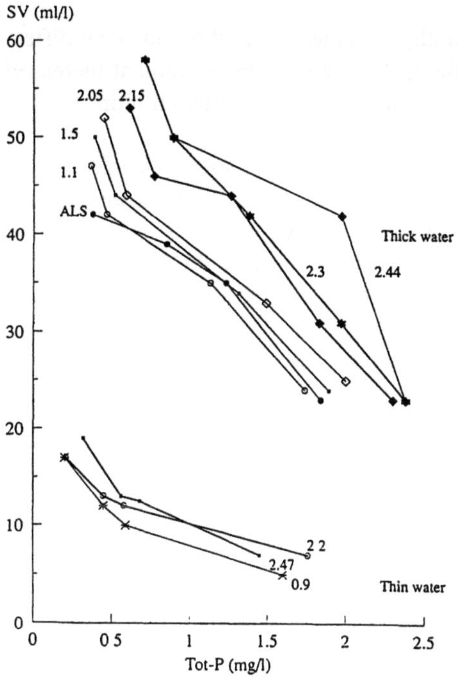

Fig. 13. The influence of the residual phosphorus concentration on the sludge volume produced when the two wastewaters were treated with aluminium sulphate and polyaluminium chlorides with different OH/Al ratios

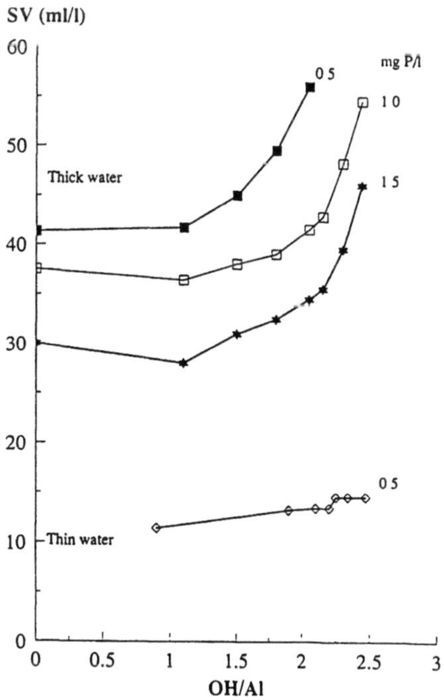

Fig. 14. The influence of the OH/Al ratio of the coagulant used on the sludge volume formed when cleaning the wastewaters to a constant phosphorus content of 0.5 mg P/l and the thickest water to a residual phosphorus concentration of 1.0 and 1.5 mg P/l, respectively

In the more concentrated water the sludge volume followed a somewhat different pattern. It decreased slightly when the OH/Al ratio of the coagulant increased from 0 to 1.1. At higher OH/Al ratios the sludge volume increased more significantly with increasing OH/Al ratios.

In Denmark and central Europe a total residual phoshorus concentration of 1.5 mg P/l is often required. Figure 14 shows that, in this case, the lowest sludge volume was produced by the coagulant with an OH/Al ratio of 1.1, the basicity of the standard coagulants Kemwater PAX-14 and Kemwater PAX-18.

Figure 15 illustrates the influence of the OH/Al of the coagulants when the waters were treated to a constant residual turbidity of either 3 or 6 FTU. It can be seen that sludge volume decreased with increasing basicity of the coagulant used. There was one exception, and that was when the thick water was treated to a turbidity of 3 FTU.

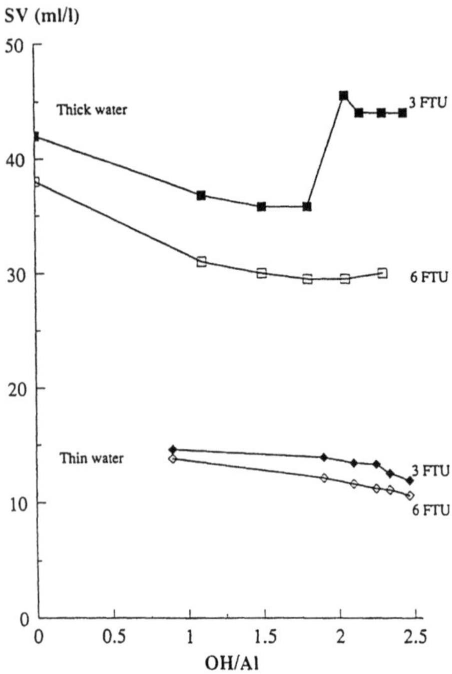

Fig. 15. The influence of the OH/Al ratio of the coagulant used on the sludge volume produced when cleaning the wastewaters to a constant turbidity of 6 (open symbols) and 3 (filled symbols) FTU, respectively

Summary

The influence of the basicity of polyaluminium chlorides when cleaning two wastewaters was studied.

Turbidity and TOC were removed to an increasing degree with increasing basicity of the coagulant. However, when the salts had OH/Al ratios of around 2 and higher the increase in basicity of the coagulant only had a marginal effect on removing turbidity and TOC.

Aluminium sulphate and polyaluminium chlorides with OH/Al ratios up to 1.8 were found to precipitate ortho-P and particles simultaneously. Polyaluminium chlorides with basicities of 2.1 and higher were found to precipitate particles preferentially at low doses. At higher doses ortho-P was precipitated to an increasing degree.

The optimal basicity of the coagulant reducing the total phosphorus concentration to the greatest extent was influenced by the thickness of the water, as determined by the total original concentration of phosphorus in the water. The coagulant with an OH/Al ratio of 1.1 was the most effective in precipitating phosphorus from the thickest water. The coagulant with an OH/Al ratio of 1.9 was the most effective from precipitating phosphorus in the thinnest water.

The alkalinity consumed strongly decreased with increasing basicity of the coagulant used. The substitution of a high basicity coagulant for a low basicity coagulant thus decreases the alkalinity consumed substantially.

Both for the volume of sludge formed and the alkalinity consumed, there seems to be a threshold concentration of coagulant for each water. If the coagulant is dosed in an amount up to the threshold concentration, the water is cleaned of turbidity and phosphorus without a substantial alkalinity consumption and without a substantial production of sludge. However, if the threshold concentration is exceeded, the alkalinity consumed as well as the amount of sludge produced will increase significantly with the dose used.

The volume of sludge produced depends of the thickness of the water, the basicity and the dose of coagulant added.

When waters were cleaned to a constant turbidity, the sludge volume decreased somewhat with increasing basicity of the coagulant.

When thin water was cleaned to a phosphorus concentration of 0.5 mg P/l, the basicity of the coagulant seemed to play only a small role in the sludge volume formed.

When the thick water was cleaned to a phosphorus concentration of 0.5 mg P/l, the lowest sludge volume was obtained with coagulants having OH/Al ratios in the range 0–1.1. The sludge volume then increased with increasing basicity of the coagulant.

When the thick water was cleaned to a phosphorus concentration of 1.5 mg P/l, the lowest sludge volume was produced by the coagulant with an OH/Al ratio of 1.1.

Aluminium sulphate gave a somewhat higher sludge volume and the sludge volume increased with OH/Al ratios above 1.1.

Table 4 summarizes the effects of the coagulants on the two waters:

Table 4. Optimal OH/Al ratios of coagulants for various parameters

	Range of optimal OH/Al ratios	
	Thin water	Thick water
Removal of		
turbidity	2.1–2.5	2.05–2.44
SS	–	1.5–2.44
TOC	2.34	–
Ortho-P	Lowest tested	Lowest tested
Non-ortho-P	2.25–2.47	1.8–2.05
Total P	1.9	1.0
Lowest alkalinity consumption	Highest tested	Highest tested
Lowest sludge volume at a residual P conc.		
of 0.5 mg P/l	2.2 or less	1.1 or less
of 1.5 mg P/l	–	1.1
Low sludge volume at a constant residual turbidity of		
6 FTU	Highest tested	1.8–2.1
3 FTU	Highest tested	1.5–1.8

Acknowledgements. Skillful laboratory assistance by Pia Ryrfors, Oliva Lhota, Adina Dumitresco and the VEAS analytical laboratory, and help in the drawing of all the figures by Fredrik Arthursson, is gratefully acknowledged.

References

Bottéro, J.Y., Poirier, J.E., Fiessinger, F.: Study of Partially Neutralized Aqueous Aluminium Chloride Solutions: Identification of Aluminium Species and Relation Between the Composition of the Solutions and Their Efficiency as a Coagulant. Prog. Wat. Tech. Toronto *12* (1980) 601–612

Bottéro, J.Y., Tchoubar, D., Cases, J.M., Fripiat, J.J., Fiessinger, F.: New Development in Knowledge of Aluminum Colloids, Interfacial Phenomena. In: Biotechnology and Materials Processing, Ed. by Y. Attia, B.M. Moudgil, and S. Chander. Elsevier Science Publisher, 1988, pp. 459–479

Bottéro, J.Y., Bersillon, J.L.: Aluminum and Iron (III) Chemistry Some Implications for Organic Substance Removal. Adv. Chem. Ser., Aquatic Humic Substances *219* (1989) Chap. 26, pp. 425–442

Brosset, C., Biedermann G., Sillén, L.G.: Studies of the Hydrolysis of Metal Ions; XI. The Aluminium Ion. Acta Chem. Scand. *8* (1954) 1917–1926

Fiessinger, F., Bersillon, J.L.; Prépolymérisation de l'hydroxyde d'aluminium pour la coagulation des eaux. CEBEDEAU-BECEWA *399* (1977) 52–68

Ødegaard, H., Fettig, J., Ratnaweera, H.C.: Coagulation with Prepolymerized Metal Salts. In: Chemical Water and Wastewater Treatment, H.H. Hahn and R. Klute (Eds.). Springer, Berlin Heidelberg New York 1990, pp. 189–220

Lars Gillberg
Kemira Kemi AB
Box 902
S-251 09 Helsingborg
Sweden

Inorganic Polymer Flocculant Polyferric Chloride, Its Properties, Efficiency and Production

H. X. Tang, B. Z. Tian, Z. K. Luan, and Y. Zhang

Abstract

This paper summarizes a series of research projects in State Key Laboratory of Environmental Aquatic Chemistry (SKLEAC).

The chemical species in polyferric solutions were examined by timed complexation spectrophotometry to classify them as Fe_a, Fe_b and Fe_c. The hydrolysity B^* taken as a principal parameter is emphasized. The classification and evolution patterns of hydrolyzed Fe(III) solutions are examined.

Short term precipitation in polyferric chloride solutions can be retarded by addition of a sort of phosphate stabilizer which contributes to this flocculant both a commercial possibility and a higher efficiency.

To examine the reuse of municipal waste water, besides jar tests in the laboratory, pilot tests of advanced treatment after the biological oxidation process were carried out. The polyferric chloride flocculant was used and compared with other coagulants. A high efficiency of removal of turbidity and color was obtained.

Introduction

Inorganic polymer flocculants have been developed over several decades and can be substituted for the traditional kinds of inorganic coagulants. They include mainly cationic polyaluminium chloride or sulfate, polyferric chloride or sulfate, anionic polysilicate and even composites of these types. Among them, polyferric chloride still cannot be produced on an industrial scale as a commercial product, though its high efficiency in water treatment has long been demonstrated in laboratories. The limitation may be caused by its instability, which makes it precipitate after a few weeks. In fact, the speciation and reaction mechanisms of Fe(III) in solution through continual complex processes have not yet been studied thoroughly, especially under concentrated flocculant production conditions.

This paper summarizes a series of research projects, carried out in our laboratory on the principal processes in polymeric ferric solution, polyferric chloride production and its technological application in pilot scale tests. A few points of view are given as well.

R. Klute / H.H. Hahn (Eds.)
Chemical Water and Wastewater Treatment III

Principal Parameters

The polynuclear hydroxy complexes of iron or polyferricpolymers are a system of intermediate compounds produced in the Fe(III) solution by the continual processes of hydrolysis, complexation, polymerization, solation, precipitation and gelation. The common formula of the species may be written as $Fe_x(OH)_y^{(3x-y)+}$, where x is the polymerization of the species, y is the number of OH groups bound in structure and $3x - y$ is their charge. However, their actual forms and their distribution in unsaturated and mainly oversaturated ferric hydroxide solution are still a big problem to be solved.

The formation function of complexes can be used as the principal constitutional parameter of hydroxy ferric species:

$$F = y/x\,. \tag{1}$$

For any single species, the definition of this parameter is the ligand number of the ferric complexes or the OH groups per Fe atom combined in the structure of polymeric species. It can also be used to consider all the species in solution as a whole [1]. Then the parameter F is only an average apparent value because the ferric species in solution are quite different in form. This was suggested by Tang and Stumm [2] as the principal parameter to characterize the changing processes for hydrolysis and polymerization of hydroxy complexes and could be expressed as:

$$B^* = [OH]_b/\,[Fe_T]\,. \tag{2}$$

In the molar concentration ratio and on average, the $[Fe_T]$ refers to all the Fe(III) atoms in solution, and $[OH]_b$ means all the hydroxy groups combined into the structure of Fe(III) compounds. The parameter can also characterize the progressive extent of hydrolysis of ferric ions in solution and may be called hydrolysity. Its value varies from 0 to 3 as the ions are progressively hydrolyzed from Fe^{3+} to $Fe(OH)_3$.

In conventional studies of hydroxy polymers, the operational parameter basicity B is often used as the principal parameter,

$$B = [OH]_a/[Fe_T]\,, \tag{3}$$

where $[OH]_a$ is the molar concentration of base added into solution in operation. Because the spontaneous hydrolysis processes already present in metal solution before the basic reagents are added and relatively decrease as more base is added, there is a difference in value between basicity B and hydrolysity B^*. This difference is significant and must be considered for Fe(III) solutions in low pH, because their spontaneous hydrolysis tendency is very strong.

Tang and Stumm [2] suggested the formula for calculating hydrolysity as:

$$B^* = B_H + B - A = \left(10^{-pH} + C_B - C_A\right)/[Fe_T]\,, \tag{4}$$

where $B_H = 10^{-pH}/[Fe_T]$ is taken as the part of spontaneous hydrolysity before or remaining after the addition of base, and C_B and C_A are base and acid, respectively, added in the operation.

In the above expressions, the complexation of the ferric ion with other anions besides OH is neglected. That should be considered where sulfate or phosphate are concerned.

The priority of hydrolysity B^* over basicity B is not only a more accurate calculation which includes the part of spontaneous hydrolysis, it is a conceptual transform. Basicity B is an operational fixed parameter while hydrolysity B^* in the sense of an average formation function is a structural and dynamic parameter usable at any time in a much broader sense.

(1) In the pure Fe(III) salt solution of various concentrations the spontaneous hydrolysis process progresses already before any base is added. The low pH value of the concentrated solution is the result of the release of H^+ in the hydrolysis reaction. While $B = 0$, hydrolysity B^* can characterize the extent of the spontaneous hydrolysis (Fig. 1). Dilution of the Fe(III) solution changes the extent of hydrolysis, having an effect similar to adding a very weak base.

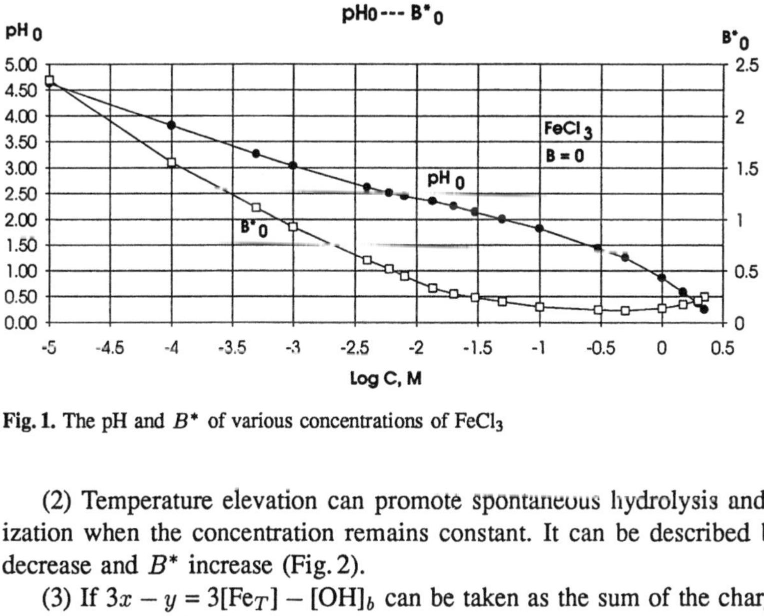

Fig. 1. The pH and B^* of various concentrations of $FeCl_3$

(2) Temperature elevation can promote spontaneous hydrolysis and polymerization when the concentration remains constant. It can be described by the pH decrease and B^* increase (Fig. 2).

(3) If $3x - y = 3[Fe_T] - [OH]_b$ can be taken as the sum of the charges for all the hydroxy species in solution, then the specific charge

$$R = (3x - y)/x = 3 - F = 3 - B^* \tag{5}$$

would be the average charge per Fe atom contained in ferric species. Obviously, as the hydrolysity B^* increases, the specific charge R will decrease correspondingly, though the apparent charge of most single polymeric species can be much higher than R.

(4) The solution oversaturated with respect to ferric hydroxide would be thermodynamically unstable. The hydrolysis and polymerization processes will continuously progress forward in time. This tendency is much stronger after adding

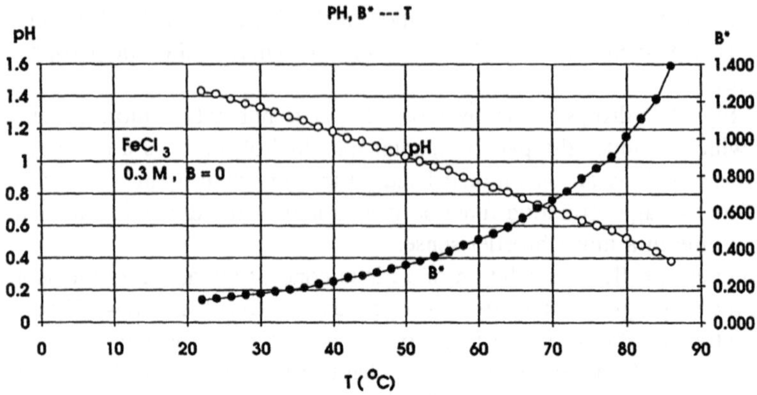

Fig. 2. The pH and B^* of 0.3 M $FeCl_3$ solution at elevated temperatures

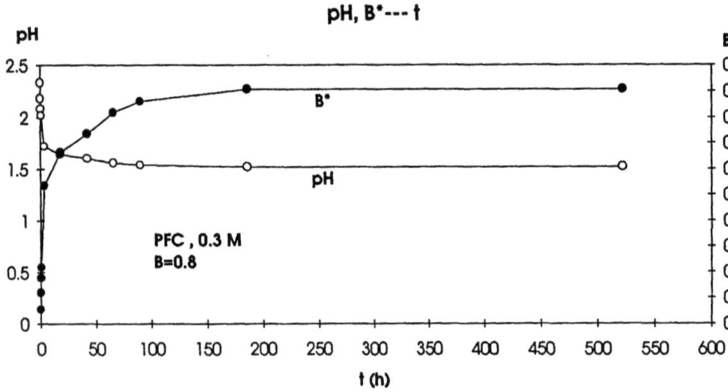

Fig. 3. The pH and B^* evolution of PFC solution with ageing time

base. The evolution can be traced by the parameter hydrolysity B^* at any time but the basicity B can only mark the initial state (Fig. 3).

In conventional literature in this field, the operational basicity B was used by many authors for various initial concentrations as the sole principal parameter. Their data are difficult to compare accurately to get a unified quantitative model.

Speciation and Pattern

The composition and transformation of hydrolyzed Fe(III) complexes and polymeric species have been investigated by a large number of authors [4–8] and summarized in many papers [9–11].

According to the suggested concept of hydrolysity B^*, Tang and Stumm divided the Fe(III) solution into five types according to the starting B_0^* for both the pure Fe(III) salt solutions and those that have been partially neutralized with base, regardless of their different preformation methods and non-equilibrium conditions. For each type of solution, the common pattern of pH evolution and its coagulating behavior was also presented [2, 3]. More data describing pH evolution in the long term ageing process are listed in detail as Table 1.

Table 1. The pH evolution of 0.3 M solution with various OH/Fe ratios

B	After preparation		Induct. time	Precipi-tation	Average rate of pH decrease (units/s)			After month	Type	B_0^*
	pH_0	B_0^*		in month	ini.	week	month	pH		
−0.34	0.88	0.102	long	no	−0.10/5 h	0	0	0.98	A	0.0–0.2
−0.1	1.22	0.101	long	no	−0.03/3 h	0	0	1.25		
0.0	1.43	0.124	long	no	−0.03/2 h	0	0	1.46		
0.1	1.65	0.175	month	no	0.01/1 h	0.0006	0	1.61		
0.2	1.79	0.254	36 h	94 h	0.01/1 h	0.027	0.007	1.49	B	0.2–0.5
0.3	1.89	0.343	18 h	46 h	0.01/1 h	0.046	0.006	1.48		
0.4	2.02	0.432	3 h	12 h	0	0.070	0.005	1.46		
0.5	2.15	0.524	0	no	0.35/3 h	0.091	0.003	1.46	C	0.5–1.0
0.6	2.31	0.616	0	no	0.46/1 h	0.113	0.001	1.50		
0.7	2.32	0.716	(–)	no	0.47/1 h	0.114	0.001	1.51		
0.8	2.33	0.816	(–)	no	0.48/1 h	0.113	0.001	1.52		
0.9	2.34	0.915	(–)	no	0.49/1 h	0.107	0.001	1.54		
1.0	2.16	1.023	(–)	no	0.46/1 h	0.084	0.001	1.56		
1.5	2.28	1.517	(–)	no	0.40/1 h	0.100	0	1.58	D	1.0–2.3
2.0	2.43	2.012	(–)	no	0.43/1 h	0.103	0	1.69		
2.3	2.85	2.304	(–)	2 weeks	0.45/1 h	0.107	0	2.04		
> 2.3	Precipitated in preparation (hour)								E	> 2.3

The speciation and structures of Fe(III) hydrolyzed species may be studied by infrared, Mossbauer, laser light-scattering spectroscopies and many other chemical and physical methods. However, the most simple and practical method is Ferron complexation timed spectrophotometry. The reagent Ferron can react with the Fe(III) monomer and different polymers in various types and rates. Depending on the progressive absorption (600 nm, pH 5) with time (3 h), the species of Fe(III) in solution can be divided into three categories: Fe_a, Fe_b, Fe_c (Fig. 4). Fe_T is determined after deep acidification of the solution.

$$Fe_c = Fe_T - Fe_a - Fe_b . \tag{6}$$

The species which reacted with Ferron immediately at $t = 0$ are considered to be Fe_a, which is assumed to include the free ion Fe^{3+}, monomers $Fe(OH)^{2+}$,

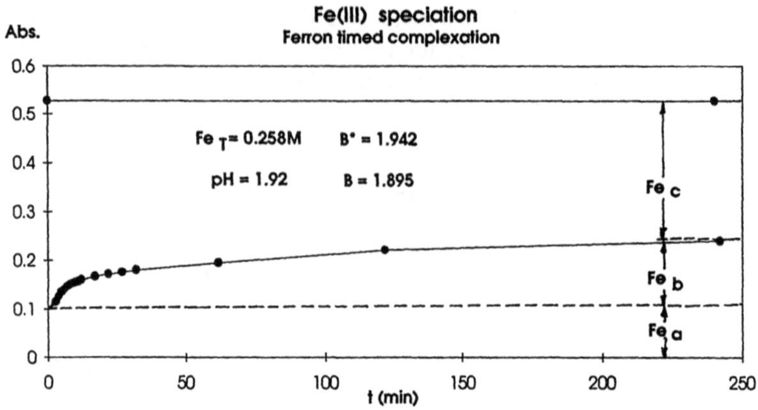

Fig. 4. Ferron timed complexation spectroscopy method for determination of Fe(III) species

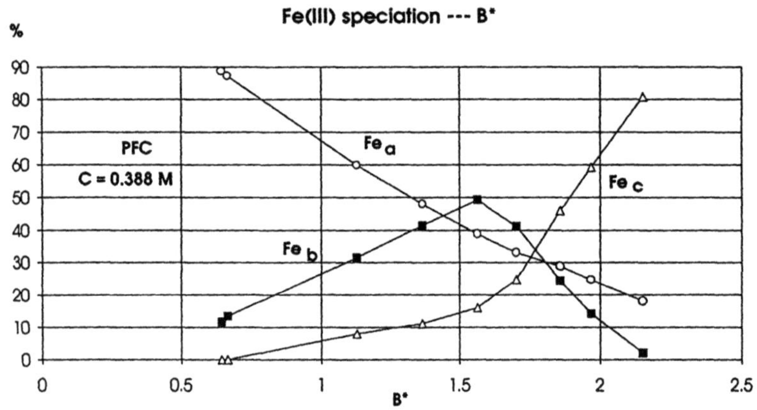

Fig. 5. Distribution of Fe(III) species as a function of B^*

$Fe(OH)_2^+$ and dimers $Fe_2(OH)_2^{4+}$. After about 2–3 hours the absorption curve levels off. The species which reacted with Ferron during this period of time are considered to be Fe_b, which is assumed to be the active part of hydroxy ferric polymers like $Fe_x(OH)_y^{(3x-y)+}$ or $Fe_xO_z(OH)_y^{(3x-2z-y)+}$ with x in the range of 3 to dozens. The species that do not react with Ferron before the limited 3 hour period are considered to be Fe_c, which is assumed to be the inert part of high polymers and fresh precipitates like $Fe_xO_z(OH)_y^{(3x-2z-y)+}$ and xFeOOH, xFe(OH)$_3$(am) with x on the order of hundreds and thousands.

The distribution of ferric species between the categories Fe_a, Fe_b, Fe_c as a function of B^* is shown in Figure 5. The transformation of the species in the ageing period in days is shown in Figure 6 and Table 2. From these tables, figures and other experimental data the following patterns can be recognized:

(1) Fe_a decreases from the dominant species to play a minor role as the B^* increases from 0 to above 2.0. However, for each B_0^* solution, the amount of Fe_a becomes constant after the initial reduction in 24 hours.

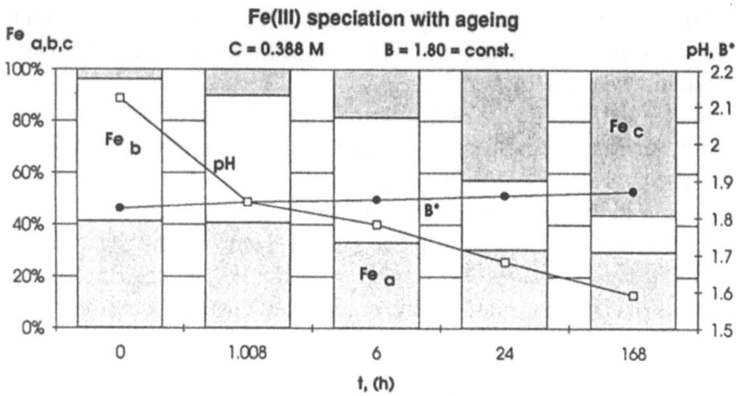

Fig. 6. The changing distribution of Fe(III) species with ageing time

Table 2. Specification of Fe(III) solution with ageing time ($C = 0.1$ M)

B	Species	Ageing time [h]				
OH/Fe	Fe *a*, *b*, *c*	1–2	7–8	24–25	120	288
0.4	*a*	95.54	78.30	76.46	–	70.71
	b	0.00	1.96	2.32	–	0.00
	c	4.46	19.65	21.22	–	29.29
1.0	*a*	66.07	57–68	56.07	53.93	56.79
	b	10.71	7.30	2.86	1.70	0.00
	c	23.22	34.82	41.07	44.28	43.21
1.5	*a*	40.18	38.39	38.21	37.50	37.50
	b	25.89	15.18	8.04	3.75	0.00
	c	33.93	46.43	53.75	58.75	62.50
2.0	*a*	25.34	21.79	20.36	19.29	19.23
	b	42.86	16.25	8.57	5.71	0.00
	c	32.80	61.96	71.07	75.00	81.71

Fe_b is the active and transition part of hydroxy ferric polymeric species. It is less stable than Al_b and more transformable to Fe_c. The Al_b portion of total Al can approach as high as 80–90 %, but the Fe_b portion can reach at most about 50 % with high B^* ratios during the initial stage of ageing. After 10–15 days almost all of these species may transform into Fe_c.

Fe_c is formed even with low B^* ratios just after the injection of base at a moderate rate that is the common procedure used in the laboratory. Its proportion increases with the B^* and ageing time and becomes dominant progressively. After 2 weeks only Fe_c and Fe_a remain in solution.

(2) In Fe(III) solutions of Type A ($B_0^* = 0.0$–0.2), Fe_a is absolutely dominant. It can be stable and near equilibrium for a long time without significant pH change and transformation of species.

In solutions of Type B ($B_0^* = 0.2$–0.5), Fe_a is also dominant. However, with higher hydrolysity the monomer species tend to transform directly to yellow precipitates after a short introduction period (without pH decrease) in several days or even hours. Solution in this range is most unstable.

In solutions of Type C ($B_0^* = 0.5$–1.0), Fe_b starts to increase, and together with Fe_c as polymeric species, becomes comparable with the Fe_a monomeric species. Solutions begin the pH decline just after preparation and change their color to red and later brown. They are rather stable in ageing but all the Fe_b will transform to Fe_c in two weeks, and then the solution turns turbid with a brown precipitate after several months.

In solutions of Type D ($B_0^* = 1.0$–2.3), Fe_b and Fe_c are dominant to Fe_a. Fe_b species, though constitute a higher proportion than in type C, still transform to Fe_c rapidly. The pH decline starts even before the end of preparation but it tends to be stable after a week and persists for months.

When B_0^* is over 2.0, solutions become less stable and precipitatation occurs in several weeks.

When B_0^* is higher than 2.3, solutions begin to precipitate before completion of preparation. These solutions may be classified as Type E.

(3) In the hydrolysis and polymerization processes of Fe(III) solutions there are two conditions that lead them to lose their pseudo-equilibrium state. One is in the range of $B_0^* = 0.2$–0.5, where a yellow precipitate is produced from the Fe(III) monomer, the other is in the high B_0^* range above 2.0 or in long term ageing, where a brown precipitate is produced. Figure 7 shows the X-ray diffraction patterns of these two precipitates treated by freeze drying. The yellow precipitates (a) are mainly lepidocrocite γ-FeO(OH) and akaganeite β-FeO(OH) and the brown ones (b) are amorphous $Fe(OH)_3$(am) [12]. The same results were obtained by infrared spectra [13]. Both methods showed results that are consistent with those found in the literature [5, 9].

Stabilization of Polyferric Chloride Flocculants

The patterns found in the present experiments show no difference in principle with the results of other authors, but the principal parameter hydrolysity B^* together with basicity B allows the species transformation process to be seen more clearly overall. Though these patterns were worked out in the low and moderate concentrations (< 0.5 M) and with the chloride anion, they can provide a necessary basic model for the production of polyferric chloride flocculants. However, the conditions between production in the laboratory and on an industrial scale differ somewhat.

(1) Concentrations of raw material $FeCl_3$ and produced reagent can be higher than 3.0 M with a very low pH value. In such supersaturated solutions, there must be influences from some kinetic effects which were not considered.

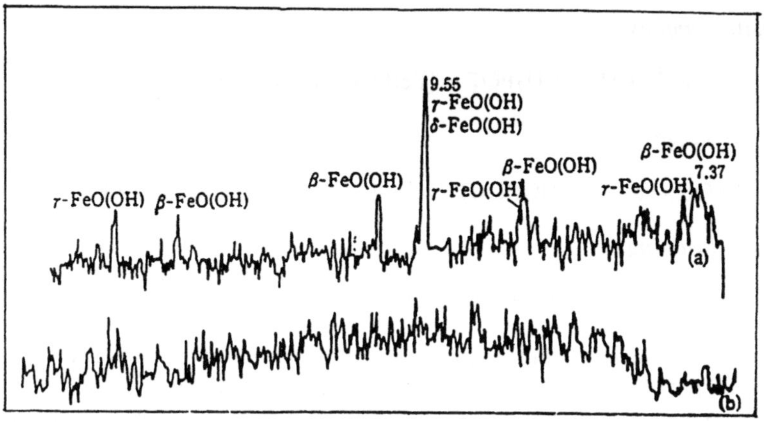

Fig. 7a,b. The X-ray diffraction patterns of Fe(III) precipitates at different OH/Fe ratios. (a) Low B_0^*, yellow precipitate, (b) High B_0^*, brown precipitate

(2) The addition of base in the industrial scale cannot occur at as slow a rate as in the laboratory. According to observation, the sharp pH gradients at the base-bulk solution boundary cause Fe_c or even fresh precipitates to form initially, which at a later time may be decomposed to polymeric species.

These operational conditions inevitably enhance further the precipitation tendency of $FeCl_3$ and obstruct the commercial production of high quality reagents. In order to produce more efficient PFC flocculants, our strategies for improving the procedure are: (1) Control the concentration so as not to be too high. (2) Apply a moderate base such as $NaHCO_3$ and dose under intense mixing. (3) Design the pertinent parameter B^* and B. (4) Search for an effective stabilizer that is indispensable for a long term stable product.

There are many organic and inorganic compounds that possess the ability to retard the hydrolysis and polymerization processes of Fe(III). The principle of screening would be: (1) the appropriate affinity to Fe(III) comparable to OH; (2) non-toxic and unproblematic in water and wastewater treatment; (3) cheap and capable of being produced industrially.

Among 24 kinds of possible compounds we chose 15 kinds to test experimentally. Phosphate was found to be among the most suitable and was then used in practice.

The interaction between Fe(III) and phosphate has been studied by many authors. The affinity of Fe(III) for phosphate is inferior to that for the hydroxy molecule but much stronger than that for other anions:

$$OH^- > PO_4^{3-} > F^- > SO_4^{2-} > Cl^- > NO_3^- > ClO_4^- .$$

Phosphate can thus compete with the hydroxy ion for part of the binding sites on Fe(III) species to form composite complexes or polymers by means of several possible mechanisms such as:

Composite complex:

$$Fe^{3+} + H_2O + H_2PO_4^- = [Fe(OH)(PO_4)]^- + 3H^+$$

Ligand exchange:

$$\text{Fe-O(OH)-Fe-OH} + HOPO_3H^- = \text{Fe-O(OH)-Fe-OPO}_3\text{H} + H_2O$$

Phosphate bridging:

$$\text{Fe-O(OH)-Fe-(OH)}_2\text{-Fe-(PO}_4)_2\text{-Fe-(OH)(PO}_4)\text{-Fe}\,.$$

No matter which mechanism is used, the phosphate groups are composited into the structure of Fe(III) species and substituted a part of the hydroxy groups. This was also demonstrated by our experiments involving base titration, pH relaxation, absorption spectrum and solution ageing, and the retardation of hydrolysis and polymerization processes were observed. Adding phosphate with a P/Fe ratio of 0.04–0.15 will make the 3.0 M $FeCl_3$ solutions stable without precipitation for over more than one year [14].

Another important contribution of the phosphate stabilizer is to improve significantly the presence of Fe_b, which is considered to be the active and effective species for coagulation and flocculation. Table 3 shows the difference in species distribution among the 0.1 M $FeCl_3$ solutions with and without phosphate after ageing for half a year. It is obvious that the addition of phosphate stabilizer preserved Fe_b in high proportions even at higher B ratios after long term ageing.

The superior efficiency of phosphate stabilized polyferric flocculants PFCP4 (3.0 M, $B_0^* = 0.5$) and PFCP6 (3.0 M, $B_0^* = 0.7$) compared with $FeCl_3$ for turbidity removal was shown in Figure 8 [15].

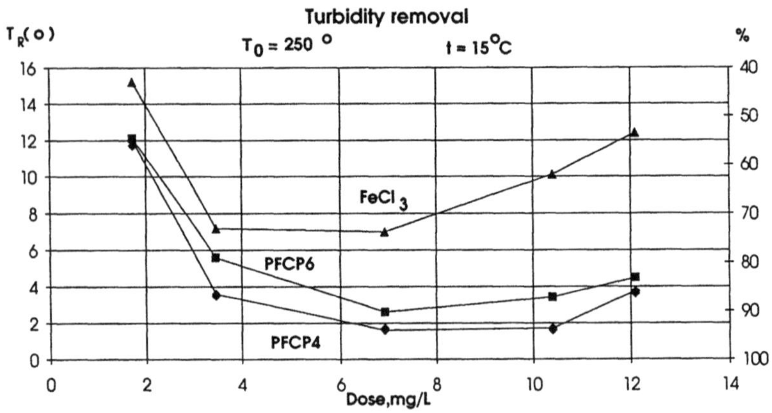

Fig. 8. The turbidity removal efficiency of PFCP

Table 3. Changes in Fe(III) solution after phosphate stabilization ($C = 0.01$ M)

B	P/Fe	1 day		24 days		80 days		Half year		
		pH	B^*	pH	B^*	pH	B^*	Fe_a	Fe_b	Fe_c
0	0.128	3.25	0.056	2.70	0.20	2.13	0.74	87	13	0
	0	3.46	0.035	2.47	0.34	1.95	1.12	45	7	48
0.3	0.128	3.49	0.33	2.93	0.42	2.32	0.78	76	17	7
	0	3.61	0.32	2.41	0.69	2.01	1.28	25	3	72
0.5	0.128	3.62	0.52	3.05	0.59	2.41	0.89	70	30	–
	0	3.42	0.44	2.46	0.85	2.03	1.43	31	–	69
1.0	0.128	3.59	1.03	3.11	1.08	2.42	1.38	39	56	5
	0	3.41	1.04	2.57	1.27	2.15	1.71	16	2	82
1.5	0.128	3.60	1.53	3.02	1.60	2.40	1.89	18	54	28
	0	3.44	1.54	2.73	1.69	2.27	2.04	8	3	89
2.0	0.128	3.64	2.02	3.18	2.07	2.57	2.27	10	39	51
	0	3.80	2.02	3.32	2.05	2.66	2.22	7	7	86

Pilot Experiments for Advanced Treatment of Wastewater

In the city of Tianjin in China a modern municipal sewage plant was constructed in 1987 and it has operated well continuously. A new project was set up for advanced treatment of the effluent after the secondary biological oxidation process of active sludge, so that the water could be reused as cooling water in a power station. A pilot test system of micro-filtration using a multi-layer filter with a flow of 5 M^3/h was established and the stabilized polyferric flocculant was chosen for use in pretreatment.

The water quality after the secondary sedimentation tank is shown as follows: Turbidity: 1.4–70°, Color: 2.9–55.8°, COD: 17.7–115.7 mg/L, BOD: 1.4–15.2 mg/L, Total P: 1.3–4.4 mg/L, pH: 6.3–7.8. The main aim of advanced treatment is removal of turbidity, color and organics. The color was caused mainly by wastewater containing dye. It was difficult to measure and remove.

A system of facilities for the production of flocculants on the scale of tons was established. The principal parameters of polyferric product in batches are: $Fe_T = 2.8$–3.0 M, $B = 0.1$-0.4, pH = 0.180–0.182, $B^* = 0.6$, P/Fe = 0.04.

Before the pilot test a series of screening jar tests was carried out. Many sorts of coagulants and flocculants such as aluminium sulfate, ferric chloride, polyaluminium chloride, polyferric sulfate and cationic, nonionic, anionic organic polymers were compared with polyferric chloride. The polyaluminium produced in our laboratory came close to the performance of polyferric chloride, while other inorganic reagents were inferior in performance, and the organic polymer cannot compete in cost.

The pilot test lasted more than a half year, in which 30 filtration cycles were completed. Figure 9 summarizes the removal efficiency of pollutants in the treated water. The result was completely satisfactory in terms of the demands of reuse [16].

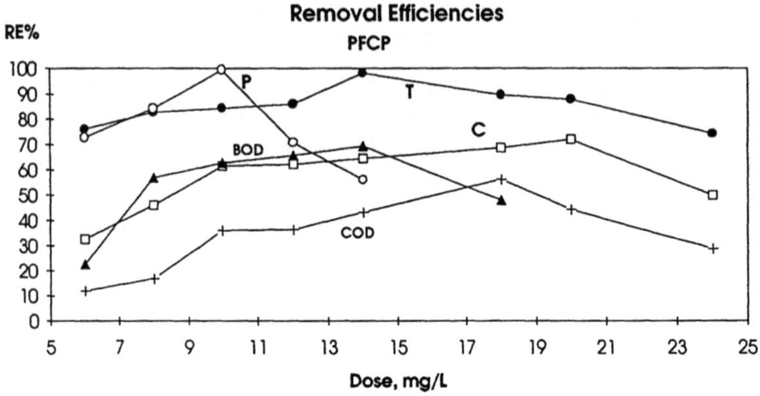

Fig. 9. The removal efficiences of PFCP in advanced treatment of municipal sewage

Conclusions

Hydrolysity B^* should be used together with basicity B as the principal parameters for studying and producing inorganic polymer flocculants. It is also useful to consider data in the literature in terms of B^* to establish a unified model.

Phosphate stabilized polyferric chloride may be commercially produced as a kind of highly efficient inorganic flocculant.

For the sake of developing further inorganic polymer flocculants, many aspects, such as their chemical speciation, special mechanisms of coagulation and flocculation, and optimum production technology still remain to be researched.

References

[1] Stumm, W., Morgan, J.J.: Aquatic Chemistry, 2nd ed. Wiley, New York 1981, p.336

[2] Tang, H.X., Stumm, W.: The Coagulating Behaviors of Fe(III) Polymeric Species, I: Preformed Polymers by Base Addition. Water Research *21* (1987) 115–121

[3] Tang, H.X., Stumm, W.: The Coagulating Behaviors of Fe(III) Polymeric Species, II: Preformed Polymers in Various Concentrations. Water Research *21* (1987) 123–128

[4] Spiro, T.G., Allerton, S.E., et al.: The Hydrolytic Polymerization of Iron(III). J. Am. Chem. Soc. *88* (1966) 2721–2726

[5] Hsu, P.H., Ragone, S.E.: Ageing of Hydrolysed Iron(III) Solutions. J. Soil Sci. *23* (1972) 17–31

[6] Murphy, P.J., Posner, A.M., Quirk, J.P.: Characterization of Partially Neutralized Ferric Chloride Solutions. J. Colloid Interface Sci. *56* (1976) 284–297

[7] Dousma, J., de Bruyn, P.L.: Hydrolysis-precipitation Studies of Iron Solutions. J. Colloid Interface Sci. *56* (1976) 527–539
[8] Van der Woude, J.H.A., de Bruyn, P.L.: Formation of Colloidal Dispersions from Supersaturated Iron(III) Nitrate Solutions. Colloids Surf. *8* (1983) 55–78
[9] Flynn, C.M.: Hydrolysis of Inorganic Iron(III) Salts. Chem. Reviews *84* (1984) 31–41
[10] Schneider, W., Schwyn, B.: The Hydrolysis of Iron in Synthetic, Biological, and Aquatic Media. In: Aquatic Surface Chemistry, W. Stumm (Ed.). Wiley, New York 1987, pp. 167–196
[11] Tang, H.X.: Basic Studies of Inorganic Polymer Flocculants. Environm. Chem. *9* (3) (1990) 1–12 (in Chinese)
[12] Tian, B.Z., Tang, H.X.: Determination of Polymeric Fe(III) Species by Ferron Complexation Timed Spectrophotometric Method. Environm. Chem. *8* (4) (1989) 27–34 (in Chinese)
[13] Tian, B.Z., Tang, H.X.: Infrared Spectra and Electrical Conductivity of Polymeric Fe(III) Species. Environm. Chem. *9* (6) (1990) 70–76 (in Chinese)
[14] Tian, B.Z., Tang, H.X.: Chemical Characteristics of Phosphate in Fe(III) Polymeric Solution. (To be published)
[15] Tang, H.X., Tian, B.Z., Luan, Z.K.: The Stabilization of Polyferric Chloride Solution and its Coagulating Efficiency. (To be published)
[16] Luan, Z.K., Tian, B.Z., Tang, H.X.: Advanced Treatment with PFCP Flocculants for Municipal Sewage After Biological Oxidation Process. (To be published)

H. X. Tang, B. Z. Tian, Z. K. Luan, and Y. Zhang
State Key Laboratory of Environmental Aquatic Chemistry (SKLEAC)
Research Center for Eco-Environmental Sciences
Chinese Academy of Sciences
100085, Beijing
Chine P.R.

Preliminary Evaluation of Polyferric Sulphate As a Coagulant for Surface Water Treatment

J. Q. Jiang, N. J. D. Graham, and C. Harward

Abstract

This paper is concerned with the performance of a relatively new form of Fe(III) coagulant, Polyferric Sulphate (PFS), which contains a range of pre-formed Fe(III) hydrolysis species with both a high positive charge and medium and large molecular weights. Preliminary laboratory and pilot-plant experiments were undertaken to evaluate the performance of PFS, in comparison with conventional coagulants such as ferric sulphate (FS), for the coagulation of surface waters containing substantial concentrations of algae and/or colour, and natural organic matter.

The performance of PFS was found to be consistently superior to other coagulants in terms of the removal of turbidity, algae, colour, UV absorbance and TOC, THMFP and the residual concentration of coagulants. In addition, sludge production and headloss development in the subsequent sand filter process were both lower with PFS than FS, and the influence of low water temperatures (e.g. 4 °C) on the coagulation performance with PFS was less than that with FS and aluminium sulphate (AS). The reasons for the superior performance of PFS and the effects of pre-ozonation and low water temperatures on coagulation behaviour are discussed in the paper.

1. Introduction

Fe(III) salts are common water treatment chemicals. Direct addition of the Fe(III) salts to the water to be treated may cause extremely rapid and uncontrolled hydrolysis, and frequently, very rapid precipitation. In contrast, a controlled hydrolysis process may produce a range of Fe(III) hydrolysis products which may adsorb to colloidal surfaces to neutralize their charge or may chemically interact with dissolved components in the raw water. For these phenomena the rate of coagulant interaction with colloidal and dissolved substances must be faster than the rate of hydrolysis to low charge species and the hydroxide precipitate. In the conventional coagulation process, the actual coagulant may be formed by dilution under the prevailing raw water conditions and in competition with other reactions. Thus, the optimal coagulant species may not be formed under these conditions.

Polyferric sulphate (PFS) is a partially hydrolysed Fe(III) chemical, containing a range of hydrolysis species and polymers. These partially hydrolysed species may

R. Klute / H.H. Hahn (Eds.)
Chemical Water and Wastewater Treatment III

reduce the rate of hydroxide precipitation upon dilution and thus may diminish the effect of competition in the coagulant-solute interaction, and may expand the region of optimum coagulation with respect to raw water conditions.

The application of polyferric coagulants in the treatment of water and wastewater has been reported previously. Jiang and co-workers [1] compared the performance of PFS with other conventional coagulants to remove algae and algal-derived organic matter in laboratory tests. PFS was able to achieve a high degree of removal of algal cells and algal-derived matter. This was believed to be due to the presence of more highly charged cationic species in PFS solutions, and due to a greater rate and extent of floc development, leading to greater settling rates. Leprince et al. [2] evaluated the efficiency of polyferric chloride (PIC) in laboratory, pilot and full scale experiments to treat natural waters. In general, for a given dose of iron, the reduction in turbidity was greater with PIC than $FeCl_3$. Tenny and Derka [3] compared the treatment efficiency of hydroxylated ferric sulphate (HFS) with aluminium sulphate (AS) at two water treatment plants in the city of Highland Park, Illinois, USA. HFS was found to produce either an equal or superior treated water quality to AS. The main advantage in the use of HFS, however, was the reduction of the treatment cost, comprising a 38 % saving in the costs of the coagulant and a lower cost in sludge disposal due to a denser sludge and subsequently a faster dewatering.

Other applications of polyferric coagulants include the pre-treatment of wastewaters from various industries, such as dyes, tannery, paper and pulp, and the treatment of waste sludges [4–10]. The general results from these studies showed that polyferric coagulants have:

- a higher treatment efficiency for turbidity, colour, heavy metals and chemical oxygen demand (COD), based on the same equivalent coagulant dose;
- a superior working performance at lower water temperatures (e.g. 4 °C);
- lower iron residuals;
- lower coagulant cost to achieve the same removal efficiency;
- faster sludge settling rates and lower costs for sludge dewatering.

This paper firstly summarises the results of laboratory scale experiments to evaluate the coagulation performance of a laboratory prepared PFS coagulant in comparison with FS and AS. The performance assessment was based on each coagulant's ability to remove algae and algal-derived organic matter from one type of model water, and to remove colour and natural organic matter from another kind of model water, at a constant or various pH values and at various coagulant dosages. Algal-type model waters contained one of the following kinds of algae, *Anabaena*, *Asterionella* and *Microcystis*. These algal species are very common in UK surface water reservoirs and are believed to be associated with quality and treatment problems in water treatment works. Coloured waters were prepared from freeze-dried humic material, previously extracted from a natural upland coloured water source. Secondly, the paper summarises the results of pilot-plant experiments which were undertaken at the Alton Water Treatment Works (AWTW), using a

2 m^3/h pilot scale water treatment process receiving influent water from a lowland reservoir. The reservoir water is characterized by seasonal blooms in algae, with peak cell concentrations (around 10^5 cells/mL) occurring in August/September. In addition, the influence of pre-ozonation and low water temperatures on coagulation performance were investigated.

2. Experimental Procedures and Methods

2.1 Algal Model Waters

Stock cultures of three kinds of algae were grown under standard conditions as described elsewhere [1]. The composition of the growth media for *Anabaena* and *Asterionella* are also described elsewhere [1], and the growth medium for *Microcystis* was the same as for *Anabaena*. Coagulation experiments were undertaken with solutions of the algae in diluted growth media and examples of the algal water quality characteristics are shown in Table 1.

Table 1. Water quality of the algal model waters

Kinds of algae	pH	Total cell cells/mL	Turbidity NTU	DOC mg/L	Colloid charge meq/(L × 10^4)	Fe mg/L	Mn mg/L
Anabaena flos-aquae	8.1	6.2 × 10^4	7.2	2.55	−79	0.138	0.231
Asterionella formosa	8.3	4.8 × 10^4	4.0	2.5	−92	0.141	0.244
Microcystis aeruginosa	7.44	5.8 × 10^4	6.1	2.35	−65	0.154	0.221

2.2 Coloured Waters

110 mg of the freeze-dried product containing dissolved organic matter, previously extracted from influent water to the Chellow Heights Treatment Plant (Yorkshire Water), was reconstituted with one litre of Millipore-Q system water to produce water samples with similar quality characteristics to the original upland water. The water quality is shown in Table 2.

2.3 Lowland Reservoir Waters at Alton Water Treatment Works (Stutton, Ipswich, UK)

During the period of the pilot-plant experiments (23 Aug.–10 Sep. 1993), the reservoir waters underwent an algal blooming period with peak cell concentrations of 3.0 × 10^5 cells/mL. About 15 algal species were monitored and the dominant species were *Coelosphaorium* and *Microcystis*. The water quality characteristics are shown in Table 3.

Table 2. Characterisation of reconstituted upland coloured water

	Raw water*	Reconstituted solution
Humus extract reconstituted [mg/L]	–	110.0
Total solids, 105 °C [mg/L]	–	109.1
Total volatile solids, 500 °C [mg/L]	–	38.1
Suspended solids, 105 °C [mg/L]	–	10.2
Dissolved solids, 105 °C [mg/L]	66.9	99.8
Dissolved volatile solids, 500 °C [mg/L]	26.7	36.7
pH	6.47	7.26
Conductivity, 25 °C [μS/cm]	7.51	11.63
Total alkalinity [mg $CaCO_3$/L]	5.03	8.00
Hardness (EDTA) [mg $CaCO_3$/L]	19.85	33.95
Dissolved carbon [mg C/L]†	9.49	10.08
Non-purgeable dissolved organic carbon [mg C/L]†	7.98	8.1
UV-absorbance, 254 nm [Abs/m]†‡	37.253	38.8
UV/Vis-absorbance, 420 nm [Abs/m]†‡	4.043	5.82
Specific UV-absorbance [Abs/mg C]†‡	4.676	4.111
E_4/E_6 ratio†‡	7.19	9.29
Al [mg/L]	–	0.14
Fe [mg/L]	–	0.12

* Mean values † Filtered, 0.45μ cellulose nitrate membrance ‡ pH 7.0

Table 3. Raw water quality of lowland reservoir at Alton (Stutton, Ipswich, UK)*

Parameters	Data
UV-Abs, 254 nm [Abs/m]	11.63
Colour, 420 nm [Abs/m]	11.87
TOC [mg C/L]	6.08
Turbidity [NTU]	2.72
Total Fe [mg/L]	0.063
Total Mn [mg/L]	0.039
Total algal concentrations [cells/mL]	1.62×10^5
pH	8.10

* Mean values during the period of the pilot-plant experiments

2.4 Jar Test Experiments

A six-beaker jar test apparatus was used with each beaker containing 300 mL of solution for algal waters and 500 mL of solution for coloured waters. The fast mixing time was 3 min at a paddle speed of 300 rpm, the flocculation period was 25 min at a paddle speed of 35 rpm, and the sedimentation period was 1 hour. Supernatant samples after settling were withdrawn for analysis of algal cells, turbidity, UV absorbance, DOC, colour, residuals of Fe, Mn and Al. Total particle numbers were measured after 5, 10 and 20 min. periods of flocculation and after 10, 20 and 60 min. periods of settling, to determine the floc size development and settling rates. The ranges of coagulant dosages and pH investigated were from 2–20 mg/L as Fe, and 2–12 mg/L as Al, and pH 3.5–8, or at pH 7.5, respectively. The coagulation pH required was achieved by prior addition of either HCl or NaOH, and the solution pH was checked during the mixing and flocculation periods.

To evaluate the coagulation performance at lower water temperatures (4 °C), a special constant temperature bath was used with the jar test beakers. The low temperature of 4 °C (±0.2) was maintained throughout the whole coagulation process and the one hour settling period.

2.5 Pilot-Plant Experiments

The pilot-plant experiments were undertaken at Alton Water Treatment Works (AWTW), Anglian Water Services LTD (Stutton, Ipswich). A schematic of the pilot plant is shown in Figure 1, which comprises the processes of pre-ozonation, coagulation, flocculation, dissolved-air flotation (DAF), sand filtration, intermediate ozonation and either GAC or BAC. The three flocculation tanks have a capacity of 1800 litres each and the rotational speeds of the mechanical stirrers are 16, 10 and 6 rpm, respectively, in the flow direction. Treated water from the flotation unit is returned at a 10 % recycle ratio to the air saturator unit which operates at a pressure of 3.5 bar. The desludging scraper operates constantly during flotation operation. The sand filter is washed at the beginning of every run condition, carried out by draining the level of water to the top of the media and then applying an air scour for 3 min (max rate > 100 L/min), followed by a backwash of water at 35 L/min for about 7 min.

In the test period of about four weeks, the pilot plant was operated under the conditions of with and without application of ozone, with coagulants of either PFS or FS for three doses (1, 3, 5 mg/L as Fe), and at a constant pH of 7.5. Raw water flow rate was 2 m^3/hr. The pre-ozone dose and the contact time were: 1.0 mg/L and 7.0 min; and the intermediate ozone dose was set to achieve a residual O_3 concentration of 0.25 mg/L after a contact time of 8.0 min. The test programme is shown in Table 4.

For every run condition the samples were not collected until the pilot plant had stabilized under the set conditions (i.e. at the given dose and pH). Samples of raw water and from points following flotation, sand filtration, BAC filter, GAC filter 1 and 2 were collected and analysed for their water qualities, including turbidity, UV absorbance (254 nm), colour, pH, residual Fe and Mn, TOC and THM formation potential. Flotation solids were collected to measure both the volume produced and their dry weight. For the ten running conditions, a separate set of samples was sent to Colchester Laboratory (Anglian Water) to identify the kinds of algal species and to measure the total algal concentrations before and after treatment.

Flotation sludge produced in a period of one run of scraping (6 min) was collected to measure its net dry weight. A set of 26 mL columns was used to

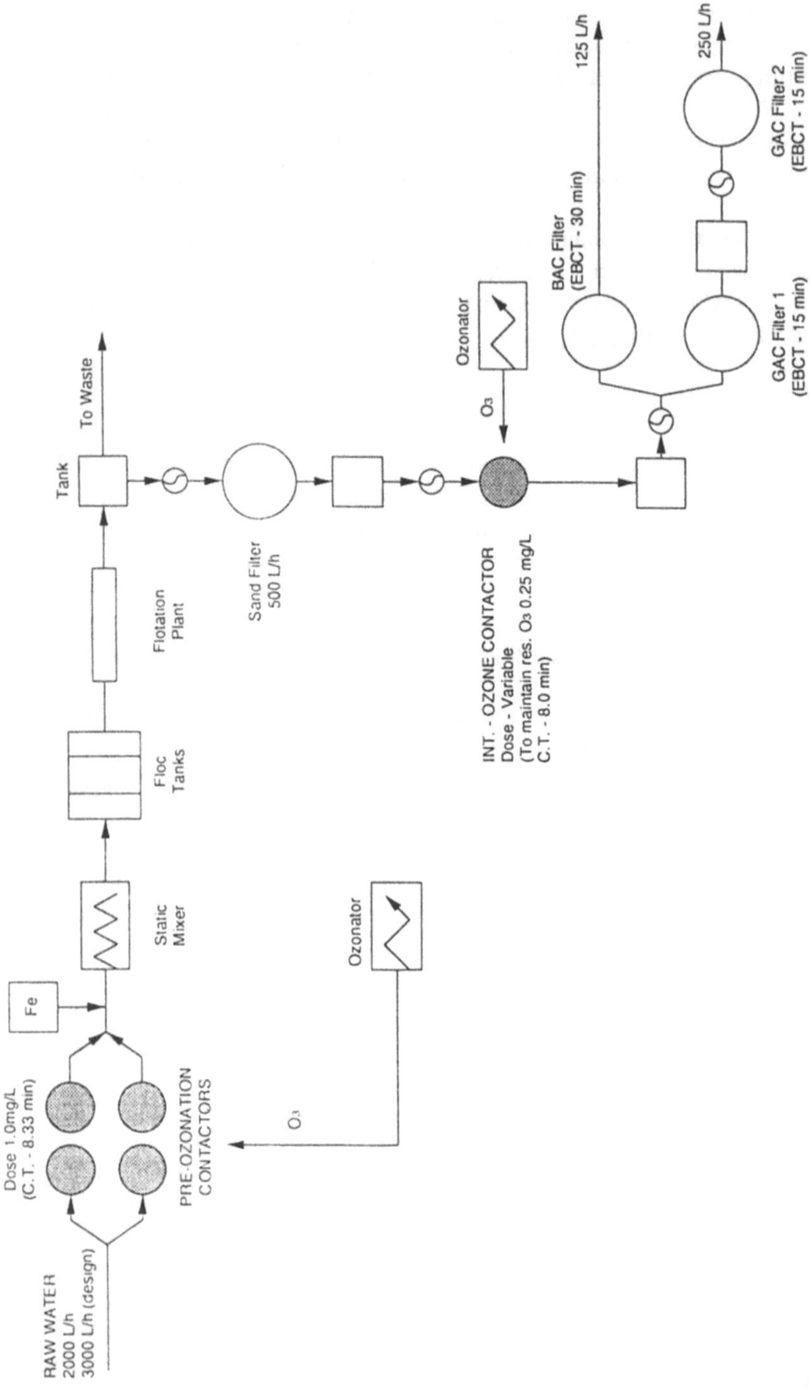

Fig. 1. Schematic of pilot plant

Table 4. Operational programme for pilo-plant experiments

Test run	Operational conditions			
	$Fe_2(SO_4)_3$, no O_3	$Fe_2(SO_4)_3 + O_3$	PFS, no O_3	PFS + O_3
1	1 mg/L as Fe			
2		1 mg/L as Fe		
3	3 mg/L as Fe			
4		3 mg/L as Fe		
5	5 mg/L as Fe			
6		5 mg/L as Fe		
	Cleaning all the facilities for the operation of PFS			
7			1 mg/L as Fe	
8				1 mg/L as Fe
9			3 mg/L as Fe	
10				3 mg/L as Fe
11			5 mg/L as Fe	
12				5 mg/L as Fe

collect the sludge to determine its density and water content, in terms of the net dry and net wet weight of sludge for the volume of 26 mL. Based on its density and the net dry weight of the sludge, the total volume produced for one run of scraping was calculated. The total volume of sludges for one run of scraping was divided by the time of one run of scraping (6 min), the surface area of the flotation tank (0.3588 m^2), and the water flow rate (2 m^3/hr), to give the expression for the sludge produced associated with the capacity of the flotation tank and m^3 of flow, which may be useful information for practical purposes. Sludges collected were firstly filtered by 0.45 μm filters, then oven dried (100°C) overnight and weighed after cooling. The average dry weight of sludge for every running condition was from the mean values of 5 measurements.

2.6 Coagulants

The PFS used in the experiments was made in the laboratory, and the FS and AS were commercial grade (E & A West Ltd and Alcan Ltd, respectively, UK).

The analytical procedures for determining algal concentration, floc size, turbidity, DOC and floc charge have been described elsewhere [1]. Residual concentrations of Fe, Al and Mn were determined by atomic absorbance spectrometry (Model 5100 PC, The Perkin-Elmer Corporation, UK). Colour and UV-Abs were measured in terms of absorbance at 420 nm and 254 nm, respectively, using a scanning UV-visible spectrophotometer (Model SP8-100, Pye Unicam Ltd, Cambridge, UK). Aqueous samples were pre-filtered with 0.45 μm membrane filters before the measurement of colour and UV-Abs. Procedures for the determination of chlorine concentration and demand, and for the THM formation potentials are described as follows:

2.6.1 Chlorine Concentration. Chlorine concentrations were determined using the colorimetric diethyl-p-phenylene diamine(DPD) method described by the UK Standing Committee of Analysis [11]. A Residometer (Wallace and Tiernan, Tonbridge, Kent, UK) was employed for the colorimetric determination of residual concentrations.

2.6.2 Chlorine Demand. At the end of the stipulated 4 day storage period (see THMFP method), measurements of the free and total chlorine residuals were made. Calculation of the 4 day chlorine demand was then carried out by subtracting the total chlorine residual from the known dose added to each of the samples.

2.6.3 THM Formation Potentials (THMFPs). THMFPs were determined using a version of a standard method described by the AWWA [12]. The 'Simulated Distribution System Trihalomethane Concentration (SDSTHMC)'was considered to be a suitable method for the Alton pilot-plant conditions. The principle of this is that samples of the treated water are stored under conditions representative of those occurring in the distribution systems, so that estimations may be made of the probable concentration of THMs which would ultimately reach consumers. The conditions of storage selected for the pilot plant samples were the following: storage time 4 days; storage temperature to be the same as that of the water in the pilot plant; chlorine dose was selected based on the chlorine dose being applied at the AWTW at the time of sampling.

THM analysis was by extraction into pentane followed by gas chromatography (GC) usinga standard method (septum vial extraction) recommended by the UK Standing Committee of Analysis [13]. Extraction and GC analysis were carried out on the day of sampling reaction mixtures in order to minimise any changes occurring during storage of samples or extracts.

GC was generally carried out using a Pye UNICAM series 204 gas chromatograph fitted with an electron capture detector (ECD) and 5 % FFAP on chromosorb HPW (100–120 mesh) packed glass column (2.7 m × 6 mm OD, 3 mm ID) (Phase Separations, Clwyd, North Wales). Instrument operating conditions were as given by the Standing Committee of Analysis [13]. The output was linked to a MEGA series integrator.

3. Results

3.1 Jar Test Experiments for Algal Model Waters

Preliminary results for the coagulation of algal model waters of *Anabaena* and *Asterionella* with PFS and other coagulants have been published elsewhere [1]. Table 5 presents comparative results for the treatment of *Microcystis* with PFS and FS at 18 °C and at pH 7.5, for a given dose of 5 mg/L as Fe. It can be seen that PFS achieved a higher removal percentage of total cells, turbidity and DOC, and lower Fe and Mn residuals, than FS; this was also the case at other coagulant doses.

Table 5. Jar test results for the treatment of *Microcystis*

Coagulants	Total cell removal %	Turbidity removal %	DOC removal %	Manganese residuals mg/L	Iron residuals mg/L
FS	60	59	20	0.18	0.188
PFS	78	81	24	0.15	0.175

The effects of water temperature on the coagulation performance of FS and PFS, using model waters of *Asterionella* and *Microcystis* are shown in Table 6. For a given dose of 5 mg/L as Fe and at pH 7.5 for comparison, the treatment efficiency (particularly of turbidity) achieved by FS appeared to be more substantially influenced by temperature than that of PFS.

Table 6. Influence of water temperature on the coagulation performance

Coagulant	Algae	Temperature °C	Total cell removal %	Turbidity removal %	DOC removal %
FS	*Asterionella*	18	77	68	24
	Asterionella	4	72	56	18
	Microcystis	18	62	60.5	15.5
	Microcystis	4	53	50	12.5
PFS	*Asterionella*	18	84	72.5	30.5
	Asterionella	4	78.5	70	27.5
	Microcystis	18	81.6	69.5	24.5
	Microcystis	4	75.0	68	22.4

3.2 Jar Test Experiments for Re-Constituted Coloured Waters

The coagulation of coloured re-constituted water with PFS, FS and AS was studied by varying pH and coagulant doses. Some typical examples are given in Figures 2–4 and in Table 7.

Comparison of PFS with FS and AS for the removal of DOC, colour and UV-Abs for a given dose (0.107 m mol/L as Fe or Al) and at various pH values is presented in Figure 2. The figures are plotted as pH vs removal percentage. It can be seen that the optimal pH for PFS and FS is in the range 3.5–4.5, and for AS approximately 5–6. However, in the optimal pH range for AS (pH 5–6), PFS can achieve a similar DOC removal percentage to AS (Fig. 2 (A)), suggesting that the efficient treatment pH range for PFS could be from 3.5 to 6, as compared to the traditional pH range of 4–5 for Fe(III) coagulants for treating coloured water [14–16].

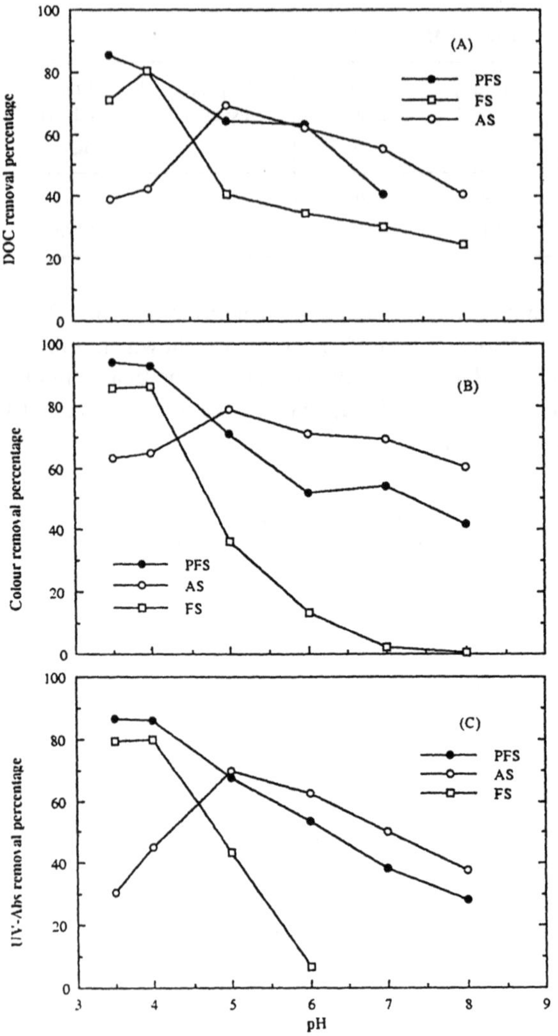

Fig. 2A–C. Comparison of the coagulation performance of PFS, FS and AS at varying pH and the given dose of 0.107 m mmol/L as Fe or Al. **(A)** pH vs. DOC removal, **(B)** pH vs. colour removal, and **(C)** pH vs. UV-Abs. removal

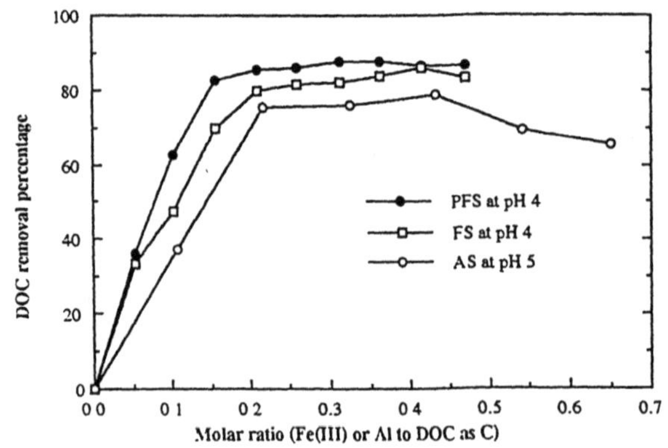

Fig. 3. The variation of DOC removal with coagulant dose for each coagulant at optimal pH

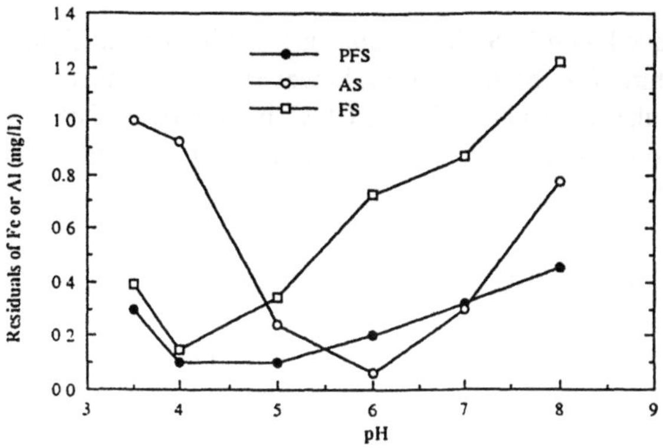

Fig. 4. Residuals of Fe and Al for a coagulant dosage of 0.107 m mmol/L as Fe or Al

Table 7. Comparison of PFS with FS and AS for colour removal at 18 °C and 4 °C

Coagulant	Dose	pH	Colour (420 nm) R %		UV (254 nm) R %		DOC R %		Residual Fe mg/l		Residual Al mg/L	
			18 °C	4 °C	18 °C	4 °C	18 °C	4 °C	18 °C	4 °C	18 °C	4 °C
PFS	6	4	92.0	91.3	83.2	82.2	80.3	78.8	0.09	0.098		
FS	8	4	91.1	89.5	84.8	80.5	80.2	76.7	0.165	0.180		
AS	5	5	92.0	91.4	82.5	75.7	75.4	68.2			0.064	0.074
Raw water	–	8.1	5.82 m^{-1}		38.8 m^{-1}		8.2 mg/L		0.05 mg/L		0.14 mg/L	

Comparison of PFS with FS and AS for the removal of DOC at various coagulant doses (expressed as molar ratio (Fe(III) or Al to DOC)) and at the optimal pH (pH 4 for PFS and FS, and pH 5 for AS), is shown in Figure 3. For the three coagulants compared, PFS achieved the highest removal percentage for DOC, colour and UV-Abs when coagulation was undertaken at their optimal pH values (Figs. 2 and 3). The removal percentage at the optimal pH ranges was higher with PFS and FS than AS, indicating Fe(III) coagulants are more effective in removing colour than AS.

Residual concentrations of Fe and Al were measured to determine the optimal pH ranges corresponding to the lowest concentrations of residuals. Figure 4 presents the residual concentrations of Fe and Al for the same coagulation conditions shown in Figure 2. For PFS and FS, the optimal pH to achieve the minimum residuals of Fe corresponded well to the pH to achieve the maximum removal percentages of DOC, colour and UV. However, the equivalent pH for AS was slightly different; the maximum removal of colour, etc. with AS was achieved at pH 5, whilst the minimum residual of Al was achieved at pH 6 (Figs. 2 and 4).

Comparison of PFS with FS and AS for colour removal at 18 and 4 °C is shown in Table 7. The coagulation pH and coagulant dosages were selected in terms of the optimal pH and minimum dose required to achieve > 90 % colour removal, > 80 % UV removal and > 75 % DOC removal at 18 °C. Under the optimal coagulation conditions selected at 18 °C, for the three coagulants compared, the removal percentages achieved with PFS did not appear to be influenced significantly by low water temperature. For the AS and FS, however, water temperature appeared to have an effect on the removal of UV and DOC, but removal of colour was only slightly reduced at low temperatures. In addition, residual concentrations for FS and AS at low water temperature were slightly increased.

3.3 Pilot-Plant Experiments for Algal-Laden Surface Waters at Alton

3.3.1 Comparison of Conventional Treatment Efficiency. The relative results for the removal percentage of TOC, colour and turbidity for the pilot plant experiments are listed in Table 8. In general, PFS achieved a consistently higher removal percentage than FS, particularly when no ozone was applied for the lower iron dosages (1 and 3 mg/L). The removal percentage of all parameters achieved by PFS was 2–4 times more than those for FS for the same dose of 1 and 3 mg/L as Fe, without the application of O_3. Moreover, treatment efficiency achieved by PFS for the dose of 1 mg/L as Fe was higher than or similar to that achieved by FS for the dose of 3 mg/L (Table 8), indicating that coagulant dosing costs may be reduced for the same treated water quality.

Table 8. Comparative results for the pilot plant experiments

Operational conditions	TOC removal percentage			Colour (420 nm) removal percentage			Turbidity removal percentage		
	DAF	Sand filter	GAC 2	DAF	Sand filter	GAC 2	DAF	Sand filter	GAC 2
(A) No ozone									
FS, [Fe] = 1 ppm	5.20	9.7	44.48	4.0	6.90	85.45	40.0	62.0	70.8
PFS, [Fe] = 1 ppm	18.1	19.70	58.27	20.97	24.20	88.71	66.6	73.3	84.0
FS, [Fe] = 3 ppm	10.8	21.90	55.43	36.67	39.83	87.3	48.0	64.8	72.4
PFS, [Fe] = 3 ppm	21.0	24.76	59.0	51.43	53.23	90.73	67.5	73.0	80.5
FS, [Fe] = 5 ppm	17.40	21.20	59.22	40.0	50.0	90.67	41.6	61.5	69.2
PFS, [Fe] = 5 ppm	22.6	26.68	61.57	48.68	55.82	89.58	62.5	69.2	75.8
(B) With ozone									
FS, [Fe] = 1 ppm	7.5	21.0	49.75	26.0	30.10	85.53	66.6	75.5	81.3
PFS, [Fe] = 1 ppm	19.8	22.90	60.28	34.21	35.86	89.06	71.4	78.57	85.7
FS, [Fe] = 3 ppm	17.6	22.45	55.90	56.67	63.33	90.0	56.0	69.6	70.4
PFS, [Fe] = 3 ppm	19.94	24.68	61.25	59.26	63.85	94.50	68.8	75.3	83.3
FS, [Fe] = 5 ppm	19.61	24.40	57.23	58.06	58.06	90.65	41.6	61.67	68.8
PFS, [Fe] = 5 ppm	21.48	25.34	58.93	66.67	73.3	92.5	62.0	68.8	77.2

Comparison of residual iron in the treated water at the different sampling points is presented in Table 9. For the conditions of iron dose of 1 mg/L and without O_3 applied, the residual iron after DAF was 0.28 mg/L for FS and 0.18 mg/L for

PFS, and were 0.07 and 0.04 mg/L, respectively, after sand filtration, indicating an overall better performance of DAF and sand filter with PFS for the dose of 1 mg/L. The better performance (lower iron residual) achieved by PFS after DAF and the sand filter was also reflected in the final iron concentration in the treated water. The iron residual after BAC was 0.05 mg/L for FS and 0.01 mg/L for PFS; and after GAC 2, was 0.04 mg/L for FS and 0.01 mg/L for PFS, respectively. For the other test conditions, the iron residuals after different sampling points were generally similar. For all the test conditions, the iron residuals in the treated water (BAC or GAC 2) were lower than the European MAC concentration (0.2 mg/L).

Table 9. Comparison of residual iron concentrations

Operating conditions	Raw water mg/L	Raw water °C	DAF mg/L	Sand filter mg/L	BAC mg/L	GAC 2 mg/L
FS, [Fe] = 1 ppm, no O_3	0.09	19.4	0.28	0.07	0.05	0.04
PFS, [Fe] = 1 ppm, no O_3	0.04	17.2	0.18	0.04	0.01	0.01
FS, [Fe] = 1 ppm, with O_3	0.07	19.4	0.17	0.05	0.04	0.03
PFS, [Fe] = 1 ppm, with O_3	0.05	17.0	0.15	0.05	0.03	0.02
FS, [Fe] = 3 ppm, no O_3	0.05	18.2	0.31	0.06	0.04	0.04
PFS, [Fe] = 3 ppm, no O_3	0.12	17.2	0.28	0.06	0.05	0.04
FS, [Fe] = 3 ppm, with O_3	0.05	18.2	0.25	0.05	0.03	0.03
PFS, [Fe] = 3 ppm, with O_3	0.05	17.2	0.20	0.03	0.03	0.02
FS, [Fe] = 5 ppm, no O_3	0.05	17.7	0.29	0.04	0.03	0.03
PFS, [Fe] = 5 ppm, no O_3	0.05	16.6	0.30	0.03	0.03	0.03
FS, [Fe] = 5 ppm, with O_3	0.04	17.2	0.23	0.03	0.03	0.03
PFS, [Fe] = 5 ppm, with O_3	0.10	17.6	0.21	0.04	0.04	0.03

3.3.2 Headloss Development in the Sand Filter. Headloss development in the sand filter under the different test conditions was compared and one set of the results is presented in Figure 5. Generally, headloss development with PFS was lower than FS for the conditions of no pre-ozone dose and for the three coagulant dosages (1, 3 and 5 mg/L as Fe). With the application of pre-ozone, the difference of headloss development with PFS and with FS was not significant. It is interesting to note that pre-ozonation had no effect onthe rate of headloss development with FS.

3.3.3 Identification of Algal Species and Measurement of the Concentration of Algae. The quality of algal-laden surface waters at Alton has been described in Table 3. It has been noted that in the samples following the different sampling points (i.e. GAC 1 and 2, or BAC, etc.), the main algal species present were different from those in the raw water. Since some data of algal concentrations in the raw water (two test conditions, 25 Aug. and 1 Sep.) and in the samples from several points were unavailable, algal removal percentage, unfortunately, cannot

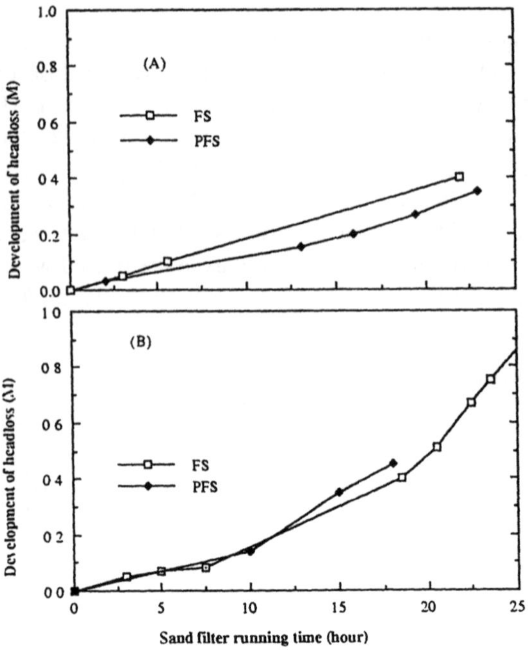

Fig. 5A, B. Development of headloss in the sand filter, for the dose of 3 mg/L as Fe(III). **(A)** without pre-ozonation and **(B)** with preozonation

Table 10. Comparative algal removal percentage

Date/time	Run conditions	Raw water conc. cells/ml	Removal percentage in Alton pilot plant							Removal percent. in Alton Water Treatment Works		
			Pre-O_3	DAF	Sand	Int-O_3	GAC 1	BAC	GAC 2	Sand	Carbon	Final
8 Sep.	PFS, 1 ppm +O_3	148 319	–	83.81	82.62	99.99	98.28	97.66	98.06	–	–	–
9 Sep.	FS, 1 ppm +O_3	118 880	–	74.45	–	–	–	–	–	–	–	–
3 Sep./ 17.00	PFS, 3 ppm, no O_3	249 220		–	96.28	–	99.55	99.60	99.98	89.14	–	–
23 Aug.	FS, 3 ppm, no O_3	79 300		–	89.86	–	95.41	97.44	99.55	–	–	99.93
3 Sep./ 7.00	PFS, 3 ppm +O_3	202 406	54.17	–	97.14	100	99.44	99.53	97.06	92.13	84.05	95.14
25 Aug.	FS, 3 ppm +O_3	–	–	–	–	–	–	–	–	–	–	–
6 Sep.	PFS, 5 ppm, no O_3	208 322		–	70.34	–	93.65	–	97.83	–	74.1	88.41
26 Aug.	FS, 5 ppm, no O_3	80 000		–	54.58	–	–	95.35	97.73	82.87	–	91.55
7 Sep.	PFS, 5 ppm +O_3	208 000	56.52		99.09	84.41	97.67	95.16	97.22	–	76.26	94.15
1 Sep.	FS, 5 ppm +O_3	–	–	–	–	–	–	–	–	–	–	–

– Date were unavailable

be calculated in these cases. Table 10 gives a comparative removal percentage of algae corresponding to the limited available data. It can be seen that PFS obtained a higher algal removal percentage than FS. This is in agreement with the previous results of algal removal, conducted by laboratory jar test experiments, using algal model waters.

3.3.4 Chlorine Demand and THM Formation Potential (THMFP). The overall results of the chlorine dose, residual concentration, the chlorine demand and THMFPs are summarized in Table 11, which compares THMFP under four test conditions: application of either PFS or FS for the doses of 3 and 5 mg/L as Fe, without ozone applied. For FS and PFS, with the dose of 3 mg/L as Fe, the chlorine dose and demand were similar, while the THMFPs at the different sample points were lower with PFS than those treated by FS, indicating that the higher removal efficiency of algae and humic substances achieved by PFS had therefore resulted in lower THMFP, under the same conditions of raw water qualities. Tables 12 and 13 summarise the raw water qualities and removal efficiency by DAF and sand filter, under the four test conditions, and confirms the analysis mentioned above. For the other two run conditions (FS and PFS for the dose of 5 mg/L as Fe, no ozone application), the raw water quality characteristics were different, causing the different Cl dose and demand. THMFPs in the final water treated by PFS were much lower than those treated by FS. This was not only caused by the different raw water quality characteristics, but also resulted from a higher removal efficiency achieved by PFS (Tables 12 and 13).

Table 11. Overall results for THMFP

Running conditions	Chlorine dose [mg/L as Cl]					Total Cl demand in the 4 days [mg/L as Cl]					Total THM Formation [μg/L]				
	Main plant	Pilot plant				Main plant	Pilot plant				Main plant	Pilot plant			
	(Final C*)	Sand filter	GAC 1	GAC 2	BAC	(Final C*)	Sand filter	GAC 1	GAC 2	BAC	(Final C*)	Sand filter	GAC 1	GAC 2	BAC
FS, no O_3 3 mg/L as Fe	1.26	1.26	1.26	1.26	1.26	1.09	1.18	1.11	1.04	1.16	105.05	49.76	70.17	72.80	58.24
PFS, no O_3 3 mg/L as Fe	1.23	1.23	1.23	1.23	1.23	1.02	1.14	1.09	1.07	1.09	109.26	43.25	58.54	64.89	51.93
FS, no O_3 5 mg/L as Fe	1.40	1.40	1.40	1.40	1.40	1.08	1.30	1.22	1.13	1.22	136.95	78.85	67.17	72.87	83.53
PFS, no O_3 5 mg/L as Fe	1.20	1.20	1.20	1.20	1.20	0.96	1.12	1.08	0.95	1.04	99.86	38.48	46.68	57.47	50.12

* Final active carbon filter

Table 12. Comparison of raw water quality characteristics and removal efficiency by DAF

Running conditions	UV (254 nm) m^{-1}		TOC mg/L		Colour (400 nm) m^{-1}		Turbidity NTU		Algae cells/mL	
	Raw water	R.%	Raw water	R.%	Raw water	R.%	Raw water	R.%	Raw water	R.%
FS, 3 mg/L, no O_3	11.0	13.64	6.17	10.8	12.0	36.67	2.5	48.0	79 300	–
PFS, 3 mg/L, no O_3	11.5	26.10	6.38	21.0	13.3	51.43	2.6	67.5	249 200	–
FS, 5 mg/L, no O_3	12.0	20.83	6.18	17.40	11.4	40.0	2.6	41.6	80 000	–
PFS, 5 mg/L, no O_3	12.5	28.0	5.54	22.6	12.54	48.68	2.5	62.5	208 322	–

– Data were unavailable

Table 13. Comparison of raw water quality characteristics and removal efficiency by Sand filter

Running conditions	UV (254 nm) m^{-1}		TOC mg/L		Colour (400 nm) m^{-1}		Turbidity NTU		Algae cells/mL	
	Raw water	R.%	Raw water	R.%	Raw water	R.%	Raw water	R.%	Raw water	R.%
FS, 3 mg/L, no O_3	11.0	18.18	6.17	21.90	12.0	39.83	2.5	64.8	79 300	89.86
PFS, 3 mg/L, no O_3	11.5	30.43	6.38	24.76	13.3	53.23	2.6	73.0	249 200	96.28
FS, 5 mg/L, no O_3	12.0	25.0	6.18	21.20	11.4	50.0	2.6	61.5	80 000	54.57
PFS, 5 mg/L, no O_3	12.5	36.0	5.54	26.68	12.54	55.82	2.5	69.2	208 322	70.34

3.3.5 Sludge Production. The results for sludge production are summarised in Figure 6. To give the rate of sludge production and to supply useful information for practical purposes, the total sludge volumes produced were expressed in alternative ways. Therefore, the comparison of sludge volume shown in Figure 6 was carried out in terms of a specific volume, which includes the factors of the time for one run of scraping (6 min), the surface area of the flotation tank (0.3588 m^2) and the water flow rate (2 m^3/hr).

Generally, lower volumes of sludge were produced with PFS than FS, especially when no ozone was applied, indicating that with PFS, lower costs and more convenience in treating sludges at water treatment works may be expected. Correspondingly, the sludge density of PFS was greater than that of FS, especially for the dose of 1 mg/L as Fe without the application of ozone, suggesting that a faster settling rate could be achieved by PFS.

When ozone was applied, there was less of a difference in the sludge volume produced with PFS and FS, compared to that without the application of ozone. This was coincident with the results of the development of headloss, in which the development of headloss in the filter with PFS was similar to that with FS with the application of ozone.

A full green colour of the sludges with PFS for the dose of 1 mg/L as Fe was speculated as being due principally to algae being neutralized by PFS

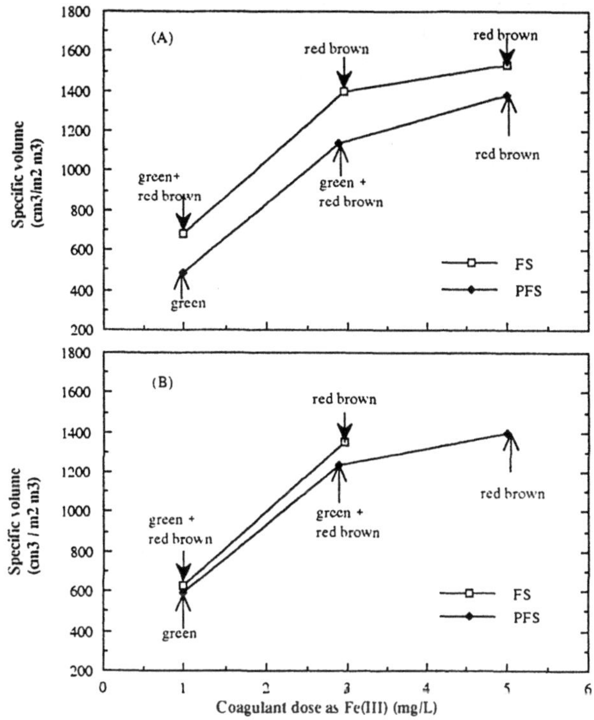

Fig. 6A, B. Sludge production vs. coagulant dose and either without application of ozone or with ozone. The definition of the specific volume of the sludge is: Total sludge volume (cm^3) produced per unit surface area of the flotation tank (m^2), per (m^3) of flow. **(A)** without pre-ozonation, **(B)** with pre-ozonation. Arrows shown in the figure indicate the colour of the sludges

cationic species, and complex precipitates between algal cells and PFS species being formed. However, sludges of a red-brown colour with FS for the dose of 3 and 5 mg/L and with PFS for the dose of 5 mg/L suggested that the algae cells were firstly absorbed onto and then enmeshed by precipitated $Fe(OH)_3$. Sludges of the colour of green together with red-brown were therefore believed to be caused by a combination of the coagulation mechanisms.

4. Discussion

It is generally claimed that prepolymerized coagulants have particular advantages over conventional metal salt coagulants, such as a wider working pH range, a lower sensitivity to low temperatures, a better overall treatment efficiency, and lower residual metal-ion concentrations. All of these advantages have been demonstrated in these studies with Polyferric Sulphate.

A superior coagulation performance with PFS for the removal of algae was shown in previous jar test experiments using two model algal waters [1]. A similar result was confirmed in this study by jar test experiments with another algal model water containing *Microcystis*. The jar test experiments were carried out at a constant pH of 7.5 in order to simulate a typical algal-laden raw water. At this pH, moderate doses of aluminium or ferric salts ($> 0.1\ mM$) lead to rapid precipitation of $Al(OH)_3$ or $Fe(OH)_3$. In this case, the removal of dissolved organic

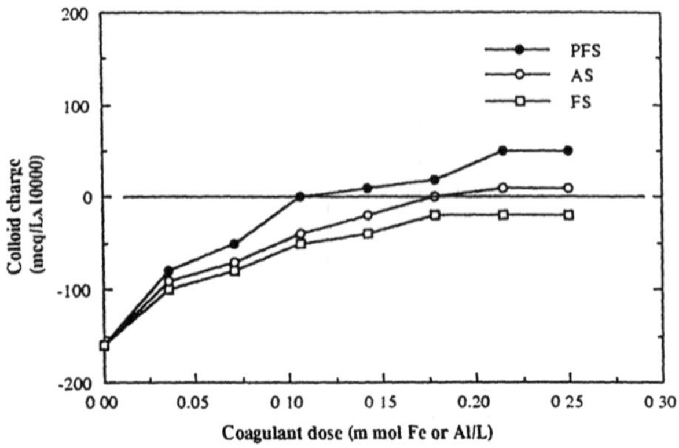

Fig. 7. Comparison of floc charges at given pH (7.5) vs. coagulant dose for model *Microcystis* waters

species is attributed to adsorption onto $Al(OH)_3$ (or $Fe(OH)_3$) particles and/or, together with algal cells, enmeshment by precipitated metal hydroxide ("sweep coagulation"). In contrast, PFS is believed to contain cationic hydrolysis species which interact principally by charge, with cationic species adsorbing onto particle surfaces (e.g. algae) and precipitating dissolved organic species (e.g. humic substances). The superior coagulation performance by PFS, for both algal cells and dissolved organic species (extracellular material and humic substances), is partly attributable to charge effects. Thus, a significantly smaller dose of PFS is required to achieve colloid neutralisation (0.1 mM as Fe) than that of the other coagulants (> 0.17 mM as Fe or Al) (Fig. 7). However, superior coagulation performance was achieved at PFS doses above that required for colloid charge neutralization, which cannot be explained by charge effects alone.

Pilot-plant experiments have also confirmed the results of previous laboratory jar test experiments concerning the removal of algae. PFS performed substantially better than FS in terms of the removal of turbidity, colour, TOC and algae (Table 8). In addition, headloss development in the sand filter and sludge production from the DAF plant were both lower with PFS than FS; this was the case especially in the absence of pre-ozonation. From Table 8 it can be seen that pre-ozonation improved the removal efficiency of UV, TOC, colour and turbidity for both coagulants, but a greater improvement was achieved in the treatment with FS at the lower doses (1 and 3 mg/L as Fe). This was generally in agreement with other studies [17–22], where coagulant doses were either very low (< 0.1 M Al/M TOC) or very high (> 1.5 M Al/M TOC) [17], and the impact of pre-ozonation on the removal of particles and TOC in the coagulation process was pronounced. Several hypotheses have been proposed to explain the observed improvement, such as:

- Ozone may polymerize natural organic matter (NOM), leading to particle aggregation via bridging reactions;
- Alterations in the number of organic carboxyl groups may affect the interactions between calcium, magnesium, Al and Fe with organic molecules, thereby causing metal humates to precipitate, or increasing removal by sorption onto Al or Fe hydroxide flocs, or both;

- Ozone may rupture complexes of iron and manganese, causing in-situ production of coagulant.

These hypotheses may explain why pre-ozonation improved the coagulation performance observed in the pilot-plant experiments. In addition, other researchers [23] have suggested this could be attributed to ozone induced pH changes or calcium carbonate precipitation during ozonation. However, in the pilot-plant experiments detailed in this paper, a constant pH (7.5) was maintained after pre-ozonation.

It is interesting to consider why, in the case of pre-ozonation, the differences of coagulation performance with PFS and FS were less pronounced at dosages of 3 and 5 mg/L (as Fe) than those in the absence of pre-ozonation. It is assumed that the pre-ozonation of NOM will decrease the MW and increase the acidic functional group content of NOM [18, 24], and these changes are expected to affect interactions between NOM and coagulants. Since ozonation reduces the average size (MW) of NOM molecules, it is expected to decrease the number of NOM molecules that are removable via precipitation reactions with coagulants. However, since ozonation increases the acidic functional group content of NOM, it seems to increase the sorption between NOM and coagulant flocs and therefore increases the removal of NOM. Thus, the above mentioned effects are expected to influence the coagulation performance achieved by both PFS and FS. Therefore, the differences in coagulation performance between FS and PFS were less pronounced than in the absence of pre-ozonation. However, since PFS contained preformed cationic species, with a range of MW sizes, the performance of PFS was still better than FS.

Several published papers [17, 24, 25] have concluded that pre-ozonation can have adverse effects on TOC removal and on coagulant residual concentrations when ozone doses are above about 0.25 mg ozone/mg TOC. This was believed to be because pre-ozonation decreased the surface charge of floc formed after coagulation with metal salts, thereby hindering adsorptive removal of the anionic organic molecules by floc surfaces and/or inducing stabilized floc formation. The results of the pilot-plant experiments, described here, where the pre-ozone dose was 0.164 mg ozone/mg TOC (i.e. < 0.25 mg ozone/mg TOC), are consistent with these earlier studies.

Conventional coagulation with Fe(III) or Al is effective in reducing DOM concentration, colour and THMFP from natural coloured waters. Actual reductions depend on the type and the dose of coagulant, pH, temperature, and raw water characteristics. Sinsabaugh III, et al. [26] concluded that the nature of the organic materials (size, charge and solubility) may directly affect the removal process. It can be seen in Table 2 that a specific UV-Abs and a E_4/E_6 ratio were used to partly characterize the NOM in the experiments. The specific UV-Abs (SUVA) is a quick and simple way of comparing the organic composition of raw waters. This is a measure of the ratio between the UV-Abs at 254 nm (m^{-1}) of a water and its DOC content (mg C/L). The SUVA of araw water therefore reflects the characteristics of the composite organic molecules in solution, independently of the organic substance concentration. Consequently, solutions containing high MW, strongly UV-Abs molecules, such as humic acids, exhibit SUVA values of between 5.0–7.5, solutions of smaller molecules, such as fulvic acids, exhibit SUVA values between 4.0 and 5.0, and those of very low MW, non-humic molecules, which are only weakly or non-UV-absorbing, have SUVA values typically less than 3 [27,28]. The E_4/E_6 ratio is a ratio of solution absorbance measured at 465 nm and 665 nm, widely used in the studies of soil humic substances. Its magnitude is primarily an inverse function of the average MW [29–31] since a greater absorbance at 665 nm exhibited larger, higher MW molecules, such as humic acids, than smaller MW molecules, such as fulvic acids. Typical values of the E_4/E_6 ratio are from 2 to 5 for solutions of humic acids and from 6 to 10 for fulvic acid solutions [31,32]. Therefore, the NOM present in the coloured waters used in this experiment consisted predominantly of fulvic acid species, the MW range of which is approximately 500–2000 [29] (SUVA values of between 4 to 5 and E_4/E_6 ratio > 6, as shown in Table 2). This is in agreement with the nature of most coloured surface waters, where fulvic acids have been shown to account for 80–90 % of the humic material present [32].

Fulvic acids have been shown to be readily removable by coagulation when the coagulationis undertaken at the optimum pH, which is 5 to 6 for AS and 3.5 to 4.5 for Fe (III) salts [33]. The optimum pH found in this study for coloured water treatment is inagreement with the literature. The optimum coagulation pH of Fe(III) or Al coagulants is believed to depend on two factors. Firstly, on the pH range which gives rise to the most favourable coagulant hydrolysis species for coagulation. Secondly, on the influence of pH on the ionization of carboxylic and phenolic functional groups associated with the humic and fulvic acids. Decreasing the pH of natural coloured waters leads to both a reduction of the net molecular charge of the humic and fulvic acids and an increase of the solution's conductivity, which leads to a greater removal of particles and humic substances by Fe(III) or Al chemicals in the coagulation process. However, for the two factors mentioned above, the first (coagulant species) appears to be the more important.

PFS contains pre-formed cationic species with a range of molecular weights. When it is added to water, interactions of the pre-formed species with colloids and dissolved substances may hinder the speed of further hydrolysis and therefore lead to a lower sensitivity to pH in the coagulation process. PFS can, thus, perform well over a wider pH range than conventional coagulants. This was demonstrated by a similar DOC removal percentage with PFS as that with AS in the optimal pH range for AS (pH 5–6), as shown in Figure 2(A). The superior performance of PFS because of its pre-formed coagulant species has also been confirmed by comparison of the DOC removal at the optimal pH for each of three coagulants and at various dosages, as shown in Figure 3. A significantly lower dose was required for PFS to achieve the same removal percentage as FS and AS. It is noted from Figure 3 that the two Fe(III) coagulants performed better than AS for the removal of DOC. This is supported by other work reported in the literature where Fe(III) coagulants performed better than AS for the removal of NOM with a MW range of between 1000 to 10000 [34], and for fulvic acids of MW range 500–2000 [29].

The adverse effect of lower water temperatures on the coagulation performance with AS and FS (shown in Table 7) is considered to be because, at lower temperatures, the favourable coagulant hydrolysis species forms more slowly and to a lesser extent than at room temperature. This, consequently, leads to higher coagulant residual concentrations (Table 7), a slow floc development (Fig. 8) and slow settling rates (Fig. 9). PFS, however, was virtually unaffected.

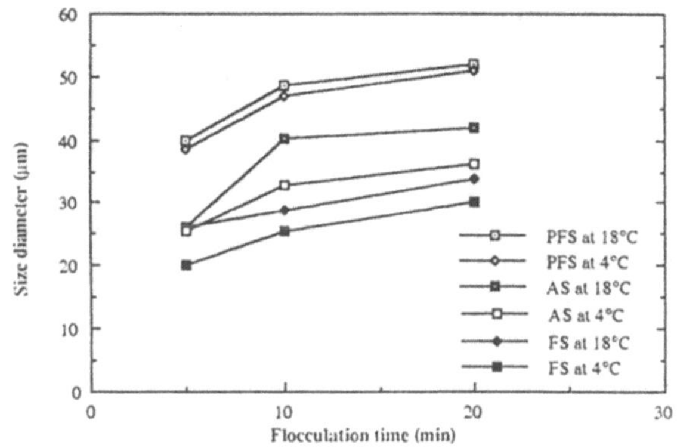

Fig. 8. Effects of temperature on the floc size development

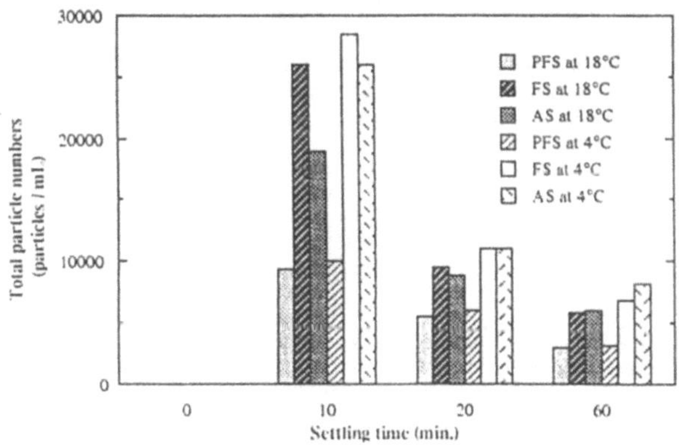

Fig. 9. Effects of temperature on the settling rates

5. Conclusions

From the experimental results detailed in this paper, the following conclusions can be made:

(i) For Algal-Laden Surface Water Treatment

- PFS performed substantially better than other conventional coagulants at both room temperature (18 °C) and at low temperature (4 °C), in terms of the removal of turbidity, algae, TOC, and THMFP. In addition, coagulant residual concentrations, headloss development in the filtration process following coagulation and sludge production were lower with PFS than with the other coagulants. The greater performance achieved by PFS occurred particularly in the absence of pre-ozonation.

- Pre-ozonation improved the coagulation performance achieved by both PFS and FS at lower coagulant doses. Improvement in the performance of FS by pre-ozonation was pronounced, but the performance of PFS was still better than FS.

(ii) For Coloured Upland Water Treatment

- The treatment efficiency of PFS for the removal of colour, UV and DOC was significantly higher than with AS and FS, under the same coagulation conditions. Alternatively, much lower doses (mg Fe or Al/L) were required for PFS to achieve the same treatment efficiency as AS and FS.
- The pH working range for PFS to treat coloured water was much wider than for conventional coagulants. The optimal pH for achieving the minimum coagulant residual concentrations corresponded to that for achieving the maximum removal percentage for both Fe(III) coagulants.
- Whilst a low water temperature (4 °C) reduced the coagulation performance of the conventional coagulants, PFS was virtually unaffected.

Acknowledgements. The authors wish to gratefully acknowledge the assistance of Dr. Paul Luckham and Dr. Jay Rajapakse of Imperial College with aspects of the experimental work. The authors also acknowledgethe financial support of Anglian Water Service Ltd for this work. Thanks are expressed to Mr. D.R.H. Price, Director of Water Quality – Anglian Water Service Ltd., for permission to publish this work. The views expressed in this paper are those of the authors and donot necessarily represent the views of Anglian Water Service Ltd.

Literature

[1] Jiang, J.Q., Graham, N.J.D. and Harward, C.: Comparison of Polyferric Sulphate with Other Coagulants for the Removal of Algae and Algal-Derived Organic Matter. In: Proc. of IAWQ/IWSA Joint Special Conference, (1992), Geneva, Switzerland, pp. 199–208

[2] Leprince, A. Fiessinger, F. and Bottero, J.Y.: Polymerized Iron Chloride: An Improved Inorganic Coagulant. J. AWWA. *76* (10) (1984) 93–97

[3] Tenny, A.M., Derka, J.: Hydroxylated Ferric Sulphate – an Aluminium Salt Alternative. Wat. Supply *10* (4) (1992) 167–174

[4] Mikami, Y. et al.: Coagulation Properties and Application of Polyferric Sulphate. PPM *11* (5) (1980) 24–32 (in Japanese)

[5] He, Z.J. et al.: Treatment of Industrial Wastewater and Sludge Dehydration with Inorganic High Molecular Weight Coagulants. Shanghai Environ. Sci. *6* (7) (1987) 32–4 (in Chinese)

[6] Mikami, Y. et al.: Application of Polyferric Sulphate on the Tertiary Treatment of Night Soil. PPM. *11* (11) (1980) 29–35 (in Japanese)

[7] Mikami, Y. et al.: Activated Sludge Treatment by Polyferric Sulphate. PPM *12* (7) (1981) 59–65 (in Japanese)

[8] Mikami, Y. et al.: Treatment of Dying Waste Water by Coagulation. PPM *15* (8) (1984) 14–19 (in Japanese)

[9] Wu, H., Liu, L.: Comparison of the Performance of Polyferric Sulphate and Other Inorganic Flocculants. Wat. Treatm. Technol. *15* (5) (1989) 290–294 (in Chinese)

[10] Du, M.: Treatment of Dying and Printing Wastes by the Combined Processes of Polyferric Sulphate Coagulation and Chlorination Oxidation. Environ. Eng. *9* (3) (1990) 1–4 (in Chinese)
[11] Standing Committee of Analysis: Chemical Disinfecting Agents in Water and Effluent, and Chlorine Demand. London, HMSO, 1980
[12] Greenberg, A.E. (APHA), Clescerii, L. (WEF), Eaton, A.D. (AWWA) (Eds.): Trihalomethane Formation. In: Standard methods for the Examination of Water and Wastewater (18th Ed.). Victor Graphics Inc., 1992, pp. 5-44–5-50
[13] Standing Committee of Analysis: Chloro- and Bromo-Trihalogenated Methanes in Water. London, HMSO, 1981
[14] Hall, E.S., Pachham, R.F.: Coagulation of Organic Colour with Hydrolyzing Coagulants. J. AWWA *57* (9) (1965) 1149–1166
[15] Van Breeman, A.N., Nieuwstad, T.J., Van der meent-Olieman: The Fate of Fulfric Acids During Water Treatment. Wat. Res. *13* (1979) 771–780
[16] Jekel, M.: Removal of Humic Substances in Ground Water Treatment. Wat. Supply *3* (1985) 61–66
[17] Edwards, M., Boller, M., Benjamin, M.M.: Effect of Preozonation on Removal of Organic Matter During Water Treatment Plant Operations. In: Proc. of IAWQ/IWSA Joint Special Conference Geneva, Switzerland, 1992, pp. 17–25
[18] Reckhow, D.A., Singer, P.C., Trussel, R.R.: Ozone As a Coagulant Aid. In: Proc. of AWWA Ann. Conf., Denver, Colo., 1986, pp. 58–69
[19] Dowbiggin, W.B., Singer, P.C.: Effects of Natural Organic Matter and Calcium Onozone-Induced Particle Destabilization. J. AWWA *81* (6) (1989) 77–84
[20] Grasso, D., Weber, W.J. Jr.: Ozone Induced Particle Destabilization. J. AWWA *80* (8) (1988) 73–81
[21] Sannier, B.M., Selleck, R.E., Trussel, R.R.: Preozonation As a Coagulant Aid in Drinking Water Treatment. J. AWWA *75* (5) (1983) 239–246
[22] Maier, D.: Microflocculation by Ozone. In: Oxidation Techniques in Drinking Water Treatment. EPA - 570/ 9 - 77 - 020, Washington D.C., 1979
[23] Edwards, M., Benjamin, M.M.: A Mechanistics Study of Ozone-Induced Particle Destabilization. J. AWWA *83* (6) (1991) 95–105
[24] Edwards, M., Benjamin, M.M.: Transformation of NOM by Ozone and Its Effect on Iron and Al Solubility. J. AWWA *84* (1992) 56–66
[25] Edwards, M., Benjamin, M.M.: The Effects of Preozonation on Coagulant-Natural Organic Interactions. J. AWWA *84* (8) (1992) 63–72
[26] Sinsabaugh III, R.L. et al.: Removal of Dissolved Organic Carbon by Coagulation with Iron Sulphate. J. AWWA *78* (5) (1986) 74–82
[27] Edzwald, J.K. et al: Surrogate Parameters for Monitoring Organic Matter and THM Precursors. J. AWWA *77* (4) (1985) 122–132
[28] Edzwald, J.K., van Benschoten, J.E.: Aluminium Coagulation of Natural Organic Matter. In: Chemical Water and Wastewater Treatment, H.H. Hahn and R. Klute (Eds.). Springer, Berlin Heidelberg New York 1990, pp. 341–359
[29] Schnitzer, M.: Humic Substances: Chemistry and Reactions. In: Soil Organic Matters, M. Schnitzer and S.U. Khan (Eds.). Elsevier Scientific Publ., London 1978, pp. 1–64
[30] Steelink, C.: Implications of Elemental Characteristics of Humic Substances. In: Humic Substances in Soil, Sediment and Water, G.R. Aiken et al. (Eds.). Wiley, New York 1985, pp. 457–476
[31] Stevenson, F.J.: Humic Chemistry – Genesis, Composition and Reactions. Wiley, New York 1982
[32] Thurman, E.M.: Organic Geochemistry of Natural Waters. Martinus Nijhoff, Boston 1985
[33] Committee Report: Organic Removal by Coagulation: A Review and Research Needs. J. AWWA *71* (10) (1979) 588–603
[34] Knocke, W.R., West, S., Hoehn, R.C.: Effects of Low Temperature on the Removal of Trihalomethane Precursors by Coagulation. J. AWWA *78* (4) (1986) 189–195

Jia-Qian Jiang and Nigel J.D. Graham
Environmental and Water Resource Engineering
Department of Civil Engineering
Imperial College
London, SW7 2BU
UK

Clive Harward
Anglian Water Services Ltd.
Chivers Way, Histon
Cambridge, CB4 427
UK

[illegible]

Stabilized Ferrates (VI): Synthesis Method and Applications

N. Neveux, N. Aubertin, R. Gerardin, and O. Evrard

Abstract

Iron (VI) ion FeO_4^{2-} possesses properties which make it potentially useful in certain areas of water purification as an oxidizing, coagulant, flocculant and bactericidal agent. Due to its high instability, ferrate (VI) has not been extensively produced and studied. In this paper, we recommend a synthesis method of alkali and alkaline-earth stable ferrates which can be easily produced industrially; these stable ferrates are characterized by the formula $M(\mathrm{Fe}, X)\mathrm{O}_4$. Applications of these compounds, in particular $K_2(Fe, S)O_4$, in water treatment have been investigated.

Introduction

Ferrate (VI) ion is a compound which has the rare quality of combining oxidating and coagulating qualities with the absence of by-products dangerous to human and environmental health [1,2].

It may be used as an oxidant to replace chlorination since it does not cause the formation of chlorinated by-products of disinfection.

In recent years, there have been studies concerning a possible link between Alzheimer's disease and intake of aluminium [3]. It is known that aluminium salts are very efficient coagulants used for drinking water treatment, and ferrate (VI) ion may be an alternative to aluminium coagulation.

Ferrate (VI) has long been known (since 1841) [4], but due to its high oxidizing power and thereby its instability and difficult preparation, it has not been produced industrially. More recently, potassium ferrate syntheses have been suggested [5,6]. They consist in oxidizing an iron (III) salt to produce FeO_4^{2-} in an alkaline aqueous solution; K_2FeO_4 is then precipitated by addition of great quantities of potash; the oxidation process can be either chemical or electrochemical.

However these ferrates having the formula $M\mathrm{FeO}_4$ are not stable, so they cannot be stored for long periods even sheltered from moisture.

In this paper, we recommend a synthesis method of alkali and alkaline-earth stabilized ferrates (VI) using a dry powders process.

R. Klute / H.H. Hahn (Eds.)
Chemical Water and Wastewater Treatment III

1. Synthesis Process

As we can see in Figure 1, pH increase improves ferrate stability in aqueous solution, but even for a pH value of 14, water is an important reducing agent of hexavalent iron.

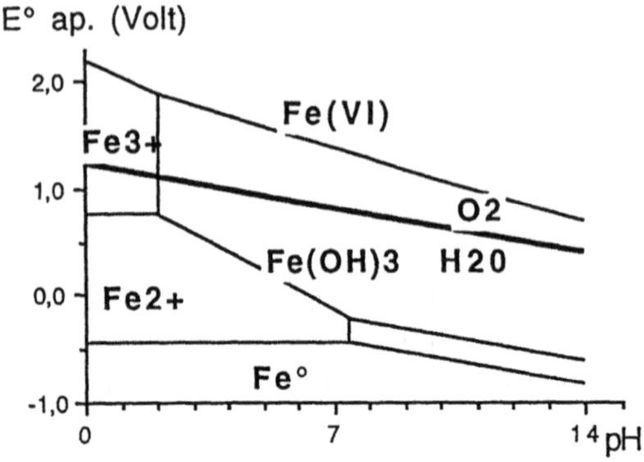

Fig. 1. Standard potentials vs. pH

Water appeared to us as the major element which may cause ferrate reduction, so we progressively eliminated the water in the reaction medium and kept only the constitution water of the different reactives to make the migration of ions, and thereby the oxidation reaction, feasible.

By mixing an iron (II) or (III) salt with an oxidizing chlorine-containing agent and a strong base like potash or soda, we produce, at room temperature, an alkali or alkaline-earth ferrate [7].

We noted that we can produce ferrate (VI) from a bivalent iron salt and that oxidation efficiency is improved when we replace a trivalent iron salt by a bivalent one, in contrast to previous processes [5, 6].

The ferrate oxidation state was determined by ^{57}Fe Mössbauer absorption spectroscopy where hexavalent iron gives a sharp single line with an isomer shift IS = -1 mm/s. As shown in Figure 2, by such a synthesis process, we oxidize whole iron to the hexavalent state.

This synthesis process produces crystallized ferrates (VI) without any aqueous solution step and thereby we achieve 100 % oxidation yield.

2. Highly Stable Ferrates

Synthesis previously described provides alkaline-earth or alkaline ferrates (VI) which can be easily produced industrially. Moreover, this process allows us to obtain highly stable ferrates characterized by the formula:

$$M(\mathrm{Fe}, X)\mathrm{O}_4,$$

where M is Ba or Ca, 2 Na or 2 K and X an element whose cation has the electronic structure of a rare gas [8].

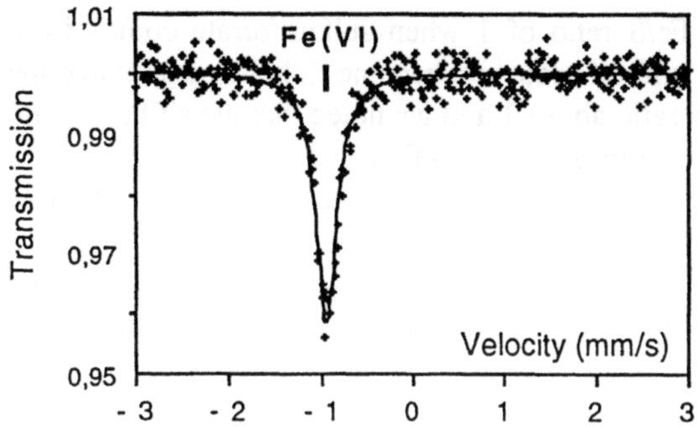

Fig. 2. Mössbauer spectrum of alkali ferrate produced by oxidation of a bivalent iron salt

MXO_4 must be an isomorph of $M\text{FeO}_4$, with close unit cell parameters in order to obtain a solid solution $M(\text{Fe}, X)\text{O}_4$. Thus highly stable ferrates are distinguished from ferrates $M\text{FeO}_4$ by the fact that some Fe atoms are replaced by X atoms.

The high stability can be explained by the fact that, in contrast to the Fe^{6+} cation, the X^{n+} cation has the electronic structure of a rare gas. Molecular orbital theory applied to a tetrahedral radical XO_4^{m-} shows that the electrons from the four O atoms come to occupy connecting orbits σ and π in the radical. Anti-connection energy orbits immediately above are not occupied when X^{n+} has the electronic structure of a rare gas but they are for FeO_4^{2-} so the tetrahedron FeO_4^{2-} is destabilized by two d electrons of Fe^{6+} which enter anti-connection orbits. It will thus be understood that the substitution of Fe^{6+} by X^{n+} stabilizes the tetrahedral system, because such a substitution involves a decreased electron concentration in the anti-connection orbits.

Cation X^{n+} can be, for example, Al^{3+}, Si^{4+}, P^{5+}, S^{6+}, Cl^{7+}, Cr^{6+}, Mo^{6+} and Mn^{7+} and in this paper we shall relate studies of what we called sulfatoferrate $M(\text{Fe}, \text{S})\text{O}_4$, where the substituted cation is S^{6+}. This sulfatoferrate and, more precisely, potassium sulfatoferrate is synthesized by oxidation of an iron (II) sulphate in the presence of potash.

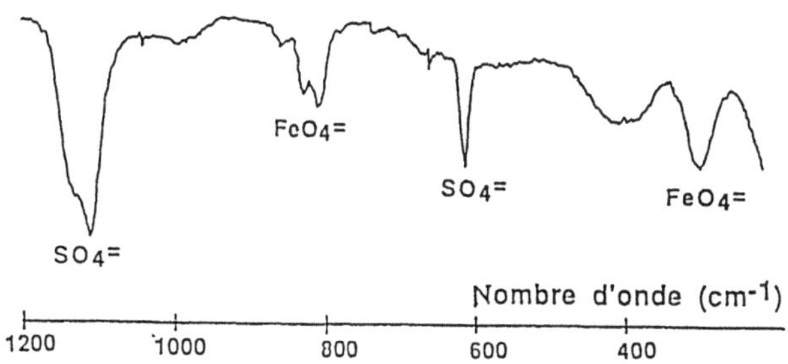

Fig. 3. I.R. absorption spectrum of potassium sulfatoferrate

As seen in Figure 3, the infrared absorption spectrum shows that FeO_4^{2-} and SO_4^{2-} coexist in the synthesized compound.

Chemical analysis gives an Fe/S ratio of 1 when sulfatoferrate comes from ferrous sulphate oxidation. This sulfatoferrate can become richer in iron when the original iron sulphate has an Fe/S ratio above 1 and we noted that the sulfatoferrate Fe/S ratio is always the same as the original iron sulphate Fe/S ratio.

By X-ray diffraction analysis, we brought to the fore the solid solution of potassium sulphate and potassium ferrate. This solid solution is an isomorph of K_2FeO_4 and K_2SO_4, it crystallizes in an orthorhombic system and we established its unit cell parameters which are compared with those of K_2FeO_4 and K_2SO_4 (Table 1).

Table 1. Unit cell parameters of K_2SO_4, K_2FeO_4 and $K_2Fe_{0.5}S_{0.5}O_4$

	a (Å)	b (Å)	c (Å)
K_2SO_4	7.48	5.77	10.07
K_2FeO_4	7.70	5.86	10.36
$K_2Fe_{0.5}S_{0.5}O_4$	7.54	5.82	10.20

Using the dry powders process, we synthesized K_2FeO_4 from iron hydroxide and we compared the stabilities of potassium ferrate and potassium sulfatoferrate. This study uses ^{57}Fe Mössbauer absorption spectroscopy where Fe(VI) reduction is shown by a doublet with an isomer shift IS = 0.3 mm/s. In Figure 4, a comparison of the Mössbauer spectra of K_2FeO_4 and $K_2(Fe, S)O_4$ shows that for a short period of time, ferrate is reduced to Fe(III) while sulfatoferrate can be stored without any problem for months.

3. Sulfatoferrate Applications

In collaboration with NANC.I.E. (International Water Center of Nancy), sulfatoferrate efficiency for water treatment was investigated. When introduced in water, hexavalent iron decomposes as follows:

$$2\,FeO_4^{2-} + 5\,H_2O \rightleftharpoons 2\,Fe(OH)_3 + 4\,OH^- + 3/2\,O_2 .$$

Moreover its oxidant power, the ferrate (VI) ion provides OH^- ions and an $Fe(OH)_3$ type gel in aqueous solution. As the pH increases due to OH^-, precipitation of metallic hydroxides will occur and $Fe(OH)_3$ acts as a flocculant.

Tests with sulfatoferrate were performed on wastewater and drinking water using jar tests and analysis following standard methods (AFNOR).

3.1 Bactericidal Efficiency

Sulfatoferrate bactericidal efficiency was studied on secondary effluent of wastewater, tests were performed in buffered solutions at pH = 6.5, 7.5 and 8.5.

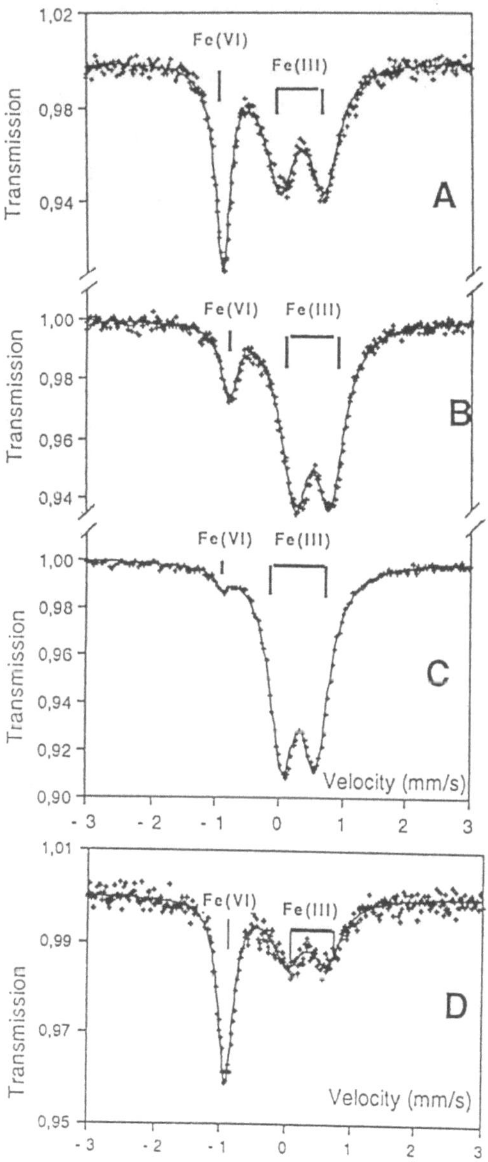

Fig. 4. Mössbauer spectra of K_2FeO_4 sheltered from moisture 1 day (A), 2 days (B), 3 days (C), $K_2(Fe, S)O_4$ 6 months after its synthesis (D)

Sulfatoferrate bactericidal efficiency improves with pH and Fe(VI) concentration, however, it has been shown that the bactericidal effect is not due to the pH increase but only to the sulfatoferrate effect. At pH = 8.5 for a concentration of 10 mg Fe(VI) per liter, total coliform removal is 99.6 % and fecal coliform removal is 99.8 %.

3.2 Flocculant Efficiency

Heavy Metals Removal. Tests were performed on Moselle river water (drinking water), and heavy metals removal was investigated and compared with results obtained when the flocculant was aluminium sulphate (commonly used flocculant).

As shown in Figure 5, sulfatoferrate possesses flocculant qualities as demonstrated by its effective removal of heavy metals (Ba and B excepted) with Fe(VI) concentrations as low as:

- 10 mg/l for Pb, Cu, Al, Ag 100 % removal;
- 20 mg/l for Ni, Cd, Zn 100 % removal.

In all cases, it is shown that sulfatoferrate is more efficient than aluminium sulphate for removing heavy metals.

Moreover, in order to compare sulfatoferrate with Fe(II) and Fe(III) flocculant effects, residual iron concentration was measured. In Figure 6, we plotted residual iron concentration vs. starting iron level for ferrous sulphate (Fe II), ferric chloride (Fe III) and sulfatoferrate (Fe VI).

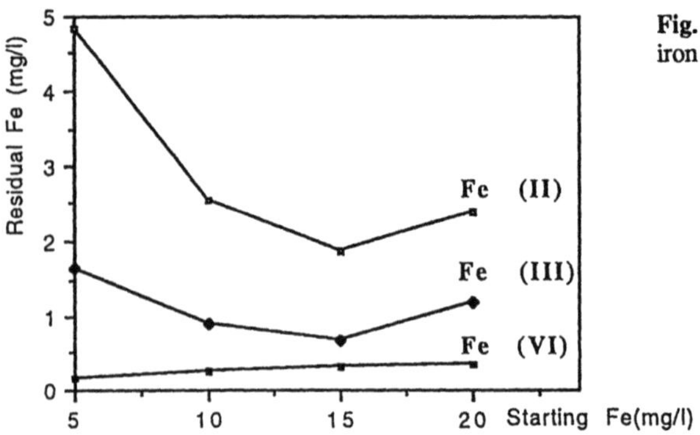

Fig. 6. Residual iron level vs. starting iron quantity

As regards Fe(VI), residual iron level increases with starting iron level; in contrast to Fe(II) and Fe(III), there is no flocculation optimum but Fe(VI) is a good quality flocculant because it results in a lower quantity of residual iron than Fe(II) and Fe(III).

3.3 Oxidizing Power

Sulfatoferrate oxidant quality was evaluated using inorganic pollutants like cyanide because of their high toxicity and nauseating sulphide odor.

Cyanide Oxidation. In Figure 7, cyanide removal by Fe(VI) oxidation is plotted; it is shown that for Fe(VI) levels as low as 10 mg/l, 100 % cyanide removal is achieved.

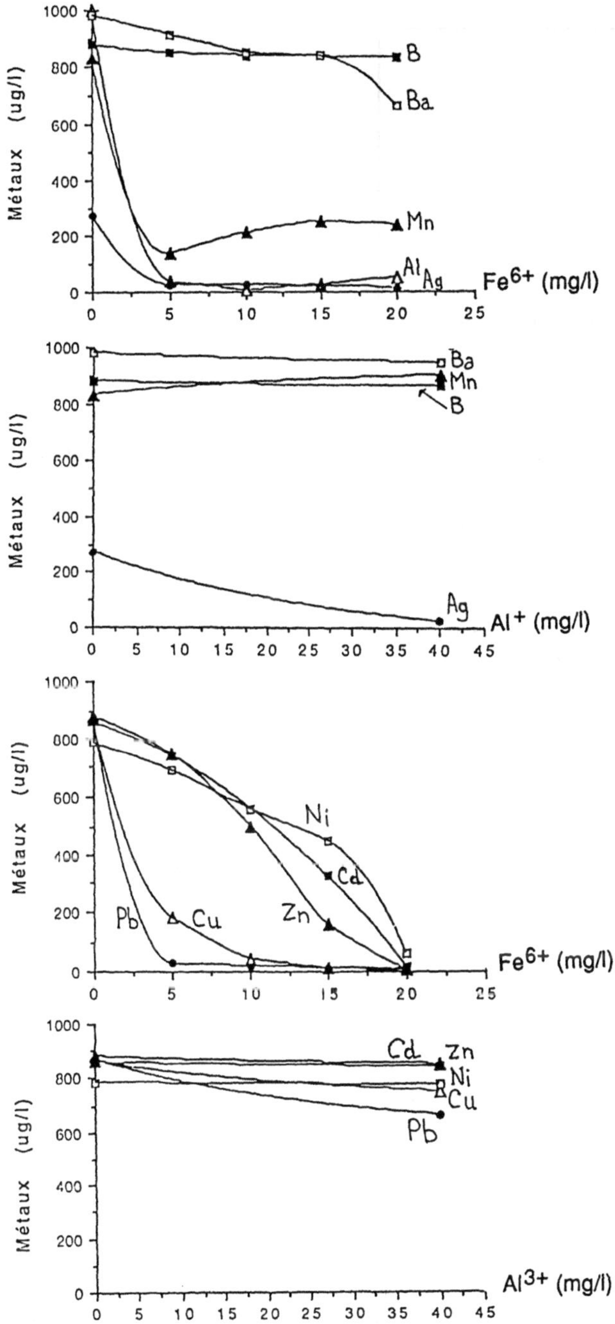

Fig. 5. Heavy metals removal vs. Fe(VI) or Al(III) concentration

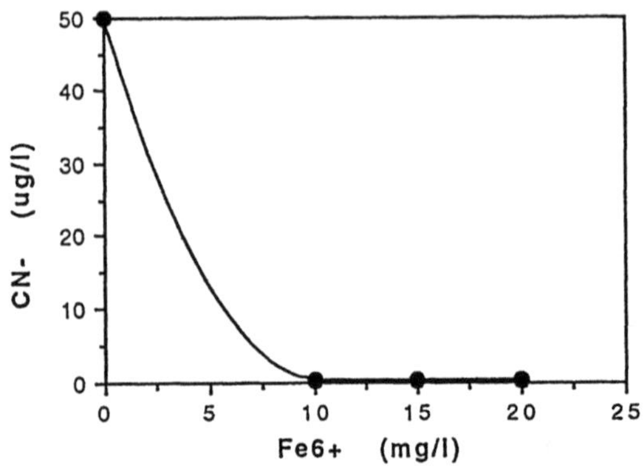

Fig. 7. Cyanide removal vs Fe(VI) concentration

Sulphide Oxidation. Both sulfatoferrate solution and sulfatoferrate powder effects on H_2S were investigated.

When bubbled in a $K_2(Fe, S)O_4$ aqueous solution (pH = 11), the complete H_2S is oxidized by Fe(VI) to SO_4^{2-}:

$$8\,FeO_4^{2-} + 3\,H_2S + 14\,H_2O \rightleftharpoons 8\,Fe(OH)_3 + 3\,SO_4^{2-} + 10\,OH^- .$$

Another important consideration is the reaction between H_2S and solid $K(Fe, S)O_4$ because it may be used for sulphided air pollution control.

By allowing H_2S to flow through solid sulfatoferrate, a chemical reaction occurs between S^{2-} and Fe(VI) to produce FeS_2 and sulphur:

$$FeO_4^{2-} + 3\,H_2S \rightleftharpoons FeS_2 + S + 2\,H_2O + 2\,OH^- .$$

In contrast, such an effect has not been observed with FeOOH where only physical adsorption occurs.

Conclusion

By the synthesis method that we described in this paper, we are capable of producing stable ferrate (VI) which can be easily made industrially and stored for a long time without decomposing.

Such a stable ferrate, which can be, for example, an aluminoferrate, a silicatoferrate, a chromatoferrate, a sulfatoferrate and so on, has oxidizing, coagulating and flocculating properties as desirable as a ferrate with $MFeO_4$ formula.

Moreover this synthesis process makes it possible to enhance an iron and steel (and titanium) industry by-product. As a matter of fact, sulfatoferrate production uses, as original synthesis salt, ferrous sulphate, whose applications are limited since millions of tons of ferrous sulphate are thrown into the sea each year.

References

[1] De Luca, S., Chao, A., Smallwood, C.: Ames Test of Ferrate Treated Water. J. Environm. Engin. *109* (5) (1983) 1159
[2] De Luca, S., Cantelli, M., De Luca, M.: Ferrate vs Traditional Coagulants in the Treatment of Combined Wastes. Wat. Sci. Techn. *26* (9–11) (1992) 2077
[3] Nilsson, R.: Aluminium in Natural Waters Toxicity to Fish and Man. In: Heavy Metals in Hydrological Cycle, M. Astruc and I.N. Lester (Eds.). Selper, London 1988
[4] Fremy, E.: Recherches sur l'action despéroxides alcalins sur les oxides métalliques. Comptes Rendus de l'Académie des Sciences *257* (1841) 23
[5] Deininger, J.P.: Process for Making Potassium Ferrate by the Electrochemical Formation of Sodium Ferrate. US Patent #4,435,256 (1984)
[6] Deininger, J.P., Dotson, R.: Process for Making Potassium Ferrate K_2FeO_4. European Patent #082 590 (1985)
[7] Schmitt, N., Evrard, O., Brice, J.F., Gerardin, R.: Procédé de préparation de ferrates alcalins et alcalino-terreux par voie solide. European Patent #89402241 (1989)
[8] Evrard, O., Gerardin, R., Schmitt, N., Evrard, J.L.: Ferrates de métaux alcalins ou alcalino-terreux, leur préparation et leurs applications industrielles. Patent #PCT/FR/ 9000775 (1990)

N. Neveux, N. Aubertin, R. Gerardin, and O. Evrard
Laboratory of Inorganic Chemistry
Nancy I University
B.P. 239
F-54506 Vandoeuvre-lès-Nancy
France

Coagulant Dosing Control
Using a Model for Wastewater Coagulation

H. Ratnaweera, H. Blom, and G. Aasgaard

Abstract

A concept for optimum coagulant dosage control based on flexible empirical models is proposed. A method for mathematical description of wastewater coagulation using multivariate calibration methods is described. The comprehensives in describing the influent quality with several parameters, rather than with one parameter, are identified to result in the high efficiency of the models. Correlation coefficients for models with good influent descriptions were over 0.90, while models like Dosage = *f* (flow, effluent turbidity) resulted in a correlation coefficient of 0.58. The latter is one of the most commonly used dosing strategies today. Calibration and cross validation coefficients for models with various structures are tabulated based on laboratory-, pilot- and full scale observations. A strategy for coagulant dosing control based on these models and instrumental error identification systems with troubleshooting functions are discussed. The high flexibility of the system to handle the changes in plant configurations, coagulant types, treatment requirements, changes in the number of real-time monitoring parameters, etc. are identified as advantages.

1. Background

The importance of carrying out the coagulation process under optimum conditions increases with the requirements for the production of safe and clean water, with minimum side effects to the environment. The inefficient control of the coagulation process may result in health hazards, high chemical costs, high sludge volumes, negative effects on consequent treatment processes, corrosion problems, etc. With the world-wide growth in the number of chemical water and wastewater treatment plants, the optimum coagulation conditions have become a considerable issue of economy, environment and health in water treatment.

For many years researchers have been investigating the optimisation of the coagulation process. The importance of physical/chemical parameters, such as the mixing of chemicals (Klute et al., 1990), the coagulation pH (Amirtharajah and O'Melia, 1991), the coagulant chemistry (Fettig et al., 1990, Ratnaweera et al., 1992), the influence of flocculant aids (Ødegaard et al., 1992), etc. are well documented. The benefits of most of these conclusions are related to the dosing of

R. Klute / H.H. Hahn (Eds.)
Chemical Water and Wastewater Treatment III

coagulants in optimum amounts required by the influent quality and the required treatment efficiency (Ratnaweera, 1991).

An investigation on the available coagulant dosing control strategies reveals that most of the wastewater treatment plants do not have the possibility either to identify the influent water quality sufficiently or to determine the optimum dosage applicable. On the other hand, the wastewater quality fluctuations during a season, a week or a day are known to be very significant (Sagberg et al., 1990), which emphasizes the necessity for an efficient coagulant dosing control. Many treatment plants monitor only the flow, while some of them also monitor pH, conductivity and temperature of the influent. However, the turbidity and the phosphates, the two major components in the municipal wastewater treatment, are seldom measured in the influent.

Unlike many other industrial processes, the coagulant dosing is difficult to control using feedback concepts, primarily due to the 2–6 hours of sedimentation times and the influent quality fluctuations during that period. The most widespread coagulant dosing strategies include the dosing proportional to flow and sometimes in combination with the over-run control of pH or conductivity in the coagulated water. Some of the large treatment plants practice dosing of coagulants based on experience curves. On the other hand, there is no simple method to determine the optimum coagulant dosage even if the influent quality is well identified. This situation forces most of the treatment plants to run either with an overdose or an under dosage of coagulants, which results in many adverse effects.

The developments in water quality measuring technology (streaming current detectors, floc characteristic monitors, automatic phosphate analysers, etc.) in the last few years indicate a promising future for real time evaluation of the wastewater quality to a sufficient level. For the optimum management of chemical treatment processes using all the important findings on coagulation process optimisation, however, it is important to have a method for their online implementation based on the real time water quality measurements. The definition of a conceptual mathematical model for wastewater coagulation would have been an ideal solution for this purpose. However, the complex nature of municipal wastewater and coagulation chemistry have obstructed the development of such a comprehensive model. The few existing empirical models address only one or two parameters and are often complicated with many experimental coefficients, sometimes also with a physical meaning.

In this paper we present a method for developing an empirical relationship between the water quality and coagulation dosage using advanced data evaluation methods. A strategy for coagulant dosing is presented. The proposed coagulant dosing control is identified to be a flexible and a reliable strategy.

2. The Concept

Coagulation is a well-defined process. The basis for this statement can be demonstrated with a simple example: if you take three identical wastewater samples and dose, coagulate, flocculate and sediment them identically, you should always get three effluents with identical quality, no matter how complex the wastewater. Any defined processes can be mathematically described. Thus, if we can define influent and effluent qualities and coagulation, flocculation and sedimentation conditions, it should be possible to describe the wastewater coagulation process mathematically, i.e. to determine the coagulant dosage based on other parameters.

The influent and effluent quality should be described sufficiently well that two influent samples with identical descriptions should result in identical effluents under the same conditions. If the description criteria are insufficient, for example, only one parameter as influent turbidity is selected, this assumption is not valid, since two samples with identical turbidities will result in different effluent qualities if they had different phosphate contents. This is also the reason for the inefficiency of flow or turbidity proportional coagulant dosing. The description criteria may vary with the application as we will demonstrate later in the different experimental phases. For all experimental phases addressed in this paper, we have excluded the specific description of flocculation and floc separation conditions, since for one and the same conditions, they are indirectly represented in well-defined influent and effluent descriptions.

Based on the above assumption, we can then construct an empirical model for determining the coagulant dosage as a function of influent and effluent quality. This is mathematically realised using multivariate statistical methods, which are further described later.

3. Experimental Methods and Procedures

We conducted experiments on the laboratory scale, pilot scale and full scale. The laboratory scale experiments were conducted using a semi-automated jar-test system in 1 l beakers and a model wastewater with very high concentration differences. We used different coagulants and dosages for the primary feasibility evaluation of the concept. The experimental conditions and the preparation method of model wastewater are reported elsewhere (Ratnaweera, 1991; Ratnaweera et al., 1994). The pilot scale experiments were conducted in a continuous reactor with tube flocculation and floc separation with a sludge blanket. The experiments were conducted on municipal wastewater following real fluctuations and at a flow of 2 l/min. The coagulants were dosed continuously using an in-line mixer (Klute, 1990) ensuring optimum mixing conditions. The wastewater quality data collection and evaluations were conducted at Lillestrøm RA-2 and Lillehamar R-2 wastewater treatment plant during the pilot and full scale experiments. The pilot scale experiments were conducted with variable coagulant dosages including under- and over-dosages. The full scale experiments were conducted with near optimum dosages.

Turbidity, ortho-phosphates, pH, conductivity, flow and temperature were measured during the pilot scale and full scale experiments. The results were registered on-line using a data logger. The remaining analyses were conducted manually using standard methods.

The complete data sets with detailed influent and effluent wastewater qualities and their coagulant types and dosages were then treated using chemometrics to construct empirical models.

4. Chemometrics and Model Construction

In the field of chemometrics, statistical methods are used in planning experiments and data collection, and for interpretation of chemical data. The methods are general and can be used in any field where large amounts of numbers are to be collected and interpreted; biometrics, econometrics, psychometrics, etc. rely on the same principle. Powerful software packages are available.

Chemometrics and the use of advanced statistical methods in data exploration are powerful tools in extracting the relevant information from large data sets. Since chemical and physical instruments have developed rapidly over the last years, we often collect many variables (parameters) from many samples resulting in extensive data matrices. It is often difficult or impossible to get an overview of such magnitudes of data. By the use of e.g. principal component analysis (PCA), the structure of relationships between variables and samples in the data set can be studied. Often a few principal components can replace many original variables without the loss of important information. This yields a considerable data compression.

In multivariate calibration, regression techniques, e.g. principal component regression (PCR) and partial least squares regression (PLSR), are used to establish relationships between variables. An introduction to multivariate calibration and analysis is given by Beebe and Kowalski, 1987, whereas a thorough discussion of the subject is presented by Martens and Næs, 1991.

In this investigation, PLSR was applied to find relationships between coagulant dosage (D), influent variables (X_{in}), and effluent variables (X_{out}) as illustrated in the equation below:

$$D = f(X_{\text{in}}, X_{\text{out}})\,.$$

There are several ways to assess the quality of the developed models. Predicted versus measured values should ideally give a straight line through the origin, with a slope = 1. In practice, however, this is normally not the case, but the deviation of the individual data points from this ideal situation is a measure of the quality of the model. The correlation coefficient between predicted and measured values is one quality parameter, the slope another. RMSEP (Root Mean Square Error of Prediction) is a useful quality parameter. The difference between a predicted and measured value is squared, the sum of squares is calculated for all samples, and then the square root of this sum is found after dividing by the number of samples. Thus, the RMSEP is a measure of the average deviation between predicted and measured values.

A model should always be validated with samples not used in the calibration procedure. The best validation method is to use a test set. First a model is developed from a calibration set and then validated with a test set. If, however, a test set is not available or the original data set is too small or not suited to be split up into a calibration and a test set, cross validation can be used. In cross validation one repeats the calibration several times, each time treating a part of the whole calibration set as prediction samples. In the end all of the calibration objects have been treated as prediction samples. Calculating e.g. the RMSEP gives an indication of the prediction capability of the model.

5. Results and Discussion

The model for laboratory scale experiments was developed using PLSR. The structure of the model is as below:

$$\text{Dosage} = f_1(TU_i, OP_i, PH_i, CO_i, COD_i, ALK_i, TU_e, OP_e, PH_e, CO_e, COD_e),$$

where TU = turbidity, OP = ortho – phosphate, PH = pH, CO = conductivity, COD = chemical oxygen demand, ALK = alkalinity and index "i" and "e" represent influent and effluent parameters, respectively. The results of the evaluation of the model are illustrated in Figure 1.

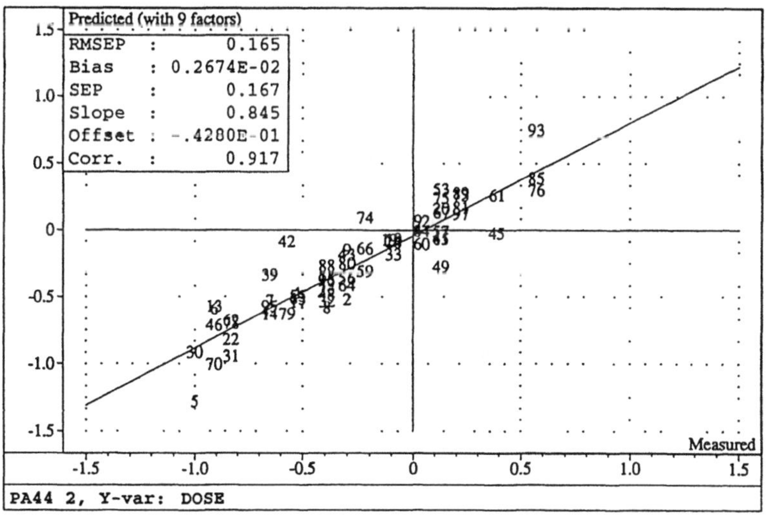

Fig. 1. Predicted vs. measured values in the laboratory scale model with all measured parameters

The values shown in the upper left corner in Figure 1 show the feasibility of the model. The calibration model seems to have a satisfactory correlation coefficient of 0.92 and a RMSEP of 0.17, which is approximately equal to 1.9 mg Al/l. Considering the cross validation (CVS) model data given in Table 1, we have the possibility of predicting of the coagulant dosage ± 2.8 mg Al/l in practice. To use this model, one has to measure the effluent turbidity, ortho-phosphates, pH, conductivity, COD and alkalinity, while the effluent qualities could be defined as the required treatment levels. Considering the wide range of influent quality parameters and the limited number of observations we suggest that the results obtained are positive, and they illustrate the feasibility of the concept.

However, in practice, it is not possible to measure all the required influent parameters on-line. That would be an expensive, if not impossible, solution. Now we have the option of reducing the number of parameters used for the description of influent quality. However, we should be careful here so that each influent sample can be uniquely identified using the selected number of parameters. Table 1 illustrates the calibration and cross validation values for different models with variable parameters in the influent description.

Table 1. Variation of calibration and cross validation parameters upon the reduction of influent quality parameters in laboratory-scale experiments

No	Model	Calibration		Cross validation	
		corr.	RMSEP	corr.	RMSEP
A_1	All influent and effluent parameters	0.92	0.17	0.82	0.24
B_1	As "A_1", excluding OP_i and COD_i	0.87	0.21	0.77	0.27
C_1	As "B_1", excluding TU_i	0.85	0.22	0.75	0.28
D_1	As "B_1", excluding CO_i, PH_i and ALK_i	0.79	0.26	0.68	0.31

The reduction of the model prediction efficiency with the decrease in the number of parameters describing influent quality is well illustrated in Table 1. The reduction in model evaluation parameters, however, does not indicate the impossibility of using models with fewer parameters describing influent quality. It illustrates the increase in the error margin in required dosage prediction, which will in practice normally require a similar overdose to ensure the required effluent quality. Thus, the poorer the influent is described, the higher the error of prediction and, consequently, the higher the waste of coagulants and the higher the other adverse effects. However, in this way, each plant can select its own level of the model considering the available instruments and expected savings versus new investments in instruments.

The pilot experiments conducted with municipal wastewater resulted in a RMSEP of 0.52 mg Al/l and a correlation coefficient of 0.92 for the preliminary model based on the structure similar to model A_1 as above, but excluding ALK_i and CO_e. These results again indicate that the concept may be realized. The relatively high error in the model in comparison to traditional mathematical models in other processes is partially caused by experimental errors.

The observations from the full scale experiments were modelled using a different approach. The model structure was similar to:

$$\text{Dosage} = f_2(Q, SED, T, TU_i, OP_i, PH_i, CO_i, TU_e, PH_e, CO_e)$$

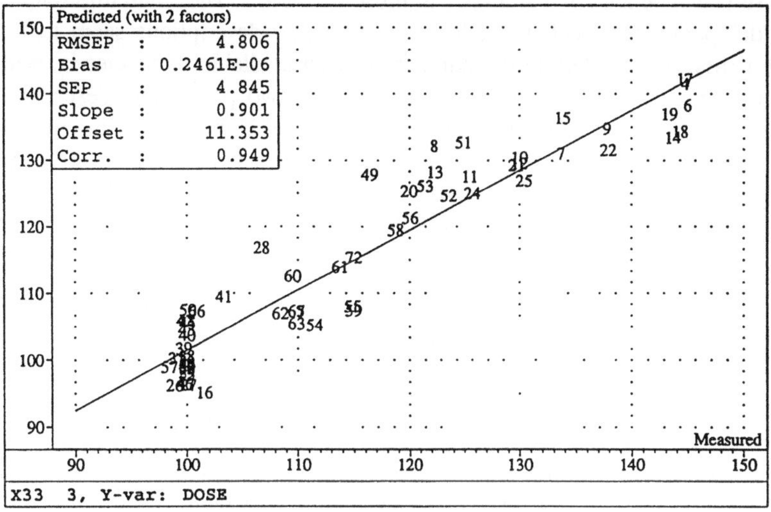

Fig. 2. Predicted vs. measured values in the full scale model with all measured parameters

where Q is flow; SED is sedimentation time and T is temperature. Figure 2 illustrates the model parameters.

The results indicate the potential of the model to predict the dosage with an error of ± 4.8 units (± 0.44 mg Al/l), and has a correlation coefficient of 0.95. Due to some practical reasons an error has occurred in the detection of the influent and effluent qualities from exactly the same wastewater portion. The maximum possible displacement is 30 minutes, and by using hourly average values we have minimised the possible error. However, this error seems to influence the model sensitivity. Our current studies are being conducted without such errors and with the averages of 10 min data registration, which is expected to increase the model efficiency considerably.

We have conducted further analysis on the full scale data, and the calibration and CVS parameters are presented in Table 2.

Table 2. Variation of calibration and cross validation parameters upon the reduction of influent quality parameters in full-scale experiments

No	Model	Calibration		Cross validation	
		corr.	RMSEP	corr.	RMSEP
A_2	All influent and effluent parameters	0.95	4.8	0.93	5.8
B_2	As "A_2", excluding OPi	0.94	5.2	0.92	6.0
C_2	As "B_2", excluding TUi	0.93	5.6	0.91	6.4
D_2	As "B_2", excluding COi, PHi, T, SED, Q	0.92	6.0	0.89	7.1
E_2	As "B_2", excluding COi, PHi, T, SED, TUi	0.90	6.6	0.87	7.4

The results can be analysed similarly to those in Table 1. The small differences in the model parameters with variable influent quality descriptions seem to be surprising at first glance. The information from each available parameter is used to construct the model. When we have only one or two parameters describing the influent quality, and a complete description of the effluent quality, the model may indirectly describe the influent quality to a certain degree by the effluent parameters. This may be the reason why model E_2, which is Dose = f_2(Q, effluent parameters), results in high apparent correlation coefficients.

In order to avoid such overfitting phenomena, we proposed another model structure, which avoids such unfavourable indirect evaluations.

$$\text{Dosage} = f_3(Q, SED, T, TU_i, OP_i, PH_i, CO_i, TU_e)$$
$$\text{Dosage} = f_3(Q, SED, T, TU_i, OP_i, PH_i, CO_i, PH_e)$$
$$\text{Dosage} = f_3(Q, SED, T, TU_i, OP_i, PH_i, CO_i, CO_e)$$

The model coefficients for these models gave much more realistic values. The coefficients for the models with TU_0 as the only effluent parameter are illustrated in Figure 3 and are also given in Table 3.

Both models D_3 and E_3 resulted in poor correlations which indicated the errors possible with coagulant dosing controls using such applications.

$$\text{Model } D_3\text{: Dosage} = f_3(TU_i, TU_e)$$
$$\text{Model } E_3\text{: Dosage} = f_3(Q, TU_e)$$

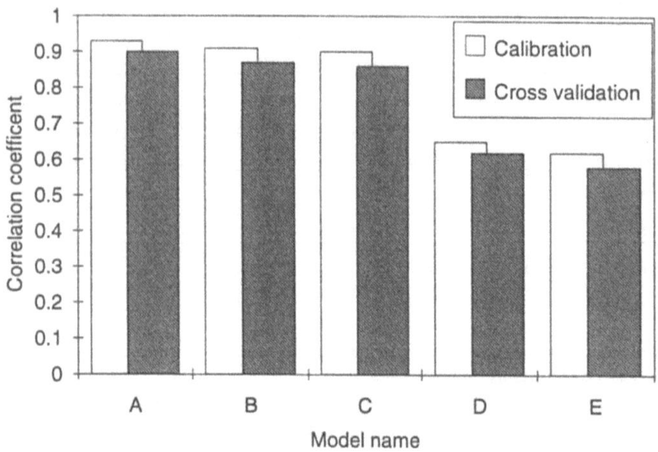

Fig. 3. Variation of calibration and cross validation parameters in different models structures

One of the most common coagulant dosing control methods is similar to model E_3 (dosing of coagulants proportional to flow in order to reduce effluent turbidity). The advantage of a proper dosing control system is clear, when this model (E_3) is compared with a model like A_3, B_3 or C_3.

Table 3. Variation of calibration and cross validation parameters upon the reduction of influent quality parameters in full-scale experiments. Effluent quality is described using only TUo

No	Model	Calibration		Cross validation	
		corr.	RMSEP	corr.	RMSEP
A_3	All influent and effluent parameters	0.93	5.7	0.90	6.7
B_3	As "A_3", excluding OPi	0.91	6.2	0.87	7.5
C_3	As "B_3", excluding TUi	0.90	6.5	0.86	7.8
D_3	As "B_3", excluding COi, PHi, T, SED, Q	0.65	11.6	0.61	12.1
E_3	As "B_3", excluding COi, PHi, T, SED, TUi	0.62	12.0	0.58	12.4

Models A_3, B_3, and C_3 seem to describe the influent well. It is the influent quality which influences the coagulant requirement most (Ratnaweera, 1991).

None of these model structures are optimised yet and therefore, we may obtain even better models in future. Further, the optimisation of the coagulation process itself and the upgrading of the water quality data monitoring system which is the basis for model construction, will indeed improve the models.

Due to the specific nature of each treatment plant's mixing, flocculation and sedimentation facilities, it will be necessary to make calibration models for each of them for the optimal use.

6. Coagulant Dosing Control Strategy

When the suitable model or the series of models is selected, it is important to consider the other factors affecting the total efficiency of a coagulant dosing control system. The concept we are using is illustrated in Figure 4.

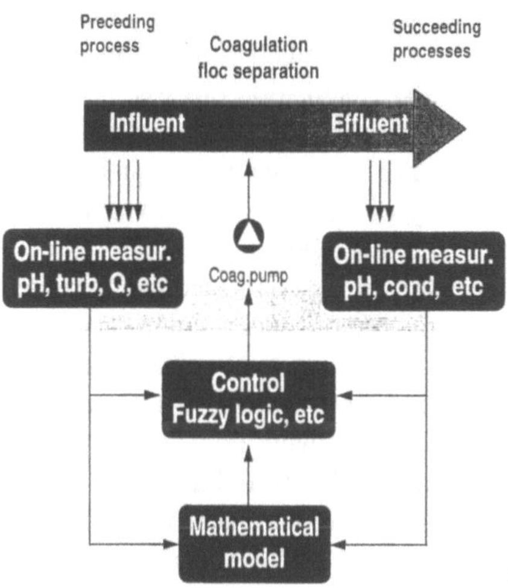

Fig. 4. Coagulant dosing control strategy

One of the most critical factors in a coagulant dosing control based on real-time measurements is the reliability of the measuring instruments. Municipal wastewater treatment plants report maintenance difficulties with such instruments. Thus, it is important to have an error identification strategy. We have studied different alternatives for this purpose, and selected to evaluate the probability of each important measurement to exist in combination with other measurements of the same sample. Investigations on the use of fuzzy technology for this purpose are being carried out now.

We have earlier demonstrated the possibility of constructing various models based on either all influent parameters or with the exception of one or another parameter. This flexibility of the concept will be used to carry out coagulant dosing functions when an instrument failure is detected. Thus, the influence of any one instrument failure will be minimum, and only a slight overdosing will be required according to the new model.

As illustrated in Figure 4, we intend to receive few signals also from the effluent water quality measurements. However, we do not expect to use these real-time effluent quality measurements for the calculation of required coagulant dosage. The effluent quality information required by models for dosage calculations will be taken from the treatment requirements. However, with the use of the same model and the selected dosage we can calculate the expected effluent quality of certain parameters. These predicted values can be compared with the real-time measured effluent qualities as indicated in Figure 4 as another control function. This possibility is also an alternative for or a supplement to the evaluation of measurements.

8. Conclusions

Coagulation of municipal wastewater is a defined process. When the parameters involved in this process (influent, effluent, etc.) are described sufficiently well, it is possible to construct an empirical model for the coagulation process.

The laboratory-scale experiments confirmed the possibility for good empirical models even for model wastewaters with very large variations in the quality. The pilot-scale experiments confirmed the possibility of obtaining improved models for municipal wastewater from one source.

The models constructed using full-scale observations confirmed the expected differences between an optimum dosing control system and a common dosing control system. A model including flow, sedimentation time, temperature, turbidity, pH, ortho-phosphates and conductivity in the influent resulted in a correlation coefficient of 0.90 in contrast to a model just based on influent flow which resulted in a coefficient of 0.58. In both models the effluent turbidity is included.

The models can calculate the minimum coagulant requirement for any given treatment efficiency of a selected effluent parameter. The final selection of coagulant dosage could be based on the treatment requirements for one or more effluent parameters. A prioritising of treatment goals (turbidity or phosphate, etc.) among them may be necessary since the dosages for the achievement of treatment goals for different parameters may not be equal.

The models should be calibrated for each application for maximum results, since flocculation and floc separation efficiencies are unique for each wastewater treatment plant.

An intelligent coagulant dosing control system based on these models and an instrument error identification system is proposed.

The system is flexible to adjust to changes in plant configuration, coagulant types and treatment requirements. It is also flexible enough to use in both large and small plants, with few or many real-time wastewater quality measuring instruments.

Acknowledgement. This research work is primarily funded by the Norwegian Institute for Water Research, the Norwegian State Pollution Authority and the "Ekspomil" program of the Norwegian Research Council. The assistance provided by the Lillestrøm RA-2 and Lillehamar R-2 municipal wastewater treatment plants is gratefully acknowledged.

References

Amirtharajah, A., O'Melia, C.R.: Water Quality and Treatment. AWWA, McGraw-Hill, 1990, 1194 p

Beebe, K.R., Kowalski, B.R.: An Introduction to Multivariate Calibration and Analysis. Analytical Chemistry *59* (17) (1987) 1007A–1017A

Klute, R.: Destabilization and Aggregation in Turbulant Pipe Flow. Coagulation with Prepolymerized Metal Salts. In: Chemical Water and Wastewater Treatment, H.H. Hahn and R. Klute (Eds.). Springer, Berlin Heidelberg New York 1990, pp. 3–18

Martens, H., Næs, T.: Multivariate Calibration. Wiley, New York 1991, 419 p

Ødegaard, H., Fettig, J., Ratnaweera, H.: Coagulation with Prepolymerized Metal Salts. In: Chemical Water and Wastewater Treatment, H.H. Hahn and R. Klute (Eds.). Springer, Berlin Heidelberg New York 1990, pp. 189–220

Ødegaard, H., Grutle, S., Ratnaweera, H.: An Analysis of Floc Separation Characteristics in Chemical Water Treatment II. In: Chemical Water and Wastewater Treatment II, R. Klute and H.H. Hahn (Eds.). Springer, Berlin Heidelberg New York 1992, pp. 97–114

Ratnaweera, H.: Influence of the Degree of Coagulant Prepolymerization on Wastewater Coagulation Mechanisms. Ph.D. dissertation, The University of Trondheim, Norwegian Institute of Technology, Norway, 1991, 143 pp.

Ratnaweera, H., Aasgaard, G., Verpe, Å.: Coagulation Control with Real-time Monitoring of Water Quality. IWSA-IWQA Workshop on Optimal Dosing of Coagulants and Flocculants, Mülheim, Germany, 1994

Ratnaweera, H., Fettig, J., Ødegaard, H.: Coagulation with Prepolymerized Aluminium Coagulants and Their Influence on Particle and Phosphate Removal. IAWPRC Biennial Conference at Washington, D.C. Wat. Sci. Techn. *26* (5–6) (1992) 1229–1237

Sagberg, P., Sæter, R., Baggerud Berge, A.: Increasing the Surface Load at a Direct Precipitation Plant, VEAS, Norway. In: Chemical Water and Wastewater Treatment, H.H. Hahn and R. Klute (Eds.). Springer, Berlin Heidelberg New York 1990, pp. 271–282

Harsha Ratnaweera and Gunnar Aasgaard
Norwegian Institute for Water Research (NIVA)
P.O. Box 173 Kjelsås
N-0411 Oslo
Norway

Hans Blom
Østlandskonsult AS
P.O. Box 123
N-1601 Fredrikstad
Norway

Coagulation of Humic Water with Fenton Reagent

P. Dolejs, N. Kalousková, H. Paillard, M. Prados, and B. Legube

Abstract

An iron (II) and hydrogen peroxide mixture, also called 'Fenton's reagent', produces an extremely strong oxidizing agent the hydroxyl radical $OH^{\bullet}$. Recent publications by several authors have shown that use of Fenton's reagent could be considered for use in water treatment, due to its very effective oxidation of specific organic substances like pesticides. Another advantage of Fenton's reagent is that it does not produce bromates. Iron (III), which is the other reaction product of Fenton's reagent, could certainly act as coagulant in the treated water.

There are some differences between coagulation with Fe^{3+} and that with Fenton's reagent. Optimal dose of iron in the case of Fenton's reagent depends on interactions between the alkalinity (pH) and H_2O_2/Fe^{2+} ratio. For the same concentration of humic substances in model humic water (5 mg/l of HS) the optimal dose varied from 40 μmol/l to 110 μmol/l of Fe, depending on the other variables. Residual colour, residual UV absorbance (254 nm) and residual iron also depend on interactions between initial alkalinity, coagulation pH, iron dose, and H_2O_2/Fe^{2+} ratio. No values of optimal pH or iron dose can be given or recommended as guide values at this time. It seems that optimal iron doses are lower when Fenton's reagent is used rather than iron (III). The H_2O_2/Fe^{2+} ratio should not drop below 0.5 but, on the other hand, there is only a slight improvement when it exceeds 0.5.

1. Introduction

Oxidation of organic substances by a mixture of iron (II) and hydrogen peroxide was published by Fenton in 1949 [1]. This paper also gave the name 'Fentons's reagent' to the ferrous iron and hydrogen peroxide ($Fe^{2+} + H_2O_2$) mixture.

The process which was studied is based on a reaction between iron (II) and hydrogen peroxide, and produces hydroxyl radicals $OH^{\bullet}$ and iron (III) according to reaction 1 [2].

$$Fe^{2+} + H_2O_2 = Fe^{3+} + OH^- + OH^{\bullet} . \qquad (1)$$

The hydroxyl radical ($OH^{\bullet}$) has a very high oxidation potential (2.80 V), which is higher than that of ozone (2.07 V). Iron (III) is suitable for subsequent coagulation in the system.

R. Klute / H.H. Hahn (Eds.)
Chemical Water and Wastewater Treatment III

The great potential advantage of using Fenton's reagent in water treatment may be that a powerful oxidation and coagulation takes place practically at the same time in one technological unit only. What may be even more interesting is the possibility that such a unit may be relatively simple to build and cheap to operate in comparison with the other oxidation processes used to date in the drinking water treatment field.

Considerable work has been conducted in the last several years with respect to the oxidation part of the use of Fenton's reagent in water treatment. The process of using hydrogen peroxide and iron (II) for oxidation and coagulation in water treatment was also patented by OTV as the 'Feroxide process' [3]. The oxidation reactions, which occur during treatment with Fenton's reagent, were studied in detail quite recently [4, 5] and may be considered relatively well understood now.

It was shown in [4, 5] that the oxidation of pesticides like simazine, atrazine, aldicarbe, izoproturon and lindane was very effective and that the Feroxide process performed better than ozonation. A good overview of the literature on Fenton's reagent can also be found in [4, 5].

On the other hand, if the iron (III), which is the second important reaction product, would not effectively perform as coagulant, the process would become much less interesting for most water treatment applications, even when we consider its excellent oxidation performance (e.g. high efficiency in oxidation of pesticides or negligible production of bromates).

The aim of this work was to assess important factors which influence the coagulation performance of the Feroxide process.

2. Materials and Methods

2.1 Waters Used for the Laboratory Experiments

Solid samples of humic substances (HS) were used in experiments reported in this paper. These samples of HS were isolated in the Czech Republic, and originated from a peatbog in a catchment area of a water supply reservoir.

The main characteristics of the HS samples are:

C = 54.5 %
H = 4.7 %
N = 3.3 %
Ash content = 0.89 %
Total $COOH^-$ acidity = 6.7 mmol/g.

Model humic water for all the experiments was prepared from distilled water, by addition of inorganic constituents and a solution of humic substances. The basic composition of this humic water was as follows:

$CaCl_2$ = 0.4 mmol/l
Na_2SO_4 = 0.4 mmol/l
Humic substances = 5 mg/l
$NaHCO_3$ = variable according to the desired coagulation pH.

This model humic water had:

A 254 = 0.524 [5 cm]
A 387 = 0.133 [5 cm]
Conductivity = 210 $\mu S/cm$.

Alkalinity is expressed in accordance with the suggested terminology in the field of aquatic chemistry, as "Acid Neutralizing Capacity to pH 4.5" – ANC.

Most chemicals used were of analytical grade. Iron (III) sulphate was a commercial water treatment chemical product, supplied by Precheza Company, Prerov, Czech Republic. Iron (III) sulphate was used instead of $FeCl_3$ to prevent any possibility of side effects on the chemistry of the system, caused by differences between sulphate and chloride concentrations in comparable Fe(III) and Feroxide experiments.

2.2 Experimental and Analytical Methods

In coagulation of humic waters, the perikinetic phase is of primary importance due to the small size of the molecules of humic substances (2–200 Å). To study coagulation (or, in other words, formation of aggregates), in such system, the method published in the previous papers [6, 7] was used for all the experiments in this study. This method is based on separation of aggregates, which are being formed in the studied system, by a centrifuge. It also enables aggregation in the system to be studied after different time intervals, i.e. times which elapsed after homogenization of coagulant with the raw water.

Before the separation of aggregates by centrifugation, rapid homogenization of sample with coagulant was achieved by injecting the raw water sample into a volumetric cylinder with the coagulant (Fe^{3+} or Fe^{2+}). In the Feroxide experiments, the raw water was dosed with the appropriate amount of H_2O_2. No mixing in cylinders took place. It is not necessary because Brownian motion is the only effective transport mechanism up to the particle size of about 300–400 nm. This particle size is, on the other hand, sufficient for their effective removal by centrifugation in a standard quality bench scale centrifuge with a g (gravity acceleration) value of more than about 3000 m/s^2.

The advantage of this method is the possibility of achieving good physical similarity of all the experiments carried out practically in every laboratory. This is based on the fact that the transport mechanism and collision frequency between molecules and small destabilized particles in the whole system are governed by only one physical factor – temperature. Experimental results can thus be compared directly as long as only two experimental parameters remain the same - temperature and centrifugation efficiency ($g \cdot t$ – i.e. product between the g value (centrifugal acceleration) and time of centrifugation – t) . Such a method also makes it possible to study early stages of aggregation, i.e. formation of particles in the first one or two minutes after the coagulant addition.

This advantage is hardly achieved with the classical jar test, where settleable flocs are usually produced after several tens of minutes of mixing and subsequent sedimentation. Therefore, from the coagulation kinetics point of view, it is not easy to apply the results of a classical jar test to the operation of direct filtration plants, where separation follows the addition of coagulant in a time of only several minutes. The development of the centrifugation method of coagulation testing was also based on the experience at several treatment plants, where the jar test results failed to optimize the dosage of coagulants.

It should also be mentioned that the reproducibility of centrifugation test results is the same or better than that of advanced jar tests which use sand filtration. Another advantage of the centrifugation test for a researcher is that consumption of the model water is reduced in comparison with jar tests by nearly one order of magnitude.

Aggregates were separated by centrifugation after 5 and 40 minutes aggregation time, respectively, i.e. time after homogenization of water with coagulant. This was done with the aim of gaining some insight into the optimum conditions at different time intervals, also to gain information about the rate of separable particles formation in order to apply the results to either direct filtration plants or classical flocculation/sedimentation plants design.

The three most important parameters in humic water treatment were determined in samples after centrifugation:

- residual coagulant (i.e. residual total Fe)
- residual absorbance at 254 nm (A 254)
- residual absorbance at 387 nm (A 387).

Absorbance at 387 nm was found in past research to be the best wavelength for calibrating spectrophotometers with inorganic colour standards [8], as this wavelength is the most accurate in determining the colour of samples with different molecular weight composition. The calculation of colour values can be made easily by multiplying the absorbance measured in 5 cm cuvettes by a factor. Factor 255 may be used to calculate the colour from values of A 387 measured in this report.

$$\text{Colour} = 255 \cdot \text{A}\,387\ (5\ \text{cm}) \qquad [\text{Pt, mg/l}]\,. \tag{2}$$

It should be mentioned that A 253 and A 387 are only surrogate parameters to estimate concentration changes of humic substances. A more detailed study of the removal of organics will be made in future experiments. DOC (dissolved organic carbon) and removal efficiency of different fractions of DOC will be determined.

Total iron was determined spectrophotometrically. The samples were acidified with HCl, iron was oxidized with H_2O_2 and colour was developed by a subsequent reaction with KSCN (which is practically an 'old' analytical method, which uses for the oxidation of iron the identical composition of reagents as the Fenton's reagent). Absorbance was measured at 480 nm. This method was tested in coagulation experiments with humic waters many times in the past. Both the precision and correctness of results were always quite satisfactory.

The centrifuge used was a Sigma, type 3–10, without cooling and with an angle rotor No. 12 159 with 6 × 85 ml buckets. The g value used in all experiments was 6000 m/s^2. Time of centrifugation was 120 seconds and the $g \cdot t$

value used throughout all experiments reported here was 676 000 m/s. A Philips PU 8625 UV-VIS spectrophotometer with 5 cm quartz cuvettes was used for all absorbance measurements.

3. Results and Discussion

To get some kind of baseline information about the coagulation behaviour of the model humic water used in the experiments, coagulation with Fe(III) was studied first. Comparison between a classical coagulant, like $Fe_2(SO_4)_3$, and the Feroxide process is important, because it can give information about the design and operational changes needed if the Feroxide process should be used in practice.

Important independent variables, which have to be studied in the field of coagulation with Fenton's reagent, were identified after a series of introductory experiments. The choice also reflected the results of the part of the Fenton's reagent studies dealing with oxidation of organic matter. They were:

- iron (II) dose [μmol/l]
- molar ratio between H_2O_2 and Fe(II)
- initial ANC (alkalinity) of the treated water [mmol/l].

The influence of pH was studied by changing the initial ANC value of the raw water, before dosing both H_2O_2 and Fe(II). After the alkali or acid addition, the raw water was held until carbonate equilibrium was reached. This is considered to be the best way to simulate the real processes which occur in treatment plants, especially in those plants which do not have any pH adjustment before dosing the coagulant. Any pH adjustment after the coagulant is dosed could influence the very important early stage reactions of coagulant in the treated water, in a way which is not quite reproducible on a full scale.

The sequence of dosing was maintained in all experiments as follows: H_2O_2 first, Fe(II) after [3]. Molar ratio between H_2O_2 and Fe^{2+} is defined as:

$$R = \left[H_2O_2\right]/\left[Fe^{2+}\right] \tag{3}$$

$R = 0.5$ was considered to be stoichiometric value in [4,5]. Authors of [4,5] studied R values up to 1.5 and concluded that for oxidation of the pesticides studied (simazine, atrazine, aldicarbe, izoproturon and lindane), values between 0.5–0.75 were quite effective in reducing pesticides concentrations below the detection limits of HPLC with a UV detector or GC with an ECD detector. Values of R down to 0.25 could also be found in other studies [9]. From the technological point of view, the higher the R value, the higher the operational costs of the Fenton's reagent oxidation.

Because the system with Fenton's reagent is, with respect to coagulation, a relatively unknown field, it would be risky to design at this phase of research a three-factor orthogonal experimental plan. Therefore, the usual selection of different combinations of variables had to be made to gain at least some basic orientation in the system in an acceptable time period.

Results are presented in the form of a group of four figures. Figures marked a) and b) show residual values at A 254, A 387 and total iron after 5 minutes of aggregation. Figures marked c) and d) show the residual values after 40 minutes of aggregation.

3.1 Coagulation of Model Humic Water with Fe(III)

An example of the results of the coagulation experiments with Fe(III) are presented in Figure 1a–d. Experiments were made at so-called "free pH", so the coagulation pH plotted in Figures 1b, d is a result of a drop in the pH of model humic water by the addition of coagulant. This simulates the treatment in waterworks, where no pH adjustment is used before or during coagulation and flocculation.

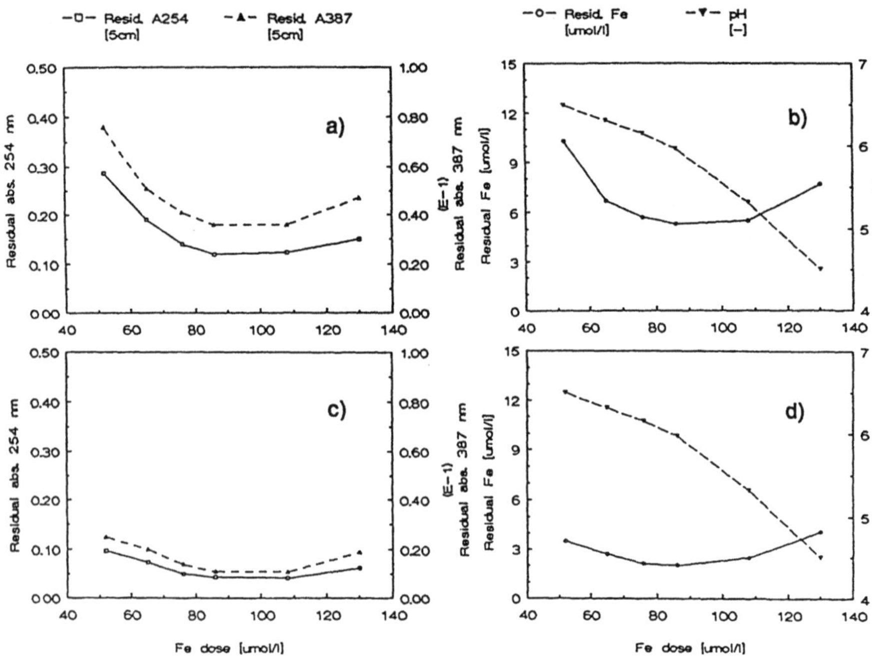

Fig. 1a–d. Coagulation experiment with iron (III) – ANC = 0.38 mmol/l. (**a**) and (**b**): aggr. time 5 min; (**c**) and (**d**): aggr. time 40 min

For 5 minutes aggregation time (Fig. 1a, b) the optimum dose of iron is about 90 μmol/l and pH is 5.8. The optimum was the same for all the three residuals, at A 254 and A 387 and Fe. Residual iron concentration was 5.3 μmol/l, which means that 94 % of dosed iron was associated with particles removable by centrifugation at the $g \cdot t$ value previously mentioned.

Comparison of the twin figures (a–c and b–d) shows the progress of aggregation into particles separable by centrifugation between 5 and 40 minutes. The zone of optimum doses is broader and there is not any significant difference between pH and iron dose optima with respect to residuals at A 254 and A 387 and iron.

3.2 Coagulation with Fenton's Reagent

3.2.1 Influence of Raw Water Alkalinity (ANC) and Iron Dose

The combined influence of both pH and iron dose on the coagulation performance of Fenton's reagent is shown in Figures 2–5.

Results of this series of four experiments are partly summarized and compared in Table 1. Values of the lowest residuals at A 254 and A 387 and Fe are presented together with the corresponding optimum iron doses and corresponding pH values.

Table 1. Values of the lowest residuals at A 254 and A 387 and Fe at different raw water ANC and their corresponding in terms of the optimum iron dose and pH at this dose. $[H_2O_2]/[Fe^{2+}] = 0.5$

Raw water ANC [mmol/l]	0.48	0.38	0.24	0.14
Residual at A 254 [5 cm]				
5 min	0.140	0.147	0.175	0.200
Optim. dose / pH	90/6.4	70/6.35	55/6.1	45/5.1
40 min	0.050	0.049	0.058	0.066
Optim. dose / pH	110/6.2	100/5.85	70/5.8	60/4.5
Residual at A 387 [5 cm]				
5 min	0.048	0.048	0.058	0.066
Optim. dose / pH	90/6.4	70/6.35	55/6.1	45/5.1
40 min	0.011	0.017	0.017	0.021
Optim. dosc / pH	110/6.2	90/6.0	65/5.9	60/4.5
Residual at Fe [μmol/l]				
5 min	6.0	5.2	6.9	13.7
Optim. dose / pH	75/6.6	70/6.35	45/6.25	42/5.3
40 min	2.1	1.7	2.1	7.5
Optim. dose / pH	75/6.6	90/6.0	45/6.25	40/5.5

It can be concluded from Table 1 that the decrease in raw water ANC increases the residual values at both A 254 and A 387 at both 5 and 40 minutes aggregation time. The optimum dose, with respect to residuals at A 254 and A 387, decreases with the drop in ANC. The optimal pH also decreases with the reduction in ANC. The difference between the two extreme ANC values (0.48 and 0.14 mmol/l) is great. An Fe dose of 90 μmol/l at pH 6.4 is the optimum at ANC = 0.48, while the optimal dose is only 45 μmol/l and pH = 5.1 at ANC = 0.14. These results illustrate how difficult it is to speak in terms of 'optimal pH value' or 'stoichiometry of coagulation' in the treatment of humic waters.

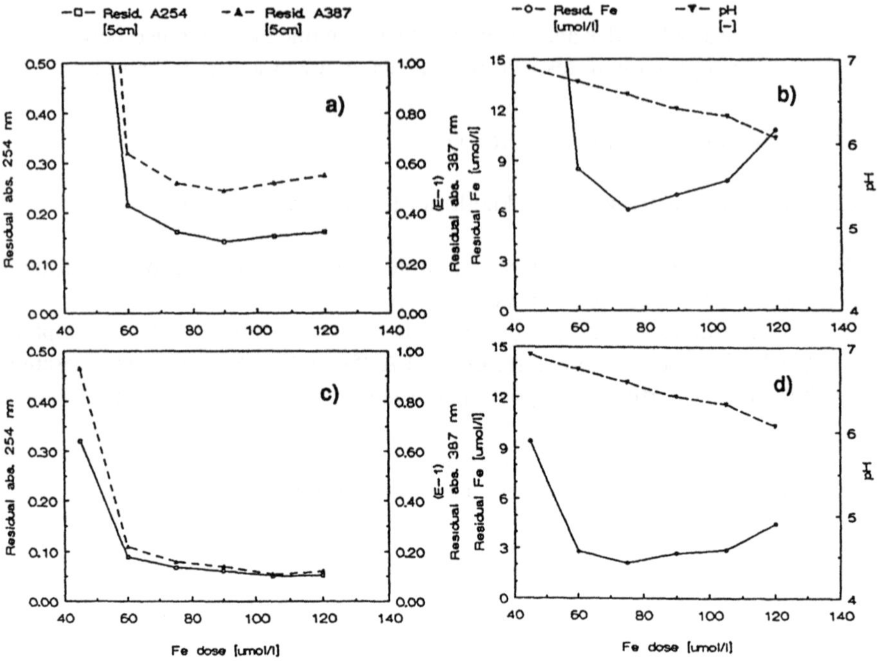

Fig. 2a–d. Coagulation experiment with Fenton's reagent at ANC = 0.48 mmol/l and R = 0.5. (**a**) and (**b**): aggr. time 5 min; (**c**) and (**d**): aggr. time 40 min

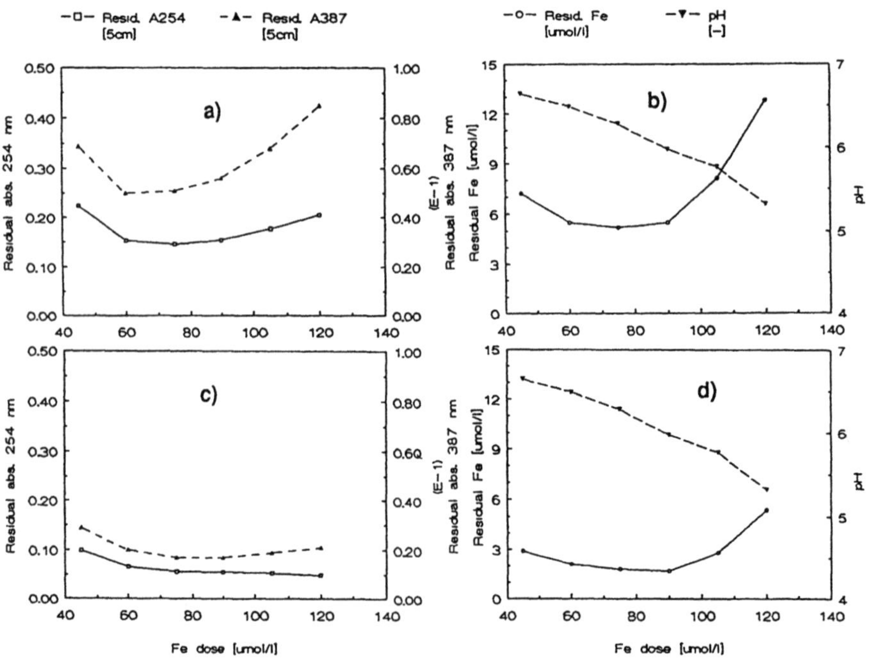

Fig. 3a–d. Coagulation experiment with Fenton's reagent at ANC = 0.38 mmol/l and R = 0.5. (**a**) and (**b**): aggr. time 5 min; (**c**) and (**d**): aggr. time 40 min

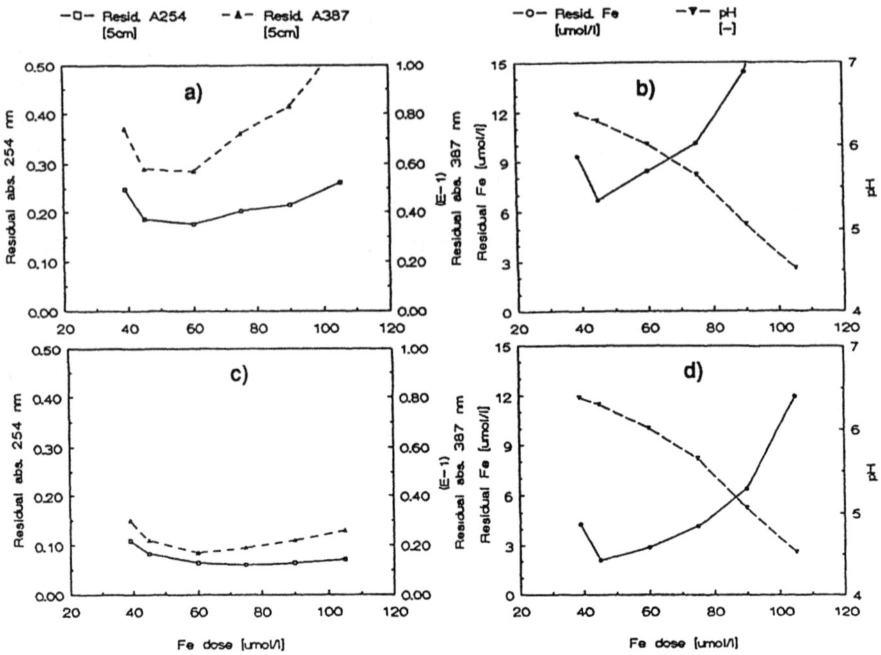

Fig. 4a–d. Coagulation experiment with Fenton's reagent at ANC = 0.24 mmol/l and R = 0.5. (**a**) and (**b**): aggr. time 5 min; (**c**) and (**d**): aggr. time 40 min

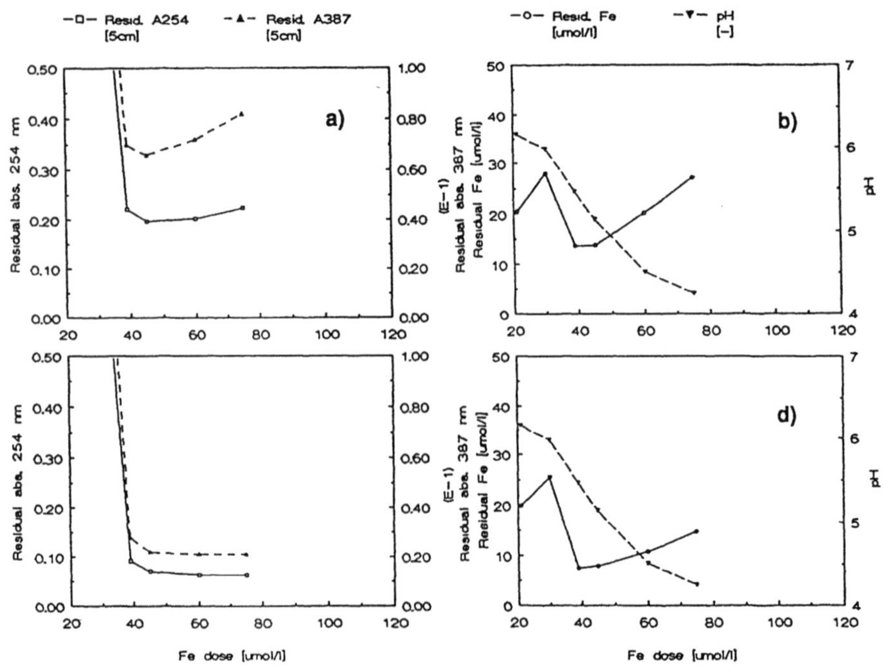

Fig. 5a–d. Coagulation experiment with Fenton's reagent at ANC = 0.14 mmol/l and R = 0.5. (**a**) and (**b**): aggr. time 5 min; (**c**) and (**d**): aggr. time 40 min

The changes in residual iron concentrations show one difference from the trends of both residual absorbances. The lowest residual concentrations of iron were found in experiments with ANC = 0.38. This could suggest that the combination of variables is very close to the optimum with respect to the residual Fe concentration, and to the characteristics of the model humic water used. Most other combinations of iron dose and pH are probably underdosed or overdosed, or the pH is too low or too high.

Comparison of data in Table 1 also shows another example of the importance of the time factor in coagulation testing. The best performing doses increased with time of aggregation at all the ANC values. This observation has already been published in previous papers [6, 7] and seems to be valid also for coagulation with Fenton's reagent. A practical consequence of this observation is that the optimal dose at a direct filtration plant will be lower than at a conventional treatment plant treating the same raw water.

A comparison between coagulation with iron (III) and coagulation with Fenton's reagent (Figs. 1 and 3), under the same chemical conditions for raw water, shows several differences.

The most apparent is that optimal dose in coagulation with Fenton's reagent (70 μmol/l) is lower than that with iron (III) (90 μmol/l). Residual iron concentrations are about 20–25 % smaller with Fenton's reagent. On the other hand, residual absorbances, especially at A 387, are higher with Fenton's reagent than with iron (III), which is quite apparent at 5 minutes aggregation time. This is rather surprising in the context of the results, and further investigation is required.

Another feature of coagulation with Fenton's reagent is apparent especially in Figures 2, 4 and 5. In these figures it is interesting that the system is very sensitive to underdosing, and instead of the usual U-type curves we more or less obtain L-type curves (see Figures 2a, c or 5a, c). While the aggregation collapses below Fe dose 60 μmol/l in Figure 2 (ANC = 0.48), we can see in Figure 4 (ANC = 0.24) that with the same dose (60 μmol/l) the system is slightly overdosed with respect to residual iron. Most similar to the curves obtained with iron (III) are the curves in Figure 3, where the optimum zone is relatively symmetrical, and at 5 minutes aggregation time also broader for residual Fe than in Figures 2, 4 and 5. This is probably because the system is close to optimum with respect to both iron dose and pH, which was already discussed above.

3.2.2 Influence of Molar Ratio H_2O_2/Fe^{2+}

The influence of the ratio between H_2O_2/Fe^{2+} (R) is shown in Figures 3, 6 and 7. Table 2 presents a combined comparison of the optima in Figures 3, 6 and 7.

We see that R values below 0.5 have a strong deteriorating effect on the residual Fe concentrations. They also slightly increase the residual values at A 254 and A 387 at optimal doses. Surprisingly, the consequences of low R value (0.4) are not so bad as far as residuals at A 254 and A 387 are concerned, in comparison with the residual iron values. At $R = 0.4$ we see that optimal doses are apparently higher with respect to the residual absorbances, which resembles the shift caused

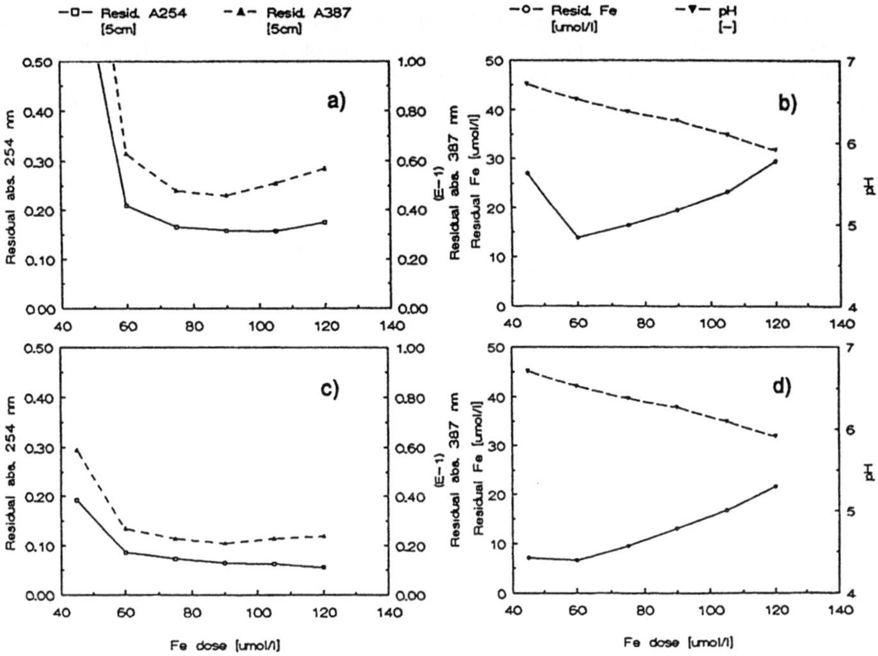

Fig. 6a–d. Coagulation experiment with Fenton's reagent at ANC = 0.38 mmol/l and R = 0.4. (**a**) and (**b**): aggr. time 5 min; (**c**) and (**d**): aggr. time 40 min

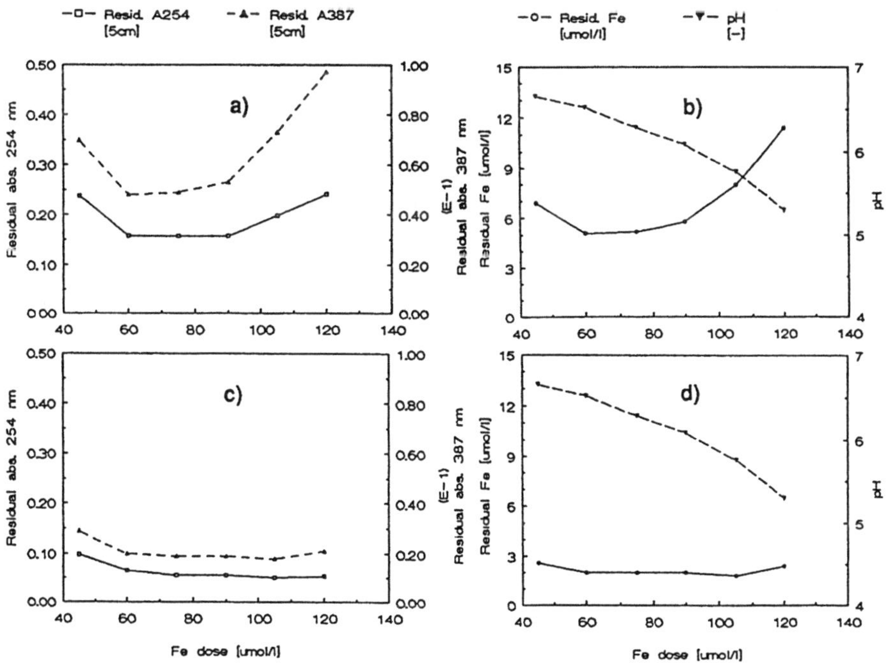

Fig. 7a–d. Coagulation experiment with Fenton's reagent at ANC = 0.38 mmol/l and R = 0.6. (**a**) and (**b**): aggr. time 5 min; (**c**) and (**d**): aggr. time 40 min

Table 2. Values of the lowest residuals at A 254 and A 387 and Fe at different H_2O_2/Fe^{2+} ratios (R) and their corresponding in terms of the optimum iron dose and pH at this dose. ANC = 0.38 mmol/l

$R = [H_2O_2]/[Fe^{2+}]$ [mol/mol]	0.40	0.50	0.60
Residual at A 254 [5 cm]			
5 min	0.160	0.148	0.150
Optim. dose / pH	90/6.3	70/6.35	60/6.5
40 min	0.052	0.049	0.046
Optim. dose / pH	110/6.0	100/5.85	75–120/6.3–5.3
Residual at A 387 [5 cm]			
5 min	0.050	0.048	0.046
Optim. dose / pH	90/6.3	70/6.35	60/6.5
40 min	0.020	0.017	0.017
Optim. dose / pH	100/6.1	90/6.0	75–120/6.3–5.3
Residual at Fe [μmol/l]			
5 min	13.9	5.2	5.2
Optim. dose / pH	60/6.5	70/6.35	60/6.5
40 min	6.6	1.7	1.7
Optim. dose / pH	60/6.5	85/6.0	60–110/6.6–5.6

by the increase in raw water ANC from 0.38 to 0.48 (see especially Figures 2a, 3a and 6a).

$R = 0.6$ has only one positive effect (with respect to coagulation) at 40 minutes aggregation time. It gives a very broad spectrum of optimal doses, which is also apparent in comparison with results of iron (III) coagulation, see Figure 1c, d.

Residual Fe concentration is the most important variable for coagulation experiments with Fenton's reagent. Optimal dose and initial ANC value must be well determined and maintained in practice.

It can be concluded that the process should never be operated below $R = 0.5$, because of resulting very high residual iron concentrations. On the other hand, if the system is not underdosed with Fe and if the pH achieved after the H_2O_2 and Fe(II) addition is favourable, higher values of R than 0.5 are not necessarily needed from the coagulation point of view. In practice, it would usually be more economical to adjust the raw water ANC to the optimum, rather than to try to overcome resulting problems by dosing higher amounts of hydrogen peroxide.

4. Conclusions

1. R values below 0.5 are not suitable from the coagulation point of view. At R below 0.5 it becomes evident that the removal becomes less efficient, especially with respect to residual iron.
2. Values of R higher than 0.5 may have some positive effect only if the other variables are not in the optimum range (e.g. pH or iron dose).

3. Optimal pH depends on interactions with other variables investigated (ANC, dose of Fe, R). No coagulation pH values can be given as recommended guide values at this time.
4. Optimal dose of iron (i.e. stoichiometry between model pollutants like humic substances and iron) also depends on the other variables studied. For the same concentration of humic substances (5 mg/l of HS) it varies from about 40 to 110 μmol/l of Fe. A very important variable influencing the optimal dose is the ANC of the treated water.

Acknowledgement. The authors would like to express thanks to M.-M. Bourbigot, Director of Anjou Recherche, for her support and interest in this work and to N. Ciba and V. Bigot for their kind assistance.

Thanks are also expressed to the French 'Ministere de la Recherche et de la Technologie', which awarded one of the authors (P.D.) the research fellowship in Anjou Recherche, Maissons Laffite, France, which made this research possible.

References

[1] Fenton, H.J.H.: J. Chem. Soc. *65* (1949) 899
[2] Merz, J.H., Waters, W.A.: J. Chem. Soc. (1949) 2427–2433
[3] Paillard, H.H., Legube, B., Carbonnier, F.D.L.: French patent 2689492, 8 October 1993
[4] Aguiar, C.A.: Ph.D. Thesis, L'Universit de Poitiers, Ecole Suprieure d'Ingenurs de Poitiers, 1992
[5] Aguiar, A., Carbonnier, F., Paillard, H., Legube B.: Water Supply *11* (1993) 129–138
[6] Dolejs, P.: Centrifugation Study of Humic Waters Coagulation by Brownian Motion. In: Chemistry for Protection of the Environment 1985, L. Pawlowski, G. Alaerts and W.J. Lacy (Eds.). Elsevier, Amsterdam 1986, pp. 297–310
[7] Dolejs, P.: A Simple Test for Determination of Optimal Doses in the Treatment of Humic Waters. In: Chemical Water and Wastewater Treatment, H.H. Hahn and R. Klute (Eds). Springer, Berlin Heidelberg New York 1990, 377–390
[8] Dolejs, P.: Spectrophotometric Determination of Colour of Humic Waters. In: Proc. 20th Conf. "Hydrochémia '83", Res. Inst. of Water Management, Bratislava 1983, 361–370
[9] Sedlak, D.L. Andren, A.W.: Environm. Sci. Technol., *25* (1991) 1419–1426

Petr Dolejs and Natasa Kalouskova
Water & Environmental Technology Team
Box 27
370 11 Ceske Budejovice
Czech Republic

Hervé Paillard
OTV-Novox, Z.I. Poitiers Chasseneuil
29, avenue des Temps Modernes
F-86360 Chasseneuil-du-Poitou
France

Maria Prados
Anjou Recherche
B.P. 76
F-78603 Maissons Laffitte Cedex
France

Bernard Legube
University of Poitiers
40, avenue du Recteur Pineau
F-86022 Poitiers Cedex
France

Drinking Water Treatment

Aspects on Future Drinking Water Treatment in Sweden

T. Hedberg

1. New Political Goals Change Water and Wastewater Technology?

Water is a strange substance with many physical and chemical peculiarities. As water is a very good solvent it is essential for all kinds of life but creates also problem for life. The phrase, "Use water wisely", is obvious and worn. But why don't we trust it? The political goal, however, is now to create a sustainable environment that creates opportunities for a variety of different species.

Our water and wastewater system is a part of the overall water cycle in which each subsidiary process affects the entire system. The vision of a sustainable society will sooner or later result in changes in the existing water supply system. In some areas it might result in localized solutions, in other areas the conventional system may continue to be used but will be maintained and operated with increased control from an environmental point of view.

The functional security of the water and wastewater system will also be paid attention to. The production of potable water has to occur at all times, even during crises. This means that the technical solutions should be simple.

Large scale centralized water supply systems have, historically, been very successful increasing the standard of living and the state of health, but have also contributed to an increased risk for waterborne diseases. Our ancient dependence on natural, local systems has been changed to a dependence on long distance, centralized systems. Such systems might be changed in the future. Information about global, regional and local environmental effects is increasingly provided today to the whole society; in schools, to the public, through political campaigns, thtough research findings etc. Examples of research activities pointing towards localized solutions or towards a careful use of resources in general are:

- research on so called alternative water-and wastewater systems.
- Multidisiplinary cooperation between different departments within universities, and with industries.
- Formations of centres of competence focused on the environment.
- research activities concerning the vulnerability of the water- and wastewater system.

R. Klute / H.H. Hahn (Eds.)
Chemical Water and Wastewater Treatment III

2. The Responsibility of the Authorities Is in Disorder?

The symptoms we notice in the water supply sector are treated by one authority, responsible for only one part of the sector. The research foundations have a similar organisation covering just a part off the water research field. This situation makes it difficult to address water research problems as water does not feel any borders. Now, in Sweden there are forces at the national, regional and local levels that act for cooperation between different organisations.

3. Water Treatment in Change

During recent years, the quality of water in reservoirs and rivers in Sweden has improved. This has also occured for rivers in Europe, e.g. the river Rhine.

The best condition for a safe drinking water is a high quality raw water source. It is important to emphasise, that simple treatment procedures and a short detention time during distribution are preferred. Modern development has, perhaps, gone in the opposite directions towards centralized treatment and large distribution systems.

Regardless, if one heads for localized solutions or keeps existing solutions, increased concern about contamination of drinking water has to be faced. It will become necessary to increase the knowledge of how different hazardous contaminates can be removed and also of what happens in the distribution network.

It will also be necessary to be able to describe and to explain that there will always be substances in drinking water but at acceptably low concentrations.

4. Does the Drinking Water Guidelines Tell Us How to Treat the Water?

In Sweden we now have new guidelines for the quality of drinking water. In the former guidelines it was not pointed out clearly enough that the water should be prepared for both human consumption and the distribution network. The new recommendations consider, for some parameters, raw water quality and give recommendations for the prevention of corrosion as well. Extra attention is paid to the presence of organic matter, both natural organic matter (NOM) and also trace organics. The microbiological status is also considered.

The consequence of these recommendations leads to the situation that several treatment plants ought to improve the removal of organic material. Beneficial measures taken at the treatment plant for organic matter removal might also be appropriate for the maintenance and operation of the distribution network, including the prevention of water quality deterioriation.

5. What Can be Done at Existing Plants, What Kind of New Solutions Can be Considered?

At some places the potential sources for water pollutions have increased. Even ground water that has been thought to be a pure source for drinking water contains substances originating from our technical society together with natural constituents such as hardness and colour. The natural constituents in groundwater and surface water may be regarded as harmless but may create water quality problems to the same extent as many compounds with anthropogenic origins. The natural environment produces substances that we do not tolerate in our drinking water e.g. taste and odour.

So, even if we are fortunate enough, to have a high quality raw water, we will always need a technical system, which in some way treats and prepares the natural water for distribution as potable water.

In addition to use as drinking water sources we want to use our lakes and rivers for many other activities, so it is a challenge to optimise the water quality, as all effective treatment procedures have draw-backs of various kinds.

One important balance to consider is the balance between the microbiological quality and the addition of disinfectants. Another important balance is to choose an optimal water quality to minimize corrosion, both chemical as well as microbiological, of different pipe materials simultaneously.

Based on knowledge of how different parameters affect removal processes in treatment plants and processes in distribution systems, one may suggest different upgrading procedures for a specific treatment plant.

Measures based only on the two balances mentioned above, may be several. Some examples follow:

- Restoration of lakes and rivers in order to prolong the detention time to favour natural biological processes.
- Increase the use of groundwater. This might be possible if the water supply system can be more local.
- Increase the use of groundwater, by using artificial groundwater or improving the natural process by methods that enhance biological processes, e.g. oxidation by O_3.
- Use of new coagulants in a flexible way, i.e. at a different pH, or at a different step.
- Integrate oxidation processes and adsorption processes in water treatment.
- Increase the possibility for biological processes to occur as pretreatment processes, as integrated processes or as final processes. The biological processes might be used at infiltration of water, in activated carbonfilters, or in slow sandfilters.
- Increase the efficiency of solid/liquid separation techniques by use of lamella separators, flotation units, multimedia filters, moving bed filters.
- Use of membrane technology to remove a number of substances.

- Combine different processes in a more efficient way in order to minimize side effects, e.g. avoid prechlorination.
- Optimise water quality from a corrosion point of view, e.g. increase alkalinity, decrease hardness.
- Try to minimize the number of different pipe materials in the distribution network.
- Try to minimize detention time in distributions systems.
- Avoid mixing of different water qualities in the distribution network. Create instead several local networks if possible.

A general conclusion is that we have moved away from natural microbiological and chemical processes occurring at ground water formation towards a use of powerful chemical methods for treating waters and now we may see an increasing use of biological processes again.

Torsten Hedberg
Department of Sanitary Engineering
Chalmers University of Technology
S-412 96 Gothenburg
Sweden

Coagulation and Corrosion Control for Soft and Coloured Drinking Water

S. W. Østerhus and B. Eikebrokk

Abstract

Soft, acidic, and coloured (humic substances) surface waters with a low turbidity are commonly used for the drinking water supply in Norway. For these waters, coagulation/direct filtration combined with a process for corrosion control is one of the most interesting water treatment alternatives.

In the experimental work described here, coagulation/direct filtration of water from Lake Jonsvatnet was investigated in a pilot plant using three different coagulants, alum (ALG), prepolymerized aluminium chloride (PAX14), and a calcium containing prepolymerized aluminium chloride (Ca-PAX). Magnafloc LT20 as a filter aid and three different combinations of chemicals for corrosion control purposes were also included in the pilot plant experiments: 1) lime/CO_2, 2) $CaCO_3$-slurry/CO_2/NaOH, and 3) CO_2/NaOH (using Ca-PAX).

A dual media (hydroanthracite on top of sand) gravity filter with a total depth of 0.85 m was used in the experiments. Headloss and effluent quality in terms of pH and turbidity were monitored continuously, and colour, Ca, alkalinity, and Al were analyzed from water samples taken throughout each filter run.

Optimum type of coagulant and combinations of chemicals for corrosion control were evaluated in terms of effluent quality, utilization of chemicals, cost of treatment, headloss, and length of filter run versus filter load.

Satisfactory treated water quality with respect to the Norwegian water quality standards was obtained with all the applied combinations of coagulants and chemicals for corrosion control. Lake Jonsvatnet is relatively low in humic content and moderate in corrosivity. The treated water quality was approximately: colour = 3 mg Pt/l, pH = 8.0, alkalinity = 1.0 meq/l, and Ca = 20 mg Ca/l. However, the utilization of chemicals, filter performance and treatment costs were largely affected by the applied combinations of coagulants and chemicals for corrosion control.

The results from the investigation show that the most cost effective choice of coagulant, coagulant dosage, and coagulation pH is largely affected by the type of corrosion control process and can be considerably different for a process without water treatment for corrosion control. The results may be used to design and optimize a more cost effective overall treatment process for these types of waters.

R. Klute / H.H. Hahn (Eds.)
Chemical Water and Wastewater Treatment III

1. Introduction

Soft and slightly acid surface waters with low turbidity and high colour due to natural organic matter (humic substances) are commonly used for the drinking water supply in Norway. These types of water are also encountered in other countries with cool and humid climate, for example, Finland, Scotland, Canada, etc. Coagulation and direct filtration has proven to be an efficient method for humus removal from these types of waters. To combat corrosion problems, dosage of calcium and chemicals to increase pH and alkalinity is normally required. However, several alternatives exist for the design of the combined coagulation/direct filtration and corrosion control processes.

2. Coagulation and Direct Filtration

The prevailing mechanisms of humus coagulation by aluminium salts can be summarized as a chemical interaction of positively charged aluminium hydrolysis products with negatively charged humic molecules. At pH < 6 and a humus concentration less than approx. 5 mg C/l, requiring a relatively low Al dose, the humus is destabilized by charge neutralization forming aluminium humate particles.The application of a higher dosage than needed for charge neutralization may result in restabilization of the precipitated aluminium humate [3,5,11]. At pH > 6, a higher coagulant dose is needed, and the humic removal occurs by sweep coagulation, where the removal of humic substances is attributed to adsorption to or enmeshment in rapidly precipitating $Al(OH)_3$-particles.

The main parameters affecting the destabilization reactions are initial humus concentration, coagulation pH, and coagulant dose. Temperature and raw water turbidity are also important variables [5,7,10]. In addition, the presence of calcium ions may increase the humic removal efficiency, reduce the required coagulant dose (and thereby the sludge production), and cause the pH range of efficient humus coagulation to shift towards slightly higher pH values [5].

3. Corrosion Control

Here, water treatment for corrosion control implies the addition of chemicals to increase the pH, alkalinity, and calcium content of the water. Several processes are available for this purpose as discussed elsewhere [9]. The perhaps most interesting corrosion control processes for typical Norwegian drinking waters are:

1) Addition of hydrated lime ($Ca(OH)_2$) and CO_2.
2) Addition of $CaCO_3$-slurry and CO_2 followed by pH adjustment using NaOH.
3) Using a calcium containing coagulant[1] and addition of CO_2 and NaOH.
4) Using alkaline filters ($CaCO_3$) and addition of CO_2 followed by pH adjustment using NaOH.

[1] Patent pending No. 922518, PCT/No. 93/00093P

At large treatment plants lime and CO_2 (process 1) is most commonly used. However, to prepare a saturated lime water feed solution requires extensive treatment of the lime water, adding high investment costs to this part of the process. Still the Ca concentration of the feed solution is relatively low. Problems like precipitation and reduced capacity of the equipment, clogging, process instability, dusty working environment, etc. commonly occur.

The lime may also be added as a slurry, which will reduce the investment costs. However, this is normally not recommended due to the risk of increased turbidity in the effluent, dosage instability, and clogging problems [9].

The investment costs can largely be reduced by using $CaCO_3$-slurry, CO_2, and NaOH (process 2) for corrosion control purposes, since $CaCO_3$ does not require any pretreatment before dosage [8,9]. The dosage is simple with few problems, and the Ca concentration of the feed solution is high. However, a third chemical is introduced (NaOH) and normally coagulation is necessary in order to avoid turbidity increase in the effluent (depending on the raw water quality and the desired effluent quality) [2,8].

The dissolution rate and the degree of $CaCO_3$ dissolution is important for the final water quality and for the operation costs of process 2. The dissolution rate of a $CaCO_3$ particle depends on water pH and the particle radius according to [13]:

$$\frac{\mathrm{d}M}{\mathrm{d}t} = 5 \times 10^{-10} \cdot \frac{M \cdot [\mathrm{H}^+]}{R_\mathrm{p}^2} \tag{1}$$

and the time required for complete dissolution of a $CaCO_3$ particle:

$$\tau = 3 \times 10^6 \cdot \frac{R_\mathrm{po}^2}{[\mathrm{H}^+]} . \tag{2}$$

The degree of $CaCO_3$ dissolution in an ideal mixed reactor can then be expressed as [8,13]:

$$\frac{[\mathrm{CaCO_3}]_T}{[\mathrm{CaCO_3}]_0} = \sum_{i=1}^{n} \frac{T}{\tau_i} \left(1 - e^{-\tau/T}\right) \cdot W_i . \tag{3}$$

$\mathrm{d}M/\mathrm{d}t$	Dissolution rate [kg/s]
M	Mass $CaCO_3$ [kg]
$[\mathrm{H}^+]$	Concentration of H^+ [mole/l]
R_p	Particle radius [m]
R_po	Particle radius [m] at time = 0
τ	Time required for complete dissolution [s]
τ_i	Time required for complete dissolution of the particle size fraction i [s]
W_i	Weight fraction of the size fraction i
T	Theoretical detention time [s] in the reactor
$[\mathrm{CaCO_3}]_T$	Concentration of $CaCO_3$ after a given detention time T
$[\mathrm{CaCO_3}]_0$	Concentration of $CaCO_3$ at time = 0

Using a calcium containing coagulant, Ca-PAX, (process 3), the problems associated with the calcium dosage and dissolution rate are avoided, and the process is very simple and flexible to operate. However, introducing Ca to the coagulant in a proper ratio reduces the Al concentration of the coagulant, thereby increasing the unit costs of the coagulant due to higher transportation costs.

The process using alkaline filters (process 4) is interesting, especially for small water works, due to simple operation and low labour demand. This process is, however, not included in this experimental work, and will not be discussed further here.

4. Experiments and Methods

Lake Jonsvatnet is the main water supply for the city of Trondheim, and some of its water quality parameters are shown in Table 1. Today there is no treatment of the raw water prior to consumption, except for straining and chlorination. Even though Lake Jonsvatnet is relatively low in humic content and moderate in corrosivity, it does not meet the Norwegian water quality standards (Table 1) with respect to the corrosion control parameters, pH, calcium, and alkalinity. In addition, colour and total organic carbon (TOC) occasionally exceed the standard limits. Pilot plant experiments were run in a coagulation and direct filtration pilotplant for the removal of humics from Lake Jonsvatnet, combined with different processes for corrosion control purposes. A total of 20 filter runs were made using three different coagulants: 1) aluminium sulphate (ALG), 2) prepolymerized aluminium chloride (PAX 14), and 3) prepolymerized calcium containing aluminium chloride (Ca-PAX) with a calcium to aluminium ratio of 7.36. Magnafloc LT20 was used as a filter aid and three different combinations of chemicals for corrosion control purposes were used: 1) lime/CO_2, 2) $CaCO_3$-slurry/CO_2/NaOH, and 3) NaOH/CO_2. The experimental matrix is shown in Table 2.

Table 1. Raw water quality during the experiments (Lake Jonsvatnet) and Norwegian standards for good drinking water quality [9]

Parameter	Lake Jonsvatnet	Standards for good drinking water quality
Calcium [mg Ca/l]	6.4–7.1	15–25
Alkalinity [meq/l]	0.24–0.27	0.6–1.0
pH	6.5–7.5	7.5–8.5
Colour [mg Pt/l]	10–20	15 (5)[1]
TOC [mg/l]	2.5–3.5	3.0 (2.3)[1]
Turbidity [NTU]	0.09–0.21	0.5 (0.3)[1]
Aluminium [mg Al/l]	0.02	(0.1)[1]

[1] Valid for effluents from coagulation/direct filtration plants

A dual media gravity filter was used in the experiments. The filter bed was 85 cm deep with 55 cm hydroanthracite (grain diameter of 0.8–1.6 mm) on top of 30 cm sand (0.4–1.0 mm). The filter bed was supported at the filter bottom by coarser sandlayers. Figure 1 shows a flow sheet of the coagulation/direct filtration pilot plant, including dosage of chemicals for corrosion control, backwash water storage tank, sludge settling tank, etc. Headloss and effluent quality in terms of pH

Table 2. Experimental matrix

Coagulation and direct filtration			Corrosion control (chemical dosages)	
Coagulant	Coagulant dose [mg Al/l]	Filter load [m/hrs]	Prior to coagulation	After filtration (pH adjustment)
ALG	1.8	10	$Ca(OH)_2$ and CO_2	$Ca(OH)_2$
ALG	2.3	7	$Ca(OH)_2$ and CO_2	$Ca(OH)_2$
ALG	2.3	10	$Ca(OH)_2$ and CO_2	$Ca(OH)_2$
ALG	2.3	15	$Ca(OH)_2$ and CO_2	$Ca(OH)_2$
ALG	2.3	7	$CaCO_3$ and CO_2	NaOH
ALG	2.3	10	$CaCO_3$ and CO_2	NaOH
ALG	2.3	15	$CaCO_3$ and CO_2	NaOH
ALG + LT20	2.3 + 0.1	15	$CaCO_3$ and CO_2	NaOH
ALG + LT20	2.3 + 0.1	22.5	$CaCO_3$ and CO_2	NaOH
PAX 14	1.8	10	$Ca(OH)_2$ and CO_2	$Ca(OH)_2$
PAX 14	2.3	7	$Ca(OH)_2$ and CO_2	$Ca(OH)_2$
PAX 14	2.3	10	$Ca(OH)_2$ and CO_2	$Ca(OH)_2$
PAX 14	2.3	15	$Ca(OH)_2$ and CO_2	$Ca(OH)_2$
PAX 14	2.3	7	$CaCO_3$ and CO_2	NaOH
PAX 14	2.3	10	$CaCO_3$ and CO_2	NaOH
PAX 14	2.3	15	$CaCO_3$ and CO_2	NaOH
Ca-PAX	1.5	10	NaOH and CO_2	NaOH
Ca-PAX	1.5	15	NaOH and CO_2	NaOH
Ca-PAX + LT20	1.5 + 0.1	15	NaOH and CO_2	NaOH
Ca-PAX + LT20	1.5 + 0.1	20	NaOH and CO_2	NaOH

and turbidity were monitored continuously, and colour, TOC, calcium, alkalinity, and aluminium were analyzed from water samples taken throughout each filter run.

The saturated lime solution was prepared by adding $Ca(OH)_2$ to the water, within the theoretical saturation limit of 950 mg Ca/l. Then the solution was allowed to settle for approx. 6 hours before it was pumped into the lime feed container. The $CaCO_3$-slurry (Setacarb from Hustadmarmor A/S) was diluted from an initial concentration of close to 560 g Ca/l to a concentration of 55,8 g Ca/l before dosage. ALG ($Al_2(SO_4)_3 \cdot 14.3\ H_2O$) was prepared at a concentration of 22.5 g Al/l in the feed solution, and PAX14 and Ca-PAX were used as delivered (undiluted) at a concentration of 104.8 and 20.3 g Al/l, respectively. Magnafloc LT20 was prepared one day before the start of each experiment at a concentration of 0.01 % by weight.

The main objectives of the investigation were to evaluate and optimize the applied coagulants and chemicals for corrosion control with respect to effluent water quality, utilization of chemicals, cost of treatment, headloss, and length of filter run versus filter load.

5. Results and Discussion

The pilot experiments aimed for a treated water quality which met the Norwegian water quality standards: Colour < 5 mg Pt/l, TOC < 2.3 mg C/l, Turbidity < 0.3 NTU, Aluminium < 0.1 mg Al/l, Calcium = 20 mg Ca/l, pH = 8.0, and Alkalinity = 0.6–1.0 meq/l.

The results from the coagulant dose and pH optimization experiments are shown in Figures 2a and b, respectively. For turbidity removal the optimum pH was slightly higher (0.1–0.2 units) than that for colour removal. Efficient coagulation could be obtained at the highest pH for PAX14, and at a higher pH for Ca-PAX than for ALG. The pH range for efficient coagulation was wider for PAX14 than for Ca-PAX and ALG. As the coagulant dose was increased, the pH range for optimum coagulation was increased (for all three of the coagulants), especially in the direction of higher pH values [1, 4, 6].

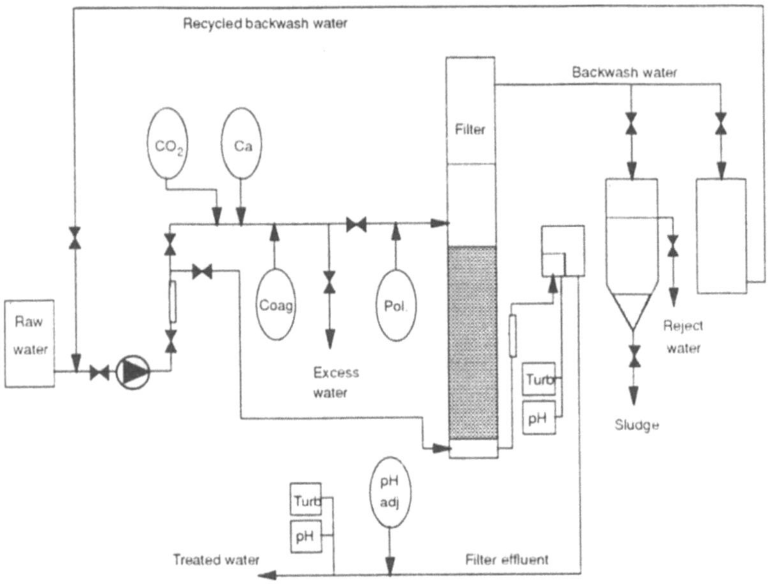

Fig. 1. Flow sheet of the pilot plant for coagulation/direct filtration and corrosion control of Lake Jonsvatnet

Satisfactory water quality as compared to the water quality standards was obtained in all the experiments. The three types of coagulants revealed no significant differences in the corresponding filter effluent water quality, except for a higher TOC concentration at high filter loads using Ca-PAX. The range of filter effluent water quality data is shown in Table 3, and the obtained water quality and applied chemical dosage data in the corrosion control treatment step are shown in Table 4 for each filter run.

Table 3. Obtained filter effluent water quality

Parameter	Concentration
Colour [mg Pt/l]	1–5
TOC [mg C/l]	1.10–2.2[1]
Turbidity [NTU]	0.05–0.10
Aluminium [mg Al/l]	< 0.02–0.11

[1] Ca-PAX at filter loads of 20 m/hr resulted in TOC of 1.7–2.2 mg C/l

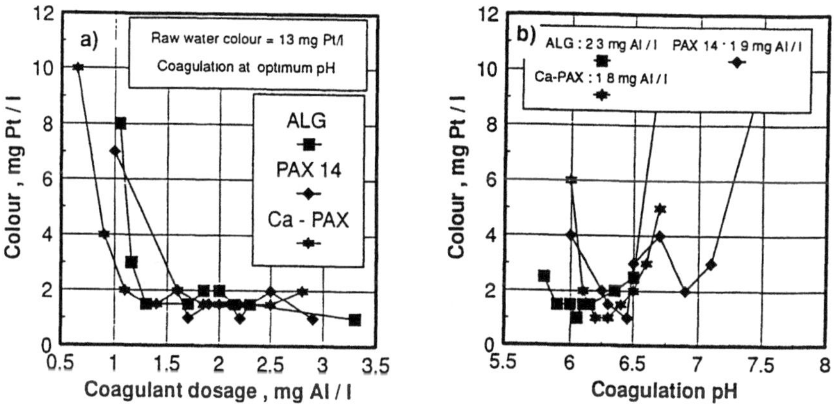

Fig. 2. Optimization of coagulant dose (a) and pH (b) with respect to colour [6]

For all the corrosion control experiments, CO_2 utilization was difficult to estimate due to some variations in the CO_2 content of the raw water. The calculated CO_2 content in the raw water during the experiments was 1.0–11.3 mg CO_2/l. In the experiments the calculated CO_2 utilization was between 60 and 95 % with anaverage of 78 % [2]. However, a considerable error may be associated with these calculations. In general, CO_2 utilization decreases with decreasing pH, as shown in Figure 3 for pH values between 5.7 and 6.2 [8]. At higher pH, a CO_2 utilization of 90–100 % is expected.

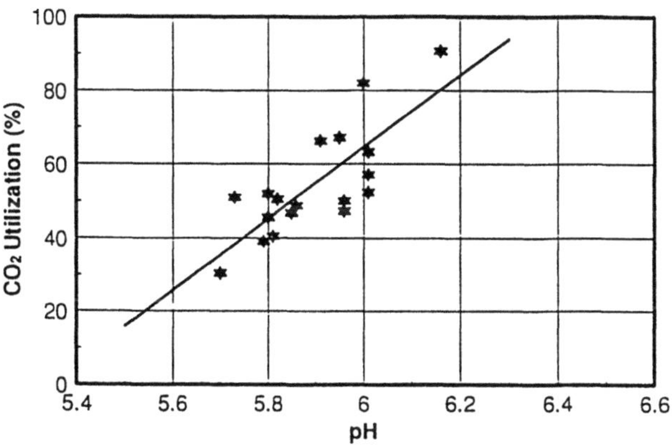

Fig. 3. Correlation between pH and CO_2 utilization [8]

Table 4. Chemical dosage and water quality data

Process (combination of chemicals)	Filter load [m/hr]	Sam-pling time [hr]	Dosage [mg/l]		Filter effluent water quality			Dosage for pH adjustm. [mg/l]	Treated water quality		
			$CaCO_3$ [2], $Ca(OH)_2$, or NaOH	CO_2	pH	mg Ca/l	Alk. [meq/l]	NaOH [2] or $Ca(OH)_2$	pH	mg Ca/l	Alkal. [meq/l]
ALG: 1.8 mg Al/l $Ca(OH)_2 + CO_2$	10	1.0	13	32	–	13	0.5	7	7.8	17	0.7
		4.5	13	32	–	13	0.4	9	7.8	20	0.8
ALG: 2.3 mg Al/l $Ca(OH)_2 + CO_2$	7	2.3	13	23	6.5	16	0.4	13	8.0	23	0.8
	10	1.5	15	22	6.5	15	0.4	13	8.0	22	0.8
	15	1.7	19	23	6.5	–	–	–	8.0	–	–
ALG: 2.3 mg Al/l $CaCO_3 + CO_2$ + NaOH	7	3.8	43	23	6.8	23	0.8	13.9	8.0	23	1.2
	10	1.8	46	23	6.8	22	0.8	13.6	8.0	22	1.1
	15	2.5	64	23	6.8	–	–	–	8.0	–	–
ALG: 2.3 mg Al/l + LT20: mg/l 0.1 $CaCO_3$ + CO_2 + NaOH	15	–	49	22	6.8	–	–	–	8.0	–	–
	22.5	–	55	21	6.8	–	–	–	8.0	–	–
PAX14: 1.8 mg Al/l $Ca(OH)_2 + CO_2$	10	1.5	15	31	–	13	0.5	9	7.9	18	0.8
		5.0	15	31	–	14	0.5	11	7.9	20	0.9
PAX 14:2.3 mg Al/l $Ca(OH)_2$ + CO_2	7	2.3	16	23	6.9	14	0.5	7	8.0	18	0.7
		6.3	16	23	6.9	15	0.5	7	8.0	19	0.7
	10	2.5	15	23	6.8	15	0.5	7	8.0	19	0.7
		4.0	15	23	6.8	16	0.5	9	8.0	21	0.7
	15	1.8	14	23	6.8	15	0.5	7	8.0	19	0.7
		3.0	14	23	6.8	15	0.5	7	8.0	19	0.7
PAX 14: 2.3 mg Al/l $CaCO_3 + CO_2$ + NaOH	7	1.8	46	19	7.2	18	0.8	4.6	8.0	18	0.9
		4.0	46	19	7.3	20	0.8	3.7	8.0	20	1.0
		8.7	46	19	7.4	22	0.9	3.1	8.0	22	1.0
	10	1.7	47	19	7.0	20	0.8	8.2	8.0	20	1.0
		4.7	47	19	6.9	22	0.9	11.7	8.0	22	1.2
	15	1.7	48	19	7.1	20	0.8	6.5	8.0	20	1.0
		3.2	48	19	6.9	22	0.9	11.7	8.0	22	1.2
Ca-PAX: 1.5 mg Al/l CO_2 + NaOH	10	4.5	28	–	–	16	–	–	8.0	16	–
	15	3.7	27	–	–	16	–	–	8.0	16	–
Ca-PAX: 1.5 mg Al/l + LT20: 0.1 mg/l CO_2 + NaOH	15	4.0	27	–	–	16	–	–	8.0	16	–
	20	2.8	21	–	–	16	–	–	8.0	16	–

[1] Sampling time (hrs) after start of a filter cycle
[2] In mg $CaCO_3$, l, mg $Ca(OH)_2$/l, or mg NaOH/l (depending on the process)

Lime and CO_2 for Corrosion Control. As expected, Ca utilization was about 100 % (with respect to the Ca content of the feed solution) in all the experiments (Table 4). However, a loss of Ca of approximately 30 % was experienced during preparation and settling of the lime solution. Typically the feed solution contained no more than 600 mg Ca/l as compared to a theoretical saturation concentration of about 950 mg Ca/l (1.7–1.8 g $Ca(OH)_2$/l was added to prepare the lime solution). Thus, the overall lime utilization was about 70 % and the loss of Ca was probably due to carbonation (CO_2 uptake from the air followed by $CaCO_3$ precipitation) [12].

When lime and CO_2 were used for corrosion control purposes, there was no significant difference in filtered water quality when ALG or PAX14 were applied as coagulants. In order to obtain 20 mg Ca/l and pH = 8.0 in treated water, there is a theoretical need for 1.7 mg more CO_2 (0.04 mmole) per mg Al when PAX14(OH/A1 ratio = 1.05) is applied, than when ALG is used. On the other hand, the treated water alkalinity is increased (0.04 meq per mg Al used) when using PAX14.

$CaCO_3$-Slurry, CO_2, and NaOH for Corrosion Control. Using $CaCO_3$ slurry usually requires that the slurry be added before the coagulation/direct filtration process in order to avoid excessive turbidity in the filter effluent [2], making the operation of this corrosion control process more complex. In order to obtain a sufficiently low pH for efficient coagulation, relatively high CO_2 doses are required. As a consequence, more NaOH is needed to obtain the desired treated water pH. Thus, even though the $CaCO_3$ dissolution rate and $CaCO_3$ utilization is reduced at increasing pH, it may be economically advantageous to apply the highest possible pH for efficient coagulation. This is discussed further below.

Figures 4a and b show the relationship between the coagulation pH and the treated water alkalinity after the required Ca increase to 20 mg Ca/l by $CaCO_3$ addition and adjustment of pH to 8.0 by NaOH addition for Lake Jonsvatnet using ALG and PAX14 as coagulants, respectively. The unreasonably high treated water alkalinities at low coagulation pH values is caused by the high CO_2 demand to maintain the coagulation pH when all the $CaCO_3$ is added before the coagulation step. This, in turn, is followed by a high NaOH demand to obtain an effluent pH of 8.0. However, another raw water quality, i.e. more acidic and lower in total inorganic carbon, would require a lower coagulation pH to obtain an alkalinity of 1.0 meq/l in the treated water.

The optimum coagulation pH for ALG at a dose of 2.0 mg Al/l was approximately 6.1 (Fig. 2b). According to Figure 4a this would result in an alkalinity of 2.8 meq/l in the treated water. In order to obtain a treated water alkalinity of 1.0 meq/l, coagulation pH of approximately 6.9 and 6.4 are required at ALG doses of 1.5 and 4.0 mg Al/l, respectively (Fig. 4a). A lower coagulation pH (requiring a higher CO_2-dose) would result in a higher alkalinity, and a coagulation pH of 6.0 would result in precipitation of $CaCO_3$ for doses of 2.0 mg Al/l or lower. Accordingly, the optimum coagulation pH for PAX14 at a dose of 2.0 mg Al/l was approximately 6.5 (Fig. 2b), which would result in a treated water alkalinity of 1.4 meq/l (Fig. 4b). At PAX14 doses of 1.5 and 4.0 mg Al/l, the required coagulation pH are 7.0 and 6.7, respectively, to obtain an alkalinity of 1.0 meq/l (Fig. 4b). At a coagulation pH of 6.0, $CaCO_3$ would precipitate at a PAX14 dose of 3.0 mg Al/l or lower.

Figures 5a and b show the CO_2 and NaOH demand versus coagulation pH for different ALG doses, and Figures 5c and d for PAX14 as coagulant. The CO_2 dose is calculated assuming the following pH dependent CO_2 utilization (Fig. 3):

$$\left((CO_2)/(CO_2)_{dosage}\right) = (-520.5 + pH \cdot 97.5)/100 \qquad 5.5 < pH < 6.3 \tag{4}$$

$$\left((CO_2)/(CO_2)_{dosage}\right) = 0.95 \qquad pH > 6.3\,. \tag{5}$$

Figure 5 shows that at a coagulation pH below 6.5–7.0, the CO_2 and NaOH demand increases dramatically, but is reduced at high coagulant doses due to the acidic behaviour of the coagulants. However, with a ALG dose of 4.0 mg Al/l it is not possible to obtain a coagulation pH higher than 6.9 even without CO_2 dosage.

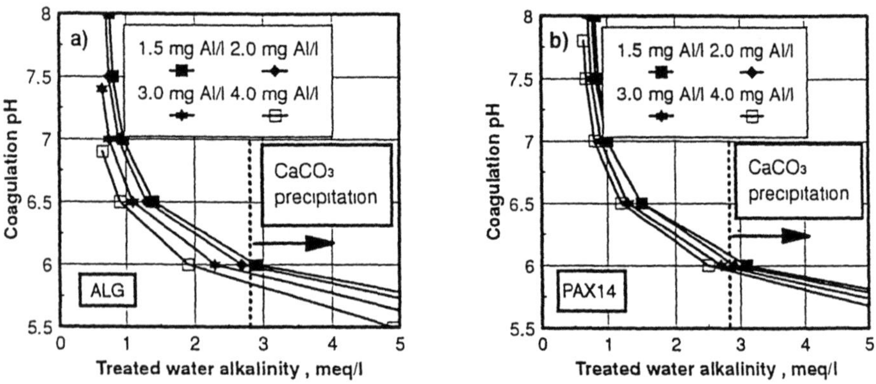

Fig. 4. Relationship between coagulation pH and treated water alkalinity for Lake Jonsvatnet using ALG (a) and PAX14 (b) as coagulants and $CaCO_3$ and CO_2 for corrosion control. Calcium concentration is increased by 13.5 mg Ca/l (to 20 mg Ca/l) by $CaCO_3$ dosage before coagulation and the treated water pH is adjusted to 8.0 by addition of NaOH

The observed degrees of $CaCO_3$ utilization were between 63 and 90 % in the experiments (Table 4), generally with the highest utilization at the lowest coagulation pH. However, there was also a tendency for the utilization to increase throughout the filter cycle, indicating that undissolved $CaCO_3$ retained in the filter bed dissolves slowly. The theoretical (empty bed) hydraulic detention time of the filter column in the pilot experiments was 5 to 25 minutes. The observed degrees of $CaCO_3$ utilization are, however, substantially higher than the calculated theoretical utilization using Eq. 3 at the given hydraulic detention times (10 to 55 %.) This confirms that the detention time for undissolved $CaCO_3$ particles is considerably higher than the hydraulic detention time, and that undissolved $CaCO_3$ is coagulated and retained in the filter bed and then subsequently partly dissolved. However, this implies that the degree of $CaCO_3$ utilization also depends on the

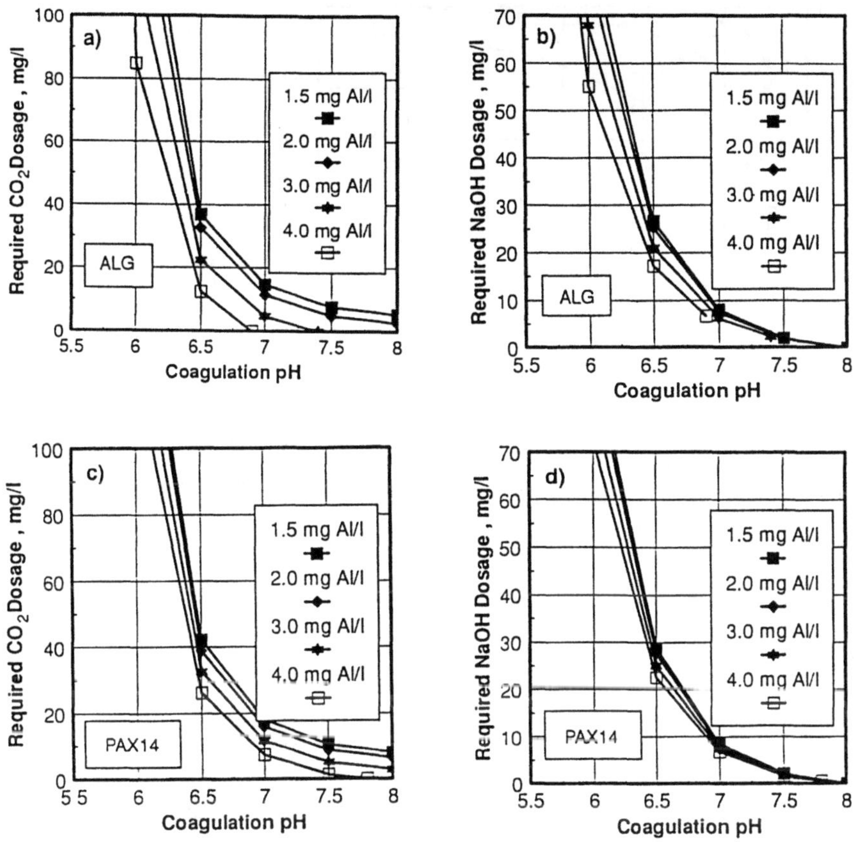

Fig. 5a–d. CO_2 and NaOH demand versus coagulation pH at different doses of ALG and PAX14 (using $CaCO_3$ slurry, CO_2, and NaOH for corrosion control) to obtain a treated water quality of pH = 8.0 and Ca = 20 mg Ca/l for Lake Jonsvatnet. CO_2 (**a**) and NaOH (**b**) demand using ALG, and CO_2 (**c**) and NaOH (**d**) demand using PAX14

length of a filter run, and that calculated degrees of $CaCO_3$ utilization from Ca analyses taken at given intervals in the filter cycle (apparent $CaCO_3$ utilization) are erroneous. In Figure 6, this is taken into account by calculating the fraction of $CaCO_3$ accumulated in the filter bed (1-(degree of $CaCO_3$ utilization)) after given filter run times using the kinetics of $CaCO_3$ dissolution [8, 13]. The apparent $CaCO_3$ utilization calculated from the theoretical Ca concentration at the given filter run times is also shown in the figure. Figure 6 shows that the degree of $CaCO_3$ utilization increases with increasing length of a filter run, and that the apparent $CaCO_3$ utilization overestimates the actual utilization.

Figure 7 shows the calculated cost of coagulants and corrosion control chemicals versus coagulation pH (with effluent pH = 8.0 and 20 mg Ca/l). The costs of CO_2 and NaOH are calculated from Figure 5, while the cost of $CaCO_3$ is calculated from Figure 6 using a filter run length of 7 hrs. This corresponds to a $CaCO_3$ utilization of 90–95 %, 80–85 %, and 55–60 % at pH 6.5, 7.0, and 7.5,

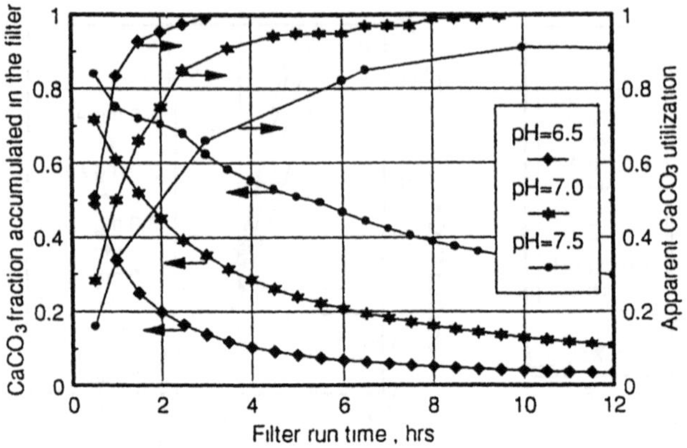

Fig. 6. Calculated fraction of $CaCO_3$ accumulated in the filter bed and calculated apparent $CaCO_3$ utilization versus filter run length

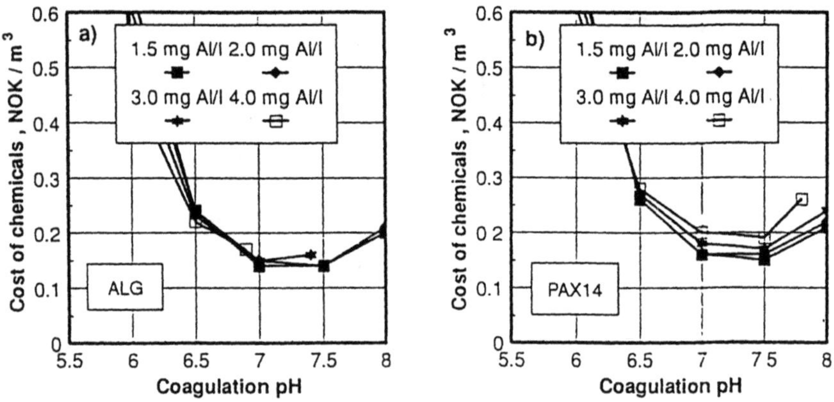

Fig. 7. Cost of chemicals versus coagulation pH for ALG (**a**) and PAX14 (**b**) as coagulants, and $CaCO_3$, CO_2, and NaOH for corrosion control. The following prices are used: CO_2 = 1.65 NOK/kg, NaOH = 3.50 NOK/kg (100% solution), $CaCO_3$ = 3.55 NOK/kgCa, ALG = 21.94 NOK/kg Al, and PAX14 = 25.61 NOK/kg Al

respectively. The experimental data agree very well with the apparent $CaCO_3$ utilization in Figure 6, indicating that the kinetics of $CaCO_3$ dissolution as described in Eqs. 1–3 and shown in Figure 6 are in good agreement with the experimental results. However, this also indicates that coagulation does not retard the $CaCO_3$ dissolution significantly, i.e. due to the formation of larger $CaCO_3$-humic particle aggregates or coverage by precipitated $Al(OH)_3$. Figure 7 shows that the lowest chemical costs are obtained at a coagulation pH of 7.0–7.5 for both coagulants. However, provided a $CaCO_3$ dosage of 13.5 mg Ca/l, it was not possible to obtain such a high coagulation pH with ALG as coagulant if the dose was higher than 3.0 mg Al/l (Fig. 5a and 7a). At a lower pH the costs increase rapidly due to higher CO_2 and NaOH demand, and at higher pH the costs increase slowly due to a low $CaCO_3$ utilization. For a given coagulation pH the costs are somewhat lower using ALG than PAX14, since PAX14 has a higher cost per unit Al and a higher CO_2 demand. Due to the acidic behaviour of the coagulants there is a coagulation pH below which the overall chemical costs are not further reduced when decreasing the coagulant dose. For Lake Jonsvatnet this pH is 6.8 and 6.4 for ALG and PAX14, respectively (where the curves intersect in Figure 7a and b). This pH is higher for ALG because ALG is more acidic than PAX14. Only above this pH is it possible to reduce the cost of chemicals by reducing the coagulant doses at a given coagulation pH.

However, a coagulation pH of 7.0–7.5 is normally too high for efficient colour and turbidity removal. PAX14 has a wider and higher optimum pH range for effective coagulation than ALG (Fig. 2). According to Figure 8, the optimum pH with respect to colour and turbidity removal may be increased by increasing the coagulant dose. However, this will in turn increase the cost of coagulants. In Figure 9, the cost of the chemicals is plotted versus the coagulant dose, using Figure 8 to determine the maximum allowable coagulation pH, and Figures 5 and 6 to determine the lowest demand of corrosion control chemicals. Figure 9a show that an ALG dose of 3.0 mg Al/l is needed to minimize the cost of the chemicals (0.154 NOK/m^3), and at lower coagulant doses the costs will increase rapidly. However, a coagulation pH of 7.0–7.5 with an ALG dose of 3.0 mg Al/l may be difficult to obtain (Fig. 5a), or will result in a relatively low effluent alkalinity (0.6–0.7 meq/l). The cost of the chemicals using PAX14 is less sensitive to the coagulant dose (Fig. 9b). A PAX14 dose of 2.0 mg Al/l is needed to minimize the costs (0.160 NOK/m^3), which will result in an effluent alkalinity of 0.8–1.0 meq/l.

From this, it can be concluded that for coagulation and direct filtration with the application of $CaCO_3$, CO_2 and NaOH for corrosion control purposes, PAX14 should be used as the coagulant at a dose of 2.0 mg Al/l, and a coagulation pH slightly above 7.0. This analysis does not include sludge treatment costs (which will increase as the coagulant dose is increased) nor the effect of shorter filter runs at higher coagulant doses. However, this will not alter the main conclusions regarding process costs, since the recommended PAX14 dose is relatively low.

The above example shows that it is important when considering coagulation/direct filtration with the application of $CaCO_3$, CO_2, and NaOH for corrosion control purposes, to take into account also the corrosion control process when optimizing the overall process. Traditional thinking, i.e. trying to minimize the coagulant dosage to reduce the treatment costs, may in fact result in higher costs of the total treatment. Optimum selection of coagulant, coagulant dose, and coagulation pH are very important to the overall treatment costs.

Coagulation by Ca-PAX, with CO_2 and NaOH for Corrosion Control. The process which uses Ca-PAX, CO_2, and NaOH is very simple to operate. The preparation and dosage of chemicals are simple, and any desired combination of pH and alkalinity can easily be achieved. The Ca-PAX has a basicity of 1.05, which compared to ALG will reduce the OH demand by 0.04 mole OH per mg Al used, to obtain the same treated water pH and alkalinity. However, the Ca content in the effluent is given by the coagulant dose. In these experiments, the coagulant had a Ca/Al ratio of 7.36, and a coagulant dose of 1.5 mg Al/l gave a treated water Ca concentration of 17 mg Ca/l. The coagulant may also be produced with

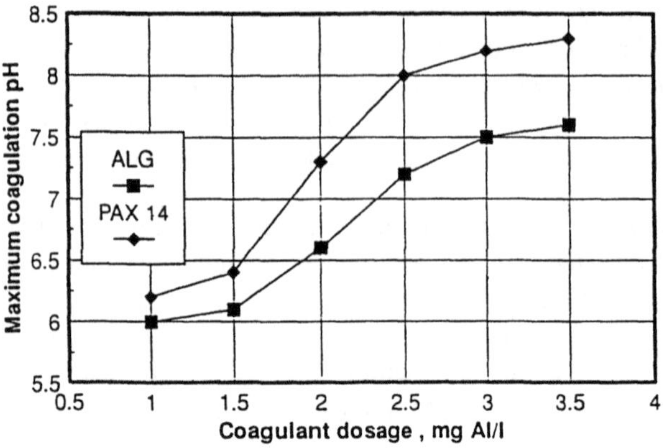

Fig. 8. Effect of the coagulant dose on the maximum allowable pH value for efficient coagulation of Lake Jonsvatnet [1, 2, 4, 6]

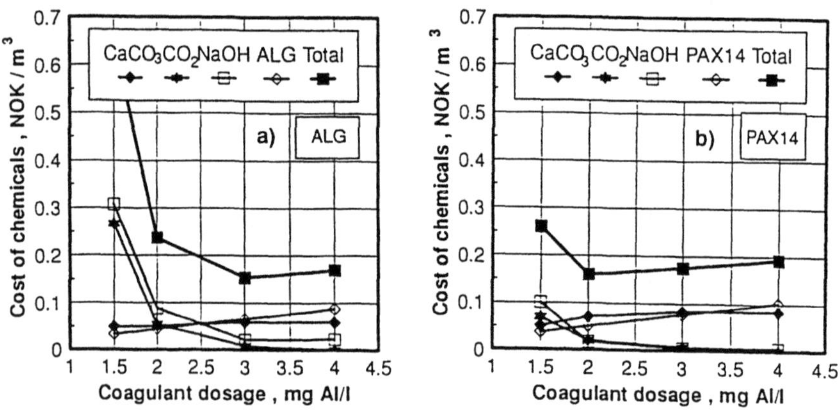

Fig. 9. Cost of chemicals versus coagulant dose for ALG (a) and PAX14 (b), and $CaCO_3$, CO_2, and NaOH for corrosion control. The unit prices are given in Figure 8

other Ca/Al ratios. The Al content of the coagulant is relatively low, 20.3 g Al/l, as compared to PAX14 (104.8 g Al/l), which probably will make it expensive per unit Al.

Filter Operation and Performance Data. The available head in the pilot filter experiments was 2 m water column. All the experiments were, however, terminated by turbidity breakthrough and not by headloss. The headloss before breakthrough was 30–50 cm for ALG, 40–60 cm for PAX14 and Ca-PAX, and 100–110 cm when polymer was used. The headloss at breakthrough is constant with respect to filter load for a given coagulant, but the rate of headloss increases with increasing filter load. No effect on the rate of headloss increase due to $CaCO_3$ dosage or $CaCO_3$ accumulation in the filter was observed.

Table 5 shows the length of filter runs obtained at different filter loads. The table shows that PAX14 gave slightly longer filter runs than ALG, and that the observed filter run, lengths were similar for Ca-PAX and PAX14. The addition of polymer as a filter aid prolonged the filter runs. The table also shows that when the coagulant dose is increased from 1.8 to 2.3 mg Al/l, the filter runs are reduced by 10–30 %. However, it is interesting to note that the use of $CaCO_3$ for corrosion control (instead of lime) prolonged the filter run. It should be expected that addition of $CaCO_3$ would reduce the length of the filter run due to accumulation of undissolved $CaCO_3$ in the filter. A possible explanation for the prolonged filter run could be that the $CaCO_3$ particles improve the coagulation/flocculation process, making stronger and denser particle aggregates which do not easily breake up.

Table 5. Filter run lengths obtained in the experiments at various filter loads, type of coagulants, dose of coagulants, and corrosion control processes

Coagulant	Corrosion control process	Filter run length [hrs] Filter load [m/hrs]				
		7	10	15	20	22.5
ALG: 1.8 mg Al/l	$Ca(OH)_2$	–	7.2	–	–	–
ALG: 2.3 mg Al/l	$Ca(OH)_2$	5.5	5.0	2.8	–	–
ALG: 2.3 mg Al/l	$CaCO_3$	7.4	6.1	2.8	–	–
ALG: 2.3 mg Al/l + 0.1 mg LT20/l	$CaCO_3$	–	–	7.2	–	3.1
PAX 14: 1.8 mg Al/l	$Ca(OH)_2$	–	6.9	–	–	–
PAX 14: 2.3 mg Al/l	$Ca(OH)_2$	7.7	5.2	3.8	–	–
PAX 14: 2.3 mg Al/l	$CaCO_3$	8.7	6.7	3.6	–	–
Ca-PAX: 1.5 mg Al/l	NaOH	–	6.9	4.5	–	–
Ca-PAX: 1.5 mg Al/l + 0.1 mg LT20/l	NaOH	–	–	5.8	4.0	–

6. Summary and Conclusions

Pilot plant coagulation/direct filtration experiments for humic substance removal combined with different processes for corrosion control were conducted on water from Lake Jonsvatnet. Alum (ALG), prepolymerized aluminium chloride (PAX14), and a calcium containing polymerized aluminium chloride (Ca-PAX) were used as coagulants, and three different combinations of chemicals for corrosion control purposes were included in the experimental matrix: 1) lime/CO_2, 2) $CaCO_3$ slurry/CO_2/NaOH, and 3) CO_2/NaOH (using Ca-PAX). The treated water quality standards were: colour < 5 mgPt/l, turbidity < 0.3 NTU, pH $= 8.0$, alkalinity = 1.0 meq/l, and calcium = 20 mg/l. The main conclusions from the experiments are:

1. Satisfactory water quality with respect to colour removal and corrosion control was obtained in all filter runs.
2. Efficient coagulation occurs at the highest pH for PAX14, and at a higher pH for Ca-PAX than for ALG. The pH range for efficient coagulation is wider for PAX14 than for Ca-PAX and ALG. Increasing the coagulant dose results in a wider optimum pH range, especially towards higher pH values.

3. The basicity (OH/Al ratio) of PAX14 (and also of Ca-PAX) is 1.05. Therefore, there is theoretically a need for 1.7 mg CO_2 (0.04 mmole) per mg Al used, when PAX14 is used as compared to ALG in order to obtain the same treated water Ca and pH. On the other hand, the treated water alkalinity is correspondingly higher (0.04 meq per mg Al used) when using PAX14. Similarly, using Ca-PAX instead of ALG reduces the OH demand by 0.04 mole OH per mg Al used, to obtain the same treated water pH and alkalinity.
4. The $CaCO_3$ slurry has to be added before the coagulation step in order to avoid excessive turbidity in the filter effluent. Undissolved $CaCO_3$ is removed efficiently during coagulation/direct filtration. The $CaCO_3$ utilization depends on pH and contact time. High pH reduces the degree of $CaCO_3$ utilization.
5. The overall cost of chemicals for coagulation/direct filtration and the application of $CaCO_3$, CO_2, and NaOH for corrosion control depends on the coagulation pH. Minimum costs are achieved at pH 7.0 to 7.5 (for both ALG and PAX14 as coagulant). The optimum pH range for coagulation depends on the coagulant dose. Thus, the cost of the chemicals is minimized at a coagulant dose of 2.0 mg Al/l and 3.0 mg Al/l for PAX14 and ALG, respectively, with minimum overall chemical costs of 0.15–0.16 NOK/m^3 in both cases. At a lower coagulant dose the costs increase rapidly due to a high CO_2 and NaOH demand, and at a higher dose the costs increase slowly due to higher coagulant dose and lower $CaCO_3$ utilization. Normally it is difficult to obtain the desired coagulation pH and effluent alkalinity with an ALG dose of 3.0 mg Al/l. Therefore, PAX14 should be selected as the optimum coagulant for this process, since it also requires the lowest dosage.
6. Due to the acidic behaviour of the coagulants, there is a coagulation pH below which the chemical costs are not further reduced when decreasing the coagulant dose. This is true when $CaCO_3$, CO_2, and NaOH are applied for corrosion control. For Lake Jonsvatnet this pH is 6.8 and 6.4 for ALG and PAX14, respectively. Only above this pH is it possible to reduce the cost of the chemicals by reducing the coagulant dose at a given coagulation pH.
7. The length of the filter runs before breakthrough was only slightly longer for PAX14 and Ca-PAX than for ALG. Addition of polymer as a filter aid prolonged the filter runs, allowed for a higher maximum filter load, and thereby also increased net daily filtered water production.
8. The use of $CaCO_3$ did not affect filter headloss development. The results also indicate that the use of $CaCO_3$ instead of lime increased filter run times (at filter loads below 15 m/hrs).
9. It is important when considering coagulation/direct filtration with the application of $CaCO_3$, CO_2, and NaOH for corrosion control purposes, to take the corrosion control process into account as well when optimizing the overall process. Traditional thinking, i.e. trying to minimize the coagulant dose only, may in fact result in higher costs of the total treatment process. Optimum selection of coagulant, coagulant dose, and coagulation pH is very important for the overall treatment costs.

References

[1] Eikebrokk, B.: Ph.D. Dissertation, Norwegian Inst. of Technol., Trondheim 1982
[2] Østerhus, S.W.: Report STF60 A93065, SINTEF-NHL, Trondheim 1993
[3] Eikebrokk, B., Fettig J.: Treatment of Coloured Surface Water by Coagulation/Direct Filtration: Effect of Water Quality, Type of Coagulant and Filter Aids. In: Chemical Water and Wastwater Treatment, H.H. Hahn and R. Klute (Eds.). Springer, Berlin Heidelberg New York 1990, pp. 361–376
[4] Eikebrokk, B.: Report 55/86 NTNFs Program for VAR-teknikk, Trondheim 1986
[5] Committee Report. J. AWWA *71* (1979) 588
[6] Sekse, T.: M.Sc. Diploma, Norwegian Inst. of Technol., Trondheim 1993
[7] O'Melia, C.R., Gray, K.A., Yao, C.: Polymeric Inorganic Coagulants. Research Report, AWWA, 1989
[8] Eikebrokk, B., Østerhus, S.W., Brattebø, H.: Report 36/89, NTNFs Program for Drikkevannsforskning, 1989
[9] Vik, E.A., Hongve, D.: Innvendig korrosjon av vannledninger. NTNFs Program for drikkevannsforskning, 1988
[10] Owen, D.M., Amy, G.L., Chowdhury, Z.K.: Characterization of Natural Organic Matter and Its Relationship to Treatability. Research Report, AWWA, 1993
[11] Eikebrokk, B., Fettig, J., Fløgstad, H.,Thorsen, T., Hem, L.J.: Fjerning avhumus fra drikkevann. NTNFs Program for Drikkevannsforskning, 1991
[12] Water Treatment Handbook, Vol. 1 and 2, 6th. edition. Degremont, France, 1991
[13] Bjerle, I.: Upplosning-Effekt-Ekonomi. Kalkning av Rinnande Vatten, University of Lund, Lund 1983

Stein W. Østerhus and Bjørnar Eikebrokk
SINTEF NHL
Klæbuveien 153
N-7034 Trondheim
Norway

Optimal Use of Ozonation and GAC in Direct Filtration for Drinking Water Treatment

J. E. Tobiason

Abstract

Results from pilot studies of the use of ozonation and biologically active GAC media to upgrade two existing direct filtration facilities are summarized. A comparison of preozonation followed by GAC filter media to post-filtration ozonation followed by a GAC contactor is made based on various aggregate measures of organic matter including DOC, THMFP, HAAFP, AOC and odors. The results show that ozonation and biologically active GAC media improve organic matter removal. For one plant, preozonation and post-ozonation at similar ozone to DOC ratios and GAC contact times produces very similar water quality while at the other plant, the post-ozonation option is significantly better. The choice of the location for ozonation and GAC at these facilities will ultimately depend on water quality goals and disinfection needs.

Introduction

In drinking water treatment, the use of a porous media filtration process without any prior solid/liquid separation step is known as direct filtration. Due to increasingly stringent drinking water quality requirements in the USA and other countries, many water utilities have been exploring process modifications to improve the performance of treatment facilities. Water quality objectives have emphasized the need for adequate disinfection while concurrently limiting the formation of disinfection by-products (DBPs). These objectives may be realized by increasing organic matter removal prior to disinfection, decreasing the use of free chlorine, using alternate disinfectants/oxidants and minimizing levels of residual disinfectants in distribution systems.

One potentially beneficial treatment process modification is the use of ozonation and granular activated carbon (GAC) media, a common practice in some European countries [1]. The South Central Connecticut Regional Water Authority (SCCRWA) and the University of Massachusetts (UMass) have been collaborating on extensive laboratory and pilot-scale investigations of the optimal placement of ozonation and GAC media for improving the performance of two existing direct filtration treatment plants. Both plants have dual media anthracite/sand filters; one is an in-line filtration plant (West River) with only rapid mixing and the other

R. Klute / H.H. Hahn (Eds.)
Chemical Water and Wastewater Treatment III

plant has three stage flocculation following static mixers (Lake Gaillard). Research has focussed on the effects of pre-ozonation on coagulant dosing and filter performance, the effects of pre-ozonation and replacement of anthracite filter media with GAC on organics removal and the effects of ozonation and separate GAC contactors following anthracite/sand filtration on organics removal. The objectives of this paper are to present results that illustrate the effects of pre-ozonation on coagulation and filtration performance and to examine the optimal placement of ozonation and GAC in the direct filtration facilities.

Background

Ozonation and GAC may be used for a variety of purposes in drinking water treatment. Objectives of ozonation may include: oxidation of reduced species (i.e., Fe(II), Mn(II)); improvement of taste and odor; destruction of disinfection by-product precursors; disinfection; direct oxidation of trace synthetic organics; increasing the biodegradability of natural organic matter (NOM), and; aid to coagulation, flocculation and particle removal [1]. GAC may be used for relatively shorter term adsorption objectives, i.e., adsorption of synthetic organics, taste and odor causing compounds, DBP precursors or NOM, or for longer term objectives as a porous medium, i.e., as a filter medium for particle removal, or as a media for the support of microorganism growth [2]. Because ozonation increases the biodegradability of NOM, usually measured as assimilable organic carbon (AOC) or biodegradable dissolved organic carbon (BDOC), ozonation is almost always used in conjunction with a process for biological stabilization, i.e., biologically active GAC, slow sand filtration or subsurface storage and retrieval.

Consideration of the use of ozonation to improve treatment for an existing direct filtration facility must involve choosing the placement of ozone contactors and a subsequent biological treatment process. Two logical alternatives are: 1) pre-ozonation followed by coagulation/flocculation and a biologically active filter medium and, 2) post-filtration ozonation followed by biological treatment. For the pre-ozonation option, previous work has indicated that the use of GAC as a filter medium will probably provide more effective biological treatment than use of conventional anthracite or sand media [3] while at the same time GAC is an effective medium for particle removal [4]. For the post-filtration ozone application, the use of subsequent GAC contactors to support biological treatment is the most likely choice if land area for slow sand filtration is not available or injection into and recovery from a subsurface site is not feasible. The choice between these two O_3/GAC placement strategies will be based on water quality (objectives, performance, and regulations), existing plant configuration (hydraulics, availability of space), and costs.

Advantages of the preozone and GAC filter media option may include lower capital costs and realization of benefits from preoxidation, i.e., Mn and Fe oxidation and improved particle removal. The preozone option may be limited by competing ozone demands in the raw water and limited empty bed contact time (EBCT) for

GAC in the filter beds (i.e., for replacement of existing filter media at typical depths of 0.5 to 1 m). Also, the effects of preozonation on coagulant dosing and overall filter performance (turbidity and particle removal, filter run time) must be evaluated. Advantages of using ozonation and GAC after conventional direct filtration may include use of a lower ozone dose due to prior NOM and particle removal, improved disinfection at a lower ozone dose due to a reduction in ozone demand, ability to choose longer EBCTs, and possibly better utilization of the GAC for adsorption objectives. The post-ozone option will most likely involve a higher capital cost and forgoes any of the preoxidation benefits. Use of post-ozonation may require continued use of existing preoxidants or the use of another preoxidant if post-filtration oxidation of manganese occurs, resulting in the formation of stable colloids that are not removed by the GAC contactors.

Earlier work at the SCCRWA's West River Treatment Plant, the in-line filtration facility (no flocculation tanks), focussed on the effects of preozonation on coagulant dosing and filter performance for their conventional anthracite/sand dual media filters. The results of intermittent pilot experiments indicated that preozonation did not change optimal coagulant doses but often caused better particle removal and somewhat poorer removal of DOC by the anthracite/sand filters [5]. This study was followed in 1991 by an investigation of the use of preozonation and GAC/sand dual media filters. The study involved continuous operation of two parallel process trains, one with and one without pre-ozonation. Each train had two GAC/sand dual media filters, with one receiving non-chlorinated backwash water while the other was backwashed with full-scale plant treated water containing approximately 1.3 mg/L free chlorine. The results of this work [6, 7] indicated that in comparison to the conventional anthracite sand filters, use of the GAC/sand filters improves DOC removal at steady state after exhaustion of bulk DOC adsorption capacity, most likely because of biological degradation. The use of chlorinated backwash water caused poorer removal of organic matter and led to degradation of the GAC media such that the use of chlorinated backwash water was discontinued after the first 6 months of the study. Results for DOC removal were similar for the preozone and no preozone trains, however, the preozone train resulted in lower DBP formation potentials and improved odors. Also, the AOC formed by preozonation was consistently removed by the GAC/sand filters, resulting in a treated water AOC level slightly lower than that of the conventional plant. In addition, the parallel comparisons of in-line filtration with and without pre-ozonation indicated that preozonation frequently resulted in somewhat longer filter run times due to a lower rate of headloss development.

Based on the observed benefits of ozonation and GAC media, the SCCRWA decided to compare the performance of preozonation and dual media GAC/sand filtration with the performance of ozonation and GAC contactors placed after anthracite/sand direct filtration. This comparison was made by conducting pilot studies, first (in 1992) at the in-line direct filtration West River Treatment Plant (WRTP) and subsequently (in 1993) at the direct filtration (with flocculation) Lake Gaillard Treatment Plant (LGTP). The water supplies for both plants are impoundments of upland rivers and are of similar quality (Table 1). Treatment processes at the 0.46 m^3/s capacity WRTP include preoxidation with potassium permanganate, coagulation with ferric chloride (or alum) and cationic polymer, dual media (anthracite/sand) filtration (3.7–7.3 m/hr), disinfection with free chlorine and corrosion control by pH adjustment (to 7.3) and addition of a phosphate inhibitor. Treatment at the 3.5 m^3/s LGTP consists of coagulant addition (alum and cationic polymer) prior to static mixing, 3 stage tapered flocculation (30 minutes total), addition of an anionic polyelectrolyte filter aid, dual media (anthracite/sand) filtration (6–12 m/hr), and disinfection and corrosion control as practised at the WRTP. In order to comply with US disinfection requirements, pre- or intermediate chlorination is practiced occasionally at the LGTP. Some of the results of the pilot studies are presented and discussed after a brief presentation of the experimental methods used for the work.

Table 1. Raw water quality

Characteristic	West River (1992)		Lake Gaillard (1993)	
	range	mean	range	mean
pH	6.5–7.3	6.9	6.6–7.5	7.0
Turbidity (ntu)	0.6–11.6	1.2	0.4–1.7	0.7
Color (Pt-Co units)	18–50	27	15–28	21
DOC (mg/L)	2.2–3.3	2.7	2.4–3.2	2.8
UV 254 (cm^{-1})	0.07–0.20	0.11	0.08–0.12	0.09
Alkalinity (mg/L as $CaCO_3$)	10–17	11	10–14	12
Hardness (mg/L as $CaCO_3$)	20–30	24	18–21	19

Experimental Methods

Description of Pilot Plants

The research has involved extensive use of pilot scale treatment units owned and operated by the SCCRWA and located at the full scale facilities. Schematic diagrams of the 1992 West River (WR) pilot plant and the 1993 Lake Gaillard (LG) pilot plant are shown in Figures 1A and 1B, respectively. In train #1 of the WR pilot plant, filtered water from the full scale WRTP was ozonated and then flowed through four GAC beds in series. Ozone was generated from air and applied in a 0.3 m diameter by 3 m tall counter-current contactor with a hydraulic detention time of 10 minutes at a water flow of 21 L/min. The GAC contactors were 10 cm diameter glass columns with 50 cm of GAC (8 × 30 Filtrasorb 300, Calgon) media. The hydraulic loading rate of 7.3 m/hr resulted in an EBCT of 4 minutes for each column, allowing for potential sampling at EBCTs of 4, 8, 12 and 16 minutes. Train #2 of the WR pilot plant included ozonation of raw water, coagulant addition with rapid mixing and GAC/sand filtration. The ozone contactor and water flow rate were the same as used for train #1. Coagulants were added to a 4 compartment rapid mixing tank with a total detention time of approximately

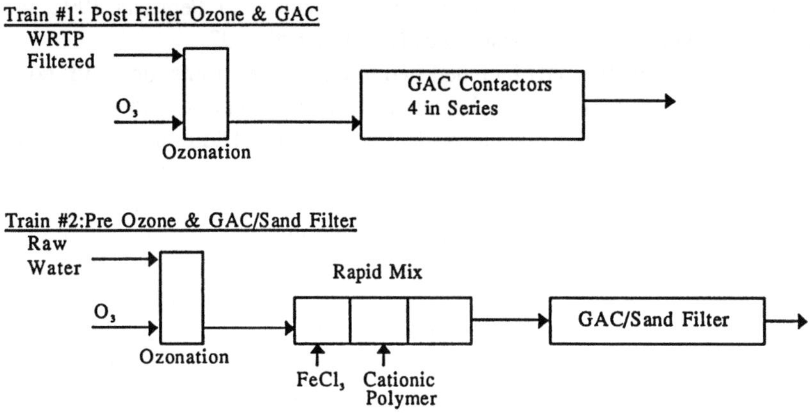

Fig. 1A. Schematic diagram of the 1992 SCCRWA West River Pilot Plant

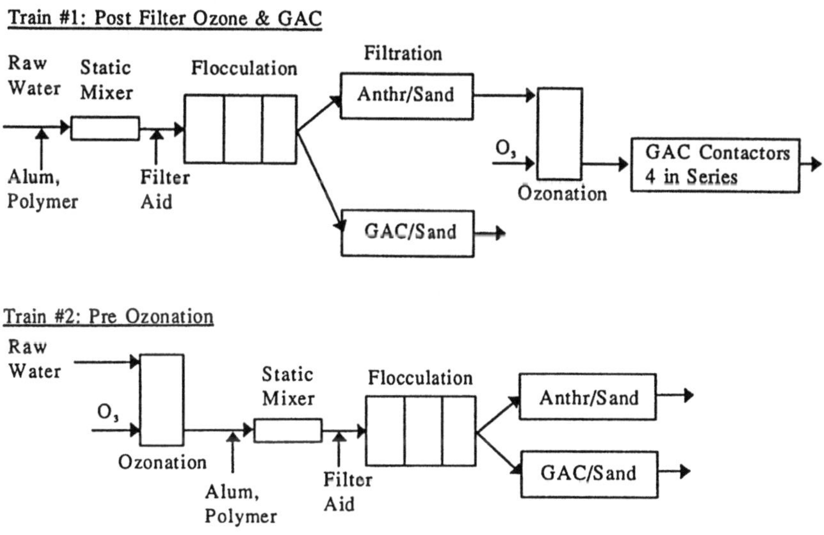

Fig. 1B. Schematic diagram of the 1993 SCCRWA Lake Gaillard Pilot Plant

10 minutes at a flow of 21 L/min. Ferric chloride and cationic polymer were added to the first and second compartments, respectively, which each contained a high speed propellor mixer. Coagulant doses were the same as used for the WRTP, approximately 6–8 mg/L $FeCl_3$ and 2–2.5 mg/L polymer (as product). The 25 cm diameter filter column contained 50 cm of GAC (8 × 20 Filtrasorb 300, approx. 1 mm effective size) and 25 cm of silica sand (0.45 mm effective size) over a gravel support layer. The filter hydraulic loading rate was either 3.7 or 7.3 m/hr and the filters were equipped with an automatic backwashing system based on filter headloss. The GAC in the filters was put into service in September of 1991

and the post-filtration GAC contactors began operation on 11 February 1992; the pilot plant was operated continuously until November of 1992.

The LG pilot plant (Fig. 1B) included two main parallel trains, one with post-filtration ozonation and GAC and the other with preozonation and GAC as a filter medium. Both pilot trains included identical chemical addition, static mixing and flocculation facilities. Alum, cationic polymer and sodium hydroxide were applied ahead of the static mixer at the doses utilized in the full scale plant: 7–12 mg/L alum, 1.2–2.5 mg/L polymer (American Cyanimid 572C), 0–2.5 mg/L NaOH. The flocculation tanks consisted of 3 compartments in series, each with a gate mixer and a detention time of 10 minutes at a flow of 38 L/min. An anionic polymer filter aid (0.05–0.08 mg/L American Cyanimid 1849A) was applied after flocculation prior to filtration. The post-ozonation train (#1) had one GAC/sand and two anthracite/sand dual media filters in parallel. Effluent from the anthracite/sand filters (50 cm anthracite, 25 cm sand) was ozonated (same contactor as used at the WR pilot plant) and applied to the same GAC columns that had been used in the 1992 WR study. The GAC medium was not changed between the pilot studies. The hydraulic loading rate for the GAC contactors was usually 2.5 m/hr, resulting in EBCTs of 5, 10, 15 and 20 minutes for the four beds, respectively. The post-ozonation pilot train for the LG study included pilot scale flocculation and filtration instead of the use of full scale filtered water because of the occasional practice of pre- or intermediate chlorination at the LGTP.

The preozonation train at LG included a 10 minute detention time countercurrent contactor for raw water ozonation prior to chemical addition. This train also had parallel GAC/sand and anthracite/sand dual media filters following flocculation. The GAC medium in the filters for both trains was the same medium that was used in the 1992 WR study. Filtered water from the pilot plant was stored and used for backwashing of the pilot filters. The pilot plant was operated continuously and a system based on a programmable logic controller and pressure transducers was used to automatically backwash filters based on headloss development.

Water Quality Analyses

Water quality analyses were conducted on samples collected from various pilot plant locations and from the full scale facilities. On-site analyses included turbidity, ultraviolet light absorbance at 254 nm wavelength (UV254), pH, particle counts, ozone transfer efficiency, aqueous ozone residual and filter headloss. The SCCRWA laboratory conducted numerous odor analyses using the flavor profile panel method. Samples were brought to the UMass laboratories for determination of dissolved organic carbon (DOC), assimilable organic carbon (AOC), keto acids, and various chlorinated DBPs (neutral extractables, including the trihalomethanes (THM); haloacetic acids (HAA); trihaloacetaldehydes, including chloral hydrate).

Samples for DOC and UV254 analyses were filtered through pre-washed glass fiber filters (Whatman GF/C). DOC samples were acidified, sparged with nitrogen and analyzed using the UV-persulfate oxidation method with a Dohrmann DC 80 instrument. AOC was determined using a modified van der Kooij method developed by Kaplan [8]. Simultaneous growth (CFU) of P17 and NOX bacterial strains was converted to equivalent carbon concentrations based on published yield values for growth on an acetate substrate [9]. By-products of ozonation and chlorination were determined using gas chromatography following appropriate sample extraction and derivatization procedures [8, 10, 11]. Formation potential (FP) conditions for determination of chlorinated DBP precursors (THMFP, HAAFP) were: pH = 7, 20 mg/L chlorine dose, 3 days incubation at 20°C. During 1993, a new HAA analytical procedure was instituted which combined a methyltertbutylether extraction step with the acidic methanol (heated) methylation step of the EPA solid phase extraction method [12]. The THMFP values reported herein are the sum of all four THM species (trichloromethane, bromodichloromethane, chlorodibromomethane and tribromomethane) while HAAFP values include only the sum of two species, trichloroacetic acid (TCAA) and dichloroacetic acid (DCAA). TCAA and DCAA consitute over 90 % of the five HAAs quantified in this work.

Results and Discussion

The presentation of results is focused on the quality of filtered waters from the existing full scale plants, the preozone trains and the post-ozone-GAC contactor trains. The effects of preozonation and GAC filter media are assessed and a comparison of the two different locations of ozonation and GAC media are made. Results from the 1992 West River study are considered first, followed by results from the 1993 Lake Gaillard study and a comparison of the two projects. The data presented are based on periodic sampling (6–10 per year) and analyses conducted by UMass and periodic odor analyses conducted by the SCCRWA. In addition, SCCWRA personnel collected daily turbidity, UV254, pH and headloss data for performance monitoring.

In-Line Direct Filtration: 1992 West River Study

The performance of the preozone (1.5 mg/L applied dose, 90–95 % absorbed), GAC/sand direct filtration train was very similar to the full scale WRTP based on traditional measures of turbidity and filter run time (UFRV), as was observed in the 1991 study [6]. Filtered water turbidity for both processes varied between 0.01 and 0.05 NTU and UFRVs varied between 90 and 200 m^3/m^2 (based on 2.4 m of headloss). Ozonation of the WRTP filtered water (0.6–0.75 mg/L applied dose, 85 % absorbed) often caused an increase in turbidity, possibly due to oxidation and precipitation of manganese. The GAC contactors usually reduced turbidity levels, however the turbidity after the 16 minute EBCT stage was typically higher than that of the WRTP filtered water.

Measurements of UV254 were used to monitor treatment performance with respect to organics removal. The UV254 results for the post-filtration GAC contactors (and the DOC results discussed below) indicate that after 30 to 60 days of operation breakthrough of the bulk organic matter occurred. After this period, UV254 values increased and decreased with changes in influent UV254 and did not display characteristic adsorption breakthrough behavior. UV254 values were consistently reduced by GAC media and the reduction increased as EBCT increased, either due to biodegradation of organic matter or adsorption of a fraction of the DOC for which the GAC has a high adsorption capacity.

The average concentrations (and range) over the study period for various measures of organic matter (DOC, AOC, DBP precursors) at several treatment process locations are summarized in Table 2. The performance of the treatment trains is

Table 2. 1992 SCCRWA West River Pilot Study – concentrations of organic matter (10 Jan.–15 Oct. 1992); mean and range for samples

Sample location	DOC		THMFP		HAAFP		AOC	
	mg/L	#	μg/L	#	μg/L	#	μg/L	#
Raw	2.71 (2.2–3.3)	6	156 (111–215)	7	220 (174–294)	7	83 (35–150)	8
Train 1								
WRTP Filtered	1.56 (1.14–2.11)	6	61 (43–85)	7	69 (57–94)	7	90 (26–181)	8
Post-ozone	1.50 (1.17–2.04)	6	50 (38–60)	6	60 (49–81)	6	154 (89–296)	7
4 min EBCT GAC	1.24 (1.13–1.40)	5	42 (33–53)	5	48 (39–61)	5	129 (69–223)	7
16 min EBCT GAC	0.76 (0.59–1.15)	5	22 (17–32)	4	27 (18–40)	4	112 (66–164)	5
Train 2								
Pre-ozone	2.65 (2.12–3.26)	6	110 (48–150)	7	182 (131–277)	7	120 (58–176)	7
GAC/sand Filtered	1.29 (1.00–1.99)	6	40 (25–58)	7	45 (28–65)	7	50 (17–126)	6

Notes:
DOC = dissolved organic carbon, nonpurgeable, passing through Whatman GF/C filter.
THMFP = trihalomethane formation potential = sum of four THM species.
HAAFP = haloaceticacid formation potential = sum of trichloracetic acid + dichloroacetic acid only.
AOC = assimilable organic carbon expressed as acetate carbon equivalents based on simultaneous growth of P17 and NOX bacteria.
# = number of samples in average.
Mean value given with (range) below.

compared in Figure 2 where the fraction of the raw water organic concentration remaining after treatment is shown. Results for full scale WRTP filtered water, preozone GAC/sand pilot filtered water, and 4 and 16 minute EBCT GAC contactors are included. For the post-filtration GAC contactors, only data collected after

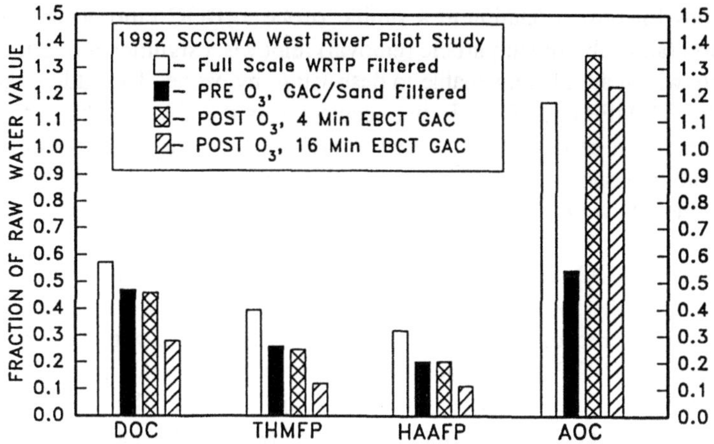

Fig. 2. Average fraction of raw water organic matter concentration remaining in treated waters for the 1992 West River pilot study

25 April 1992 are included in the averages. It is important to note that the O_3/DOC ratio (0.4–0.5) and the GAC contact time are very similar for the preozone train and for the post-ozone train at the 4 minute EBCT sampling point.

The DOC results (Fig. 2) show that the use of ozonation and GAC media resulted in an increase in DOC removal from 43 % for the WRTP to approximately 54 % for both the preozone train and the post-ozone, 4 minute EBCT GAC train. DOC removal increased further to over 72 % for the post-ozone, 16 minute EBCT GAC process. The effects of the preozone, GAC/sand filter train on DOC removal are in agreement with the results of the 1991 West River pilot study [6]. The DOC results suggest that, for similar ozone to DOC ratios and GAC contact times, the use of post-filtration ozonation followed by GAC did not result in significantly lower DOC levels than did the use of preozonation and GAC/sand direct filtration.

The results for the removal of chlorinated DBP precursors are similar to the DOC results with respect to the relative effects of the treatment processes. The WRTP achieves approximately 60 % removal of THMFP while for the preozone train and for the post-ozone, 4 minute EBCT GAC train, the removal is increased to 75 %. Similar results are found for the removal of HAAFP although absolute levels of removal are slightly higher (68 % for the WRTP and 80 % for the ozone trains). Increasing the post-filtration GAC contact time to 16 minutes results in a further increase in DBP precursor removal to nearly 90 %. As is usually found, the chlorinated DBP precursor fraction of the NOM is more amenable to removal than the bulk DOC, for both coagulation and conventional treatment and for ozone and GAC media treatment.

The results in Table 2 show that the direct effects of ozone on DBP precursor destruction are quite significant for this water. Preozonation causes a 30 % reduction in THMFP and a 17 % reduction in HAAFP while for post-ozonation the reductions are 18 and 13 % for THMFP and HAAFP, respectively. The lower percent reductions for post-ozonation at a similar ozone to DOC ratio occur because coagulation (and subsequent particle removal) removes a significant fraction of the precursors which are also amenable to destruction by ozone. The additional removal of DOC, THMFP and HAAFP that occurs across the GAC media in the post-ozonation train can be directly attributed to biological activity (and possibly adsorption). A similar conclusion is suggested by the results for the preozone train although measurements were not made immediately after flocculation prior to the GAC/sand filter.

Measurements of AOC were made to assess the effects of treatment on the relative biological stability of the water. The results in Table 2 and Figure 2 for the WRTP and for the preozonation, GAC/sand train are in accord with the 1991 study. On average, the WRTP does not remove AOC present in the raw water. Preozonation causes AOC to increase by an average of 45 % but after the GAC/sand filter AOC levels are reduced on average to 54 % of the raw water values, suggesting significant biological activity. The results for the post-ozone, GAC contactor pilot train were not encouraging. Ozonation of WRTP filtered water caused AOC to increase on average by 71 % (Table 2), similar to the preozonation effect and perhaps a result of the similar ozone to DOC ratios for the two trains. Removal of AOC across the post-GAC contactors was very poor, perhaps because of more limited biological activity. However, the consistent removal of other organic paramaters across the GAC well after the period where adsorption of bulk DOC is significant suggests that biological activity was responsible for removal of organic matter.

The SCCRWA utilizes internal flavor profile analysis panels to assess the taste and odor of water. The panel members assess several categories of odors (musty, fishy, medicinal, fruity, chlorinous, etc.) using a relative scale of 0 (no odor) to 12 (very strong). The study period average odor intensities for samples of raw water, WRTP filtered water and several pilot plant treated waters are shown in Figure 3. None of the samples were treated with chlorine. The results show that the conventional treatment process at the WRTP reduces raw water odor levels by only 30 %. Preozonation alone has the same effect as the WRTP, however, odor levels after the GAC/sand filter are reduced by over 95 %. In contrast, the direct effect of ozonation of WRTP filtered water is large and there is little additional reduction across the 4 minute EBCT GAC column following post-ozonation.

The results of the 1992 West River pilot study indicate that for similar ozone to DOC ratios and GAC contact time, preozonation and GAC filter media provide treated water quality similar to post-filtration ozonation followed by a GAC contactor. If the improvements in water quality are sufficient to meet water quality goals and regulations, then cost considerations may make preozonation and replacement of filter media with GAC the preferred option over ozonation and new GAC contactors after filtration. However, if the additional improvement in water quality realized by longer GAC contact times is needed, and if the disinfection benefits of post-filtration ozonation are desired, then the post ozonation and GAC option may be required.

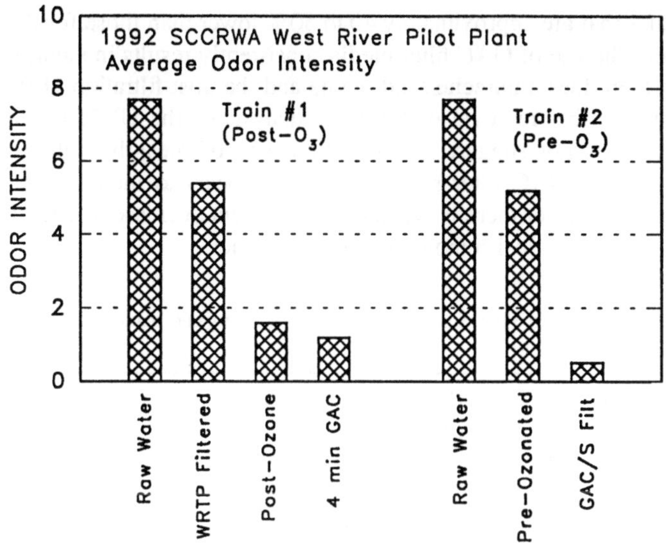

Fig. 3. Average of results of odor analyses conducted by SCCRWA for the 1992 West River pilot study

Direct Filtration with Flocculation: 1993 Lake Gaillard Study

The effects of ozonation and GAC media were also examined at the SCCRWA's other direct filtration facility, the Lake Gaillard Treatment Plant. This study was initiated in January of 1993 and is ongoing as of January 1994. In order to begin the study with GAC that was already exposed to natural organic matter and that was probably biologically active, it was decided that the GAC media from the West River study would be moved to Lake Gaillard. As described earlier, train #1 simulated the full scale plant processes and provided anthracite/sand filtered water for the post-ozonation and GAC contactor processes while the other train (#2) included preozonation prior to coagulant addition. Both trains had parallel anthracite/sand and GAC/sand dual media filters. As in the 1992 WR study, the ozone to DOC ratios for the preozonation (2.0 mg/L applied O_3, 73–86 % transfer) and the post-ozonation (0.6–0.75 mg/L applied O_3, 85–97 % transfer) trains were similar (0.58 and 0.35 an average, respectively). Preliminary work during pilot plant start up in January 1993 indicated that coagulant dosing was not altered by preozonation and that filter aid dosing for the pilot plant needed to be slightly higher than for the full scale plant to avoid premature turbidity breakthrough.

The performance of the pilot trains with respect to filtered water turbidity and filter run time was similar to the full scale plant. Filtered water turbidity was typically in the range of 0.05 to 0.12 NTU for the LGTP and the pilot trains and a further reduction of 0.01 to 0.03 NTU was achieved by the post-filtration (and ozonation) GAC contactors. Note that filtered water turbidities at the West River plant are typically lower, in the range of 0.01 to 0.05 NTU. The results of routine UV254 measurements show that the use of either pre- or post-ozone results in UV254 values that are approximately 40 to 50% lower than those for the LGTP filtered effluent. The use of GAC filter media consistently results in slightly lower UV254 values than does anthracite/sand media and the post-filtration GAC contactors also cause a further removal of UV254. Analysis of the UV254 data as a function of time does not suggest an adsorption type of breakthrough for the GAC contactors (or the GAC filter media). UV254 appears to increase and decrease with influent quality yet there is typically a consistent decrease as EBCT increases, perhaps due to biological removal of organic matter.

Table 3. 1993 SCCRWA Lake Gaillard Pilot Study – concentrations of organic matter (Jan.–Nov. 1993); mean and range for samples

Sample location	DOC		THMFP		HAAFP		AOC	
	mg/L	#	μg/L	#	μg/L	#	μg/L	#
Raw	2.76 (2.42–3.19)	10	200 (135–263)	7	229 (185–306)	7	75 (47–97)	6
LGTP Filtered	1.70 (1.50–2.00)	10	98 (75–120)	5	94 (83–127)	5	44 (39–51)	5
Train 1								
GAC/sand Filtered	1.64 (1.48–1.82)	9	96 (74–118)	6	88 (79–102)	6	37 (25–62)	6
Anthr./sand Filtered	1.71 (1.50–1.91)	9	99 (70–118)	6	92 (79–115)	5	40 (34–50)	6
Post-ozone	1.70 (1.53–1.82)	9	76 (55–97)	7	78 (57–113)	7	132 (102–158)	6
5 min EBCT GAC	1.46 (1.33–1.52)	8	63 (42–77)	6	57 (44–79)	6	76 (47–99)	6
20 min EBCT GAC	1.20 (1.09–1.47)	8	46 (30–51)	5	40 (31–55)	5	50 (36–66)	(5)
Train 2								
Pre-ozone	2.68 (2.27–3.12)	9	140 (90–171)	7	182 (142–233)	6	145 (120–162)	6
GAC/sand Filtered	1.58 (1.35–1.71)	9	80 (52–96)	6	71 (57–103)	6	46 (40–52)	6
Anthr./sand Filtered	1.66 (1.46–1.98)	9	82 (54–97)	6	74 (58–104)	6	56 (49–73)	6

Notes:
DOC = dissolved organic carbon, nonpurgeable, passing through Whatman GF/C filter.
THMFP = trihalomethane formation potential = sum of four THM species.
HAAFP = haloaceticacid formation potential = sum of trichloracetic acid + dichloroacetic acid only.
AOC = assimilable organic carbon expressed as acetate carbon equivalents based on simultaneous growth of P17 and NOX bacteria.
# = number of samples in average.
Mean value given with (range) below.

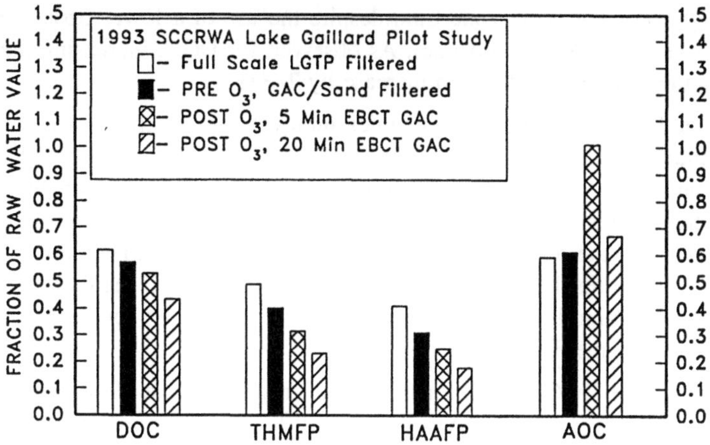

Fig. 4. Average fraction of raw water organic matter concentration remaining in treated waters for the 1993 Lake Gaillard pilot study

Results of measurements of organic matter in samples from the 1993 Lake Gaillard study are summarized in Table 3 where the average concentrations and the observed range in concentration for DOC, THMFP, HAAFP and AOC are listed. In Figure 4, a comparison of filtered water from the LGTP, the preozone GAC/sand train (#2) and the post-ozonation train (#1) GAC contactors (5 and 20 minutes EBCT) is made by considering the fraction of raw water DOC, THMFP, HAAFP and AOC remaining in the filtered waters. The DOC results show that ozonation and GAC media results in increased DOC removal. Preozonation and GAC/sand filtration decreases the fraction remaining from 0.62 for the LGTP to 0.57. However, post-ozonation and a similar 5 minute EBCT results in better DOC removal (0.53 remaining) while the 20 min EBCT further decreases the fraction remaining to 0.43.

The relative effects of the different treatment processes on THMFP and HAAFP removal follow the same pattern as for DOC. However, the increased removal for the preozone train as compared to the LGTP is more significant for the DBP precursors than for DOC, probably because of the direct effects of ozonation. The results in Table 3 indicate that on average preozonation destroyed 30 % of the THMFP and 20 % of the HAAFP. After preozone and GAC/sand filtration, removals averaged 60 % and 69 % for THMFP and HAAFP, respectively, reflecting the greater removal of HAA precursors (especially TCAA) by coagulation despite the greater destruction of THMFP by ozonation (largely because ozone has little effect on DCAA precursors). Post-filtration ozonation of anthracite/sand filtered water destroys on average 22 % of the THMFP and 17 % of the HAAFP, resulting in concentrations similar to that of the preozonated, GAC/sand filtered water (see Table 3). As a result, subsequent treatment with the 5 minute EBCT GAC contactor causes removals to increase to 69 % for THMFP and 75 % for HAAFP (Fig. 4), probably due to biodegradation. It is interesting that the overall removals are higher for the post-ozone train despite the use of a lower ozone to DOC ratio (0.35 versus 0.58 for preozone). As observed for DOC, increasing the EBCT to 20 minutes causes removals to increase by 7–8 %.

The effects of filter media type on organics removal can be considered for the 1993 LG study by comparing the results shown in Table 3 for the parallel anthracite/sand and GAC/sand dual media filters. In general, there is very little difference between the results for the two types of filters, both with and without preozonation; however, the average values are consistently 3–5 % lower for the GAC/sand filters. These results suggest that adsorption by the GAC media is not significant and that similar levels of biological activity may exist for both the anthracite and the GAC media.

A comparison of the DOC, THMFP and HAAFP results for the 1992 WRTP and the 1993 LGTP (Figs. 2 and 4) shows that overall levels of removal are quite similar although removals are consistently greater for the WR study than for the LG project. This could be due to differences in raw water quality (both general and temporally), coagulant type ($FeCl_3$ at WR versus alum at LG, both with cationic polymer) or flocculation processes. For both studies, removals increase in the order DOC $<$ THMFP $<$ HAAFP. Ozonation and GAC media result in improved removals as compared to the existing treatment for both locations, however, the relative effect of the placement of ozone and GAC appears to be different. For the WR study, levels of DOC, THMFP and HAAFP were nearly identical for preozonation and post-ozonation in combination with GAC with the same contact time. In contrast, for the LG study, the post-ozonation with a 5 minute EBCT GAC contactor resulted in significantly better organic matter removal than the preozonation option with GAC filter media (5 minute EBCT). It appears that the difference is due to relatively poorer performance of the preozone GAC train at Lake Gaillard as opposed to relatively improved performance of the post-ozonation GAC contactor train. The difference in performance could be due to a number of factors including differences in raw water quality, especially DOC characteristics, different ages of the GAC (filter media GAC put in service September 1991, post-contactor GAC put in service February 1992) or different effects of preozonation and coagulation between WR and LG.

The AOC results (Table 3, Fig. 4) show that on average the LGTP removed 41 % of raw water AOC and that preozonation caused a 93 % increase in AOC. The GAC/sand pilot filter on the preozone train reduced AOC to a level simlar to the LGTP effluent, but the reduction was not as large as was observed in the 1992 WR study. Note that the anthracite/sand filter on this train also removed much of the AOC created by preozonation, suggesting biological activity, although filtered AOC values were consistently higher than those of the GAC filter. For the other train, ozonation of anthracite/sand filtered water caused a 76 % increase in AOC, yielding values similar to the preozonated water, perhaps because of the similar (although lower) ozone to DOC ratio. AOC removal by the GAC contactors was much better than for the 1992 West River study. Note that AOC concentrations for the 5 minute EBCT were consistently higher than for the preozone GAC/sand train, in contrast to the relative performance for removal of DOC, THMFP and HAAFP at LG. After the 20 min EBCT GAC contactor, AOC was reduced to a level similar to that found for the LGTP and preozone GAC/sand filtered waters. Lack of acclimation time should not be responsible for poor AOC removal at the 5 min EBCT; further studies are being conducted in 1994.

The average odor intensity results from the flavor profileanalyses for the 1993 LG study are shown in Figure 5. The LGTP typically removes only 24 % of raw water odors, similar to the West River results (Fig. 3). Preozonation and GAC/sand filtration increases removal to 58 % and post-ozonation and 5 minute EBCT GAC results in 63 % removal. Overall removals by the pilot trains are not as good as for the West River study and, for Lake Gaillard, the post ozonation option yields the lowest odors as compared to the preozonation option for West River (for the same GAC EBCT for both options).

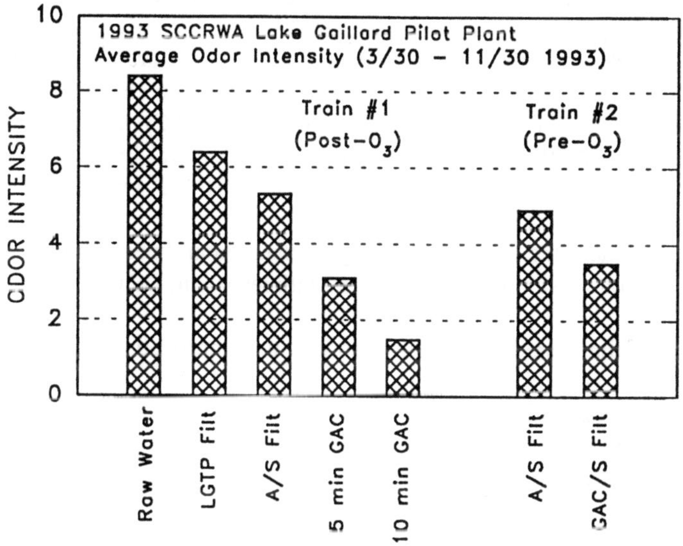

Fig. 5. Average of results of odor analyses conducted by SCCRWA for the 1993 Lake Gaillard pilot study

Overall, the results of the 1993 LGTP pilot project show that ozonation and GAC media provide improved water quality in terms of the removal of DOC, THMFP, HAAFP and odors. The results suggest that, for the same GAC EBCT, post-filtration ozonation and GAC provides significantly better water quality than does the use of preozone and GAC as a filter medium. As a result, the SCCRWA is continuing to study this option at the LGTP in 1994, with an emphasis on the effects of EBCT, startup of new GAC and optimization of coagulation in conjunction with the post-filtration ozone and GAC.

Conclusions

Results from two separate years of pilot studies of the use of ozonation and biologically active GAC media to upgrade existing direct filtration facilities have been summarized in this paper. The emphasis of the data analysis was on a comparison of the use of preozonation and GAC filter media with the use of post-filtration ozonation followed by GAC contactors. The evaluation herein was based mostly on various aggregate measures of organic matter including DOC, THMFP, HAAFP, AOC and odors. For each study, the pilot plants were operated continuously for approximately one year and each plant included parallel trains for the different processes being studied. The following conclusions are made:

1. The 1992 WRTP study showed continued (from 1991) excellent performance of the preozonation, GAC/sand direct filtration train in terms of filtration performance and removal of organic matter. DOC, THMFP, HAAFP, AOC and odor removal were significantly increased as compared to the existing plant.
2. For the 1992 West River study, the performance of the post-filtration ozonation and 4 minute EBCT GAC contactor was very similar to the preozone train (which also had 4 min of EBCT for the filter media GAC) with respect to DOC, THMFP, HAAFP and odors, however, AOC removal by the GAC contactors following post-ozonation was poor. DOC, THMFP and HAAFP removal improved with increasing GAC EBCT.
3. For the West River plant, the choice between preozonation and post-ozonation will depend on absolute water quality objectives with respect to DOC and DBP precursors.
4. The 1993 Lake Gaillard study also demonstrated the improved removal of organic matter achieved with ozonation and GAC media. However, the improvements using preozonation and GAC/sand filter media were not as large as for the West River study.
5. For the 1993 LG study, the post-ozonation, 5 minute EBCT GAC contactor results showed significant improvements in DOC, THMFP, HAAFP and odor removal as compared to the existing plant, with performance improving with increasing EBCT.
6. As compared to the WR study, the removal of AOC formed by ozonation at LG was not as good for the preozone train but was better for the post-ozonation train. However, neither biologically active train produced water with an AOC level lower than the existing LGTP (unlike the results for the 1991 and 1992 West River preozone train).

Acknowledgements. The work reported on herein was supervised jointly by the author and co-workers James Edzwald, David Reckhow and Michael Switzenbaum of the University of Massachusetts and Howard Dunn and Gary Kaminski of the SCCRWA. Sampling and analyses were conducted by graduate students James Ayers, Rebecca McEnroe, Paul Schmidt, Denise Springborg, Nagaraju Vinod, Jonathan Weiner and Qing-wen Zhu. We are very grateful for the funding provided by the SCCRWA of New Haven, CT, USA.

References

[1] Langlais, B., Reckhow, D.A., Brink, D.R., editors: Ozone in Water Treatment, Application and Engineering. American Water Works Association Research Foundation, Compagnie Geneale des Eaux, Lewis Publ., Chelsea, Michigan (1991).

[2] Sontheimer, H., Crittenden, J.C., Summers, R.S.: Activated Carbon for Water Treatment. DVGW-Forschungsstelle, Engler-Bunte-Institut, Universitt Karlsruhe (1988).

[3] Krasner, S.W., Sclimenti, M.J., Coffey, B.M.: Testing Biologically Active Filters for Removing Aldehydes Formed During Ozonation. Journal AWWA, 85:5 (1993) 62.

[4] Graese, S.L., Snoeyink, V.L., Lee, R.G.: Granular Activated Carbon Filter-Adsorber Systems. J AWWA, 79:12 (1987) 64.

[5] Tobiason, J.E., Edzwald, J.K., Schneider, O.D., Fox, M.B., Dunn, H.J.: Pilot Study of the Effects of Ozone and PEROXONE on In-Line Direct Filtration. Journal AWWA, 84:12 (1992) 72.

[6] Tobiason, J.E., Edzwald, J.K., Reckhow, D.A., Switzenbaum, M.S.: Effect of Pre-Ozonation on Organics Removal by In-Line Direct Filtration. Wat. Sci. Techn. *27* (11) (1993) 81

[7] Reckhow, D.A., Tobiason, J.E., Switzenbaum, M.S., McEnroe, R., Xie, Y., Zhu, Q., Zhou, X., McLaughlin, P., Dunn, H.J.: Control of Disinfection Byproducts and AOC by Pre-Ozonation and Biologically-Active In-Line Direct Filtration. Proceedings of 1992 American Water Works Association Annual Conference, Vancouver, B.C., 1992, p. 487

[8] Standard Methods for the Examination of Water and Wastewater, 18th Edition. American Public Health Assoc., American Water Works Assoc., Water Environment Federation, Washington, D.C., 1993

[9] van der Kooij, D., Hijnen, W.A.M., Kruithof, J.C.: The Effects of Ozonation Biological Filtration, and Distribution on the Concentration of Easily Assimilable Organic Carbon in Water. Ozone Sci. Techn. *11* (3) (1989) 297

[10] Koch, B. et al.: Analysis of Halogenated Disinfection By-Products by Capillary Chromatography. AWWA Water Quality Technology Conference 1988

[11] Glaze, W.R., Konga, M., Cancilla, D.: Ozonation By-Products. 2. Improvement of an Aqueous Phase Derivatization Method for Detection of Formaldehyde and Other Carbonyl Compounds Formed by Ozonation of Drinking Water. Environm. Sci. Techn. *23* (7) (1989) 838

[12] Xie, Y., Springborg, D., Reckhow. D.A.: Analysis of HAAs and Ketoacids Without Diazomathane. Proceedings of 1993 American Water Works Association Water Quality Technology Conference, Miami, Florida, November, 1993

John E. Tobiason
Department of Civil and Environmental Engineering
University of Massachusetts
Amherst, Massachusetts 01003
USA

Combined Design, Operation and Economic Optimization of Contact Filtration

A. Adin and S. Hatukai

Abstract

Granular filter design should rely not only on process evaluation but also on overall economic considerations. A logical model is introduced for optimizing a deep bed filter and minimizing experimentation. The model consists of two stages, a process stage and an economic stage. Conventional filtering materials as well as basalt and tuff are evaluated in single, dual, and mixed media configurations. Tests were performed in a pilot plant treating Lake Kinneret water. It is concluded that the optimization process should involve a process optimization stage followed by an economic optimization stage. The filter process parameters, run length and produced water volume per run do not provide enough data for an economic evaluation of different filter beds. Single deep bed filters of basalt or tuff can remove turbidity as efficiently as single, dual, and mixed media filters containing conventional filtering materials, and produce cheaper water. Determination of optimal media parameters should be carried out at the optimal mode of backwashing and chemical treatment. Process efficiency indicators, such as volume of produced water per filter run, are poor economic indicators and should be used carefully.

Introduction

Granular media characteristics such as filter bed type (number of layers), media materials, media grain size and its distribution affect significantly filtration efficiency, head loss development and backwash regime. The values of the abovementioned granular media variables are determined in the design process of filtration systems according to raw water characteristics, desired ultimate quality of filtrate, and type of chemical pretreatment [1]. There is more than one combination of filter media and filter operation variables that could give similar filtration efficiency. The question in this case is which combination is also the most cost effective.

Generation of data concerning filter system varaiables is carried out either by running pilot plant experiments or by applying reliable mathematical models of the filtration process. Unfortunately, the mathematical models available are either limited to specific conditions or impractical for design purposes, unless they are supported by data from pilot studies. The indispensability of pilot plant

R. Klute / H.H. Hahn (Eds.)
Chemical Water and Wastewater Treatment III

experiments for optimization of filtration plants was recognized long ago both by Mintz [2] and Camp [3]. A methodology for the application of filtration theory to pilot plant design is given by Adin et al. [1].

The ultimate objective in the development of an optimization model for granular filter beds is the determination of design and operational parameters of the filtration system which will produce the desired filtrate quality of a given water at the lowest cost. The model's goal is to provide the cost of water treatment as a function of possible combinations of design and operation parameters. It is suggested that this goal be achieved through two successive stages: The first stage would provide alternate solutions of design which produce the desired filtrate quality. The second stage would provide the solution/s which obtain/s the required quality of water at the lowest cost.

This work introduces a logical model of the above nature for the optimization of filter media utilising pilot plant experiments. Cost equations are also developed for the evaluation of the effect of filter media characteristics on filter system economy. Results of pilot plant experiments and data evaluation for process optimization and economic optimization are presented. Two new filtering materials, basalt and tuff, are considered in addition to the limited number of filtering materials commonly used in water treatment (quartz sand, anthracite, garnet and occasionally granular activated carbon).

Optimization Model

Process Optimization Model

In principle, process optimization for a filter system should provide all possible combinations capable of providing the target filtrate quality. For that purpose, the filtration system is divided into a number of subsystems, the optimization of each subsystem is carried out while parameters of the other subsystems remain unchanged. In the presented model the filtration system is divided into three subsystems which are: filter bed, chemical pretreatment, and backwashing.

Plainly speaking, the objective of process optimization of the deep bed filter run is to maximize volume of filtrate and to minimize volume of backwash water, subjected to critical filtrate quality and head loss.

Three initial conditions are determined in a system which is about to be optimized. The first is raw water quality (usually turbidity), the second is filtrate quality, and the third is total head loss.

The variables affecting the performance of the filtration process and whose values have to be determined during the optimization process are [4]:

a) Type of Media. Media types in practice are single, dual, and mixed media. Media materials commonly in use are sand as single medium or bottom layer, anthracite as single medium, intermediate, or upper layer, and granular activated carbon as upper layer. Dual and mixed media filters make high filtration rates possible, which, according to European practice, can be achieved also by coarse deep single medium filters.

b) Grain Size and Its Distribution. The smaller the grain size, the better the filtrate quality and the higher the head loss across the media. Media grain size and its distribution in dual and mixed media filters must be carefully selected. The finer grains of the lower layer and the coarser grains of the upper layer must be compatible to accomplish proper separation of the layers after backwashing.

c) Depth of Media. Increasing media depth improves filtrate quality while increasing head loss across the media.

d) Rate of Filtration. Increasing filtration rate affects filter performance by increasing head loss across the media and by deteriorating filtrate quality.

e) Backwashing. Filter bed is washed either by water alone, or by water and air. Water and air are used subsequently or simultanously. Rate of backwashing with water is determined by rate of fluidization, specific gravity and grain size of the upper layer in dual and mixed media filters. Low backwash rates are required when air scour is applied. Air scour cannot be applied simultanously with water when dual, mixed, or anthracite media are in use, to avoid intermixing of the layers and loss of media. Period of backwashing is dependent on the amount of deposits in the filter bed.

f) Chemical Pretreatment. Filter performance is affected more by chemical pretreatment than by any other parameter. The objective is to change the physical-chemical characteristics of the particles in suspension in order to increase their attachment ability to the filter grains. Chemical pretreatment generally includes pH modification, coagulation and flocculation by adding salts of hydrolizing metal ions and/or polyelectrolytes to the incoming water.

It is suggested that the approach to the design and operation of filter bed optimization should be conducted according to the following steps, which are also presented schematically in Figure 1:

1. Selecting available media materials.
2. Selecting media grain sizes that apply to single media filters on the basis of previous experience with similar raw water or literature.
3. Constructing deep single medium experimental filters, each filter containing one of the selected materials.
4. Selecting type of proper flocculant/s and determining optimal doses by performing filtration runs, all at a common predetermined rate (cycle 1 in Fig. 1).
5. Running filters at different rates to the highest possible rate with the flocculants and doses determined in step 4 (cycle 2 in Fig. 1). Refine chemical pretreatment if necessary.
6. Selecting materials for dual media experiments considering the previous results (cycle 3 in Fig. 1).
7. Performing step 5 with dual media filters.
8. Repeating steps 6 and 7 for mixed media (cycle 3 in Fig. 1).

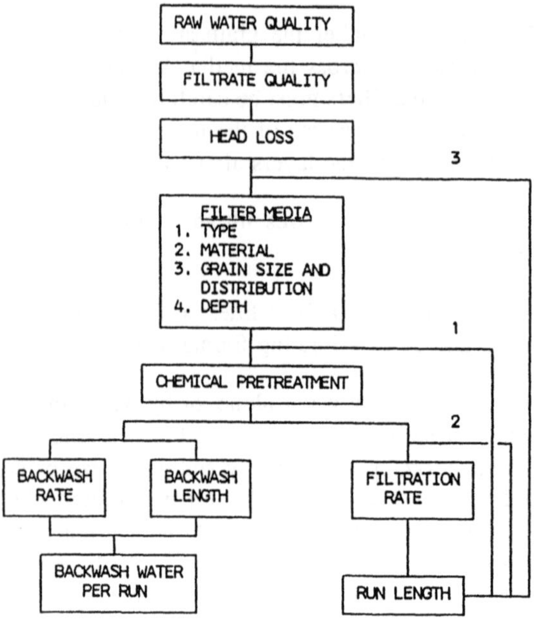

Fig. 1. Process variables affecting performance of deep bed filter

Economic Optimization Model

Parameters of filtration systems obtained in the process optimization provide the basic data for cost estimations further employed for economic optimization. Filtration cost is defined in the model as the sum of capital cost and operation and maintenance cost. Capital cost is expressed in the form of annual recovery

$$\text{ACC} = (\text{CRF})\ (M)\ (\text{DP})^X \tag{1}$$

ACC	Annualized capital cost [$/yr]	DP	Design parameter affecting the cost
CRF	Annual capital recovery coeff.	M, X	Constants determined experimentally

Filter area has been used by some investigators [5–7] as the variable affecting cost, DP, and design flowrate by others [8]. The costs of filter plant components associated with the filter area are presented in Figure 2.

The total filtration area, A (m^2), of a filter is proposed to be determined using the following expression

$$A = \frac{Q}{24}\,\frac{\text{TF} + \text{TS}}{\text{TF}\ V - \text{TB}\ \text{VB}} \tag{2}$$

Q	Design flow rate [m^3/day]	VB	Backwash rate [m^3/m^2/hr]
V	Filtration rate [m^3/m^2/hr]	TB	Time per backwash with water [hr]
TF	Run length (hr)	TS	Filter down time per backwash [hr]

A similar approach for filter area calculation is given by Letterman [6]. Increasing filtration rate will reduce filter area more than any other variable because it appears

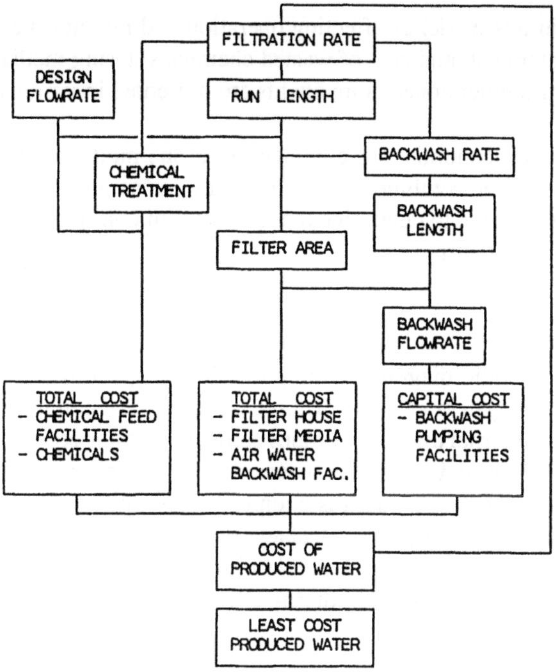

Fig. 2. Variables affecting the economics of deep bed filter

only in the denominator of the above expression. This reflects the desire of filter plant designers to reach higher filtration rates. Increasing run length will also reduce filter area but less than filtration rate. The effect of backwash rate (VB) and filter down time (TB) are less important because their values are relatively small.

Equation (1) is too general and does not demonstrate the effect of filter house, media charecteristics and backwashing regime as separate variables. The following equation is suggested instead

$$\text{TACC} = M_1 A^{X_1} + M_2(AL)^{X_2} + M_3(AV_\text{B})^{X_3} \quad (3)$$

TACC	Total annualized capital cost	L	Media depth [m]
M_1, M_2, M_3	Cost coefficients	V_B	Backwash velocity [m/hr]
A	Filter area [m^2]		

The first term on the right hand of the equation presents cost of filter house as a function of filter area; the second term presents cost of filter media as function of filter media volume; and the latter presents cost of backwash system as function of flow rate.

Coefficients and powers appearing in Eq. 3 are determined by applying linear regression on cost data provided by Gumerman et al. [9]. Original data costs were adjusted to the April 1990 US$ value (ENR = 4.38) for the economic evaluations of this study.

Costs that are considered in this model as affecting operation and maintenance costs are energy cost, wash water cost, and cost of wasted chemicals during media washes. Cost equations of the aforementioned components are presented in various forms in other studies [6, 10].

The two models, process and economic, are combined into one model as described schematically in Figure 2. The combined model links process parameters to cost components, and makes it possible to determine the filter system configuration which produces water at the lowest cost.

Experimental

Pilot Plant Description. A schematic diagram of the pilot plant is given in Figure 3. The experimental filtration system consists of three clear perspex columns, each 200 cm high, and 29 cm in internal diameter. Probes across each filter column at 15 cm intervals permit sampling and head loss measurements via differential pressure transmitters. The bottom of each column is equipped with nozzles which are covered with stainless steel sieves to prevent leakage of filter media grains. Arrangement of nozzles permits backwashing of media either by water and air separately or simultaneously. Constant flow rate during filtration run is attained by a flow rate controller.

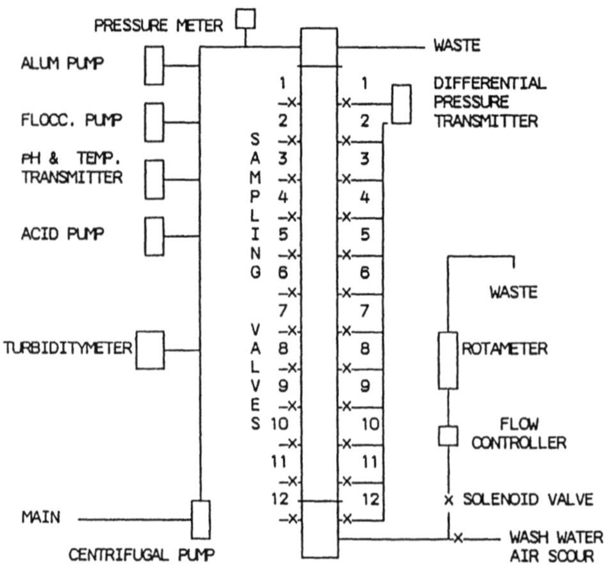

Fig. 3. The filtration pilot plant

A centrifugal pump supplies raw water to the filter columns from a main which carries Lake Kinneret water. Three dosing pumps allow pH modification, primary flocculant and flocculant aid injection. The water, after addition of chemicals, is introduced immediately to the columns without flocculation. Raw water temperature, pH, and turbidity are measured at constant intervals.

Operation of the pilot plant and data acquisition are carried out by a personal computer. A specific program has been developed for that purpose, allowing manual and automatic control and operation. Data files can be imported to a Lotus 123 worksheet for subsequent processing.

Raw Water Quality. The lake water turbidity and pH range usually between 2 to 4 NTU and 7.5 to 9.0, respectively. The alkalinity and total hardness (both as $CaCO_3$), range 78 to 126 and 200 to 270 mg/l, respectively. The total suspended solids range 4 to 6 mg/l (40–60 percent are volatile).

Media. Three different filtering materials, anthracite, basalt, and tuff, were tested in single, dual, and mixed filter beds. Basically, filter media tested include: (1) anthracite with d_{10} of 1.3, 1.5, and 1.7 mm; (2) basalt and tuff with d_{10} of 1.4 and 1.5 mm; (3) sand as bottom layer with 0.5 mm; (4) granular activated carbon as upper layer with d_{10} of 2.65 mm. Total height of each media combination extended to 150 cm. Bottom and lower layers extended to 30 cm. Media types, sizes, and depths tested are summarized in Table 1, their physical characteristics are given in Table 2.

Table 1. Media types, sizes, and depths tested in pilot study

Medium	Eff. size mm	Uniformity coefficient	D_{90} mm	Size range mm	Depth cm
Monomedium					
A1-Anthracite	1.30	1.23	1.76	1.00–2.50	120
A2-Anthracite	1.52	1.12	1.80	1.40–2.50	120
A3-Anthracite	1.70	1.30	2.90	1.40–2.50	150
B1-Basalt	1.55	1.16	2.25	1.40–2.80	120
B2-Basalt	1.42	1.20	2.15	1.40–2.80	150
T1-Tuff	1.40	1.20	2.25	1.40–2.80	150
Dual media					
D1-Anthracite	1.30	1.23	1.76	1.00–2.50	120
Sand	0.47	1.42	0.85	0.30–1.00	30
D2-Anthracite	1.52	1.12	1.80	1.40–2.50	120
Sand	0.53	1.32	0.74	0.40–1.00	30
D3-Basalt	1.55	1.16	2.25	1.40–2.80	120
Sand	0.53	1.32	0.74	0.40–1.00	30
Mixed media					
M1-GAC	2.65	1.13	4.0	2.00–5.00	30
Anthracite	1.30	1.23	1.76	1.00–2.50	120
Sand	0.47	1.42	0.85	0.30–1.00	30
M2-GAC	2.70	1.18	4.20	2.00–5.00	30
Anthracite	1.52	1.12	1.80	1.40–2.50	120
Sand	0.53	1.32	0.74	0.40–1.00	30
M3-Tuff	2.70	1.24	4.20	2.00–5.00	30
Basalt	1.55	1.16	2.25	1.40–2.80	120
Sand	0.53	1.32	0.74	0.40–1.00	30

Table 2. Physical characteristics of tested filtering materials

Material	Sp. gravity	Roundness	Sphericity	Coarseness
Tuff	2.3	sub-rounded	0.81	very coarse
Basalt	2.6	very angular	0.69	coarser
GAC	1.4	very angular	0.65	coarse
Anthracite	1.7	very angular	0.69	less coarse
Sand	2.6	well rounded	0.84	smooth

In each series of experiments three different media configurations were examined simultaneously at various filtration rates, different chemical pretreatments and backwashing regimes.

Results and Discussion

Filtration experiments were conducted according to the process optimization approach. The results are presented and interpreted below.

Media Material and Filter Bed Type. A summary of the results from pilot experiments is given in Table 3. All single medium materials (anthracite, basalt, and tuff) gave the same ultimate filtrate turbidity ($\leq$ 0.3 NTU). In filter beds with the same depth and effective grain sizes, the anthracite gave the highest volume of water produced per run at a rate of 12.5 m/hr, and the basalt gave the lowest. However, raising the rate of filtration diminished the superiority of the anthracite medium over the others. At a rate of 17.5 m/hr the anthracite produced the lowest volume of water (Fig. 4). Single medium filter beds produced higher volumes of water per run than dual and mixed media filter beds. In dual and mixed media filter beds, the main removal of turbidity occurred only at the upper layer/s, therefore the sand layer just introduced more head loss, decreasing the volume of water produced per run.

Filtration Rate. Filtration rates of 10, 12.5, 15, and 17.5 m/hr were investigated. At each rate the filter run was terminated after critical head loss was reached. Increasing filtration rate up to 17.5 m/hr did not change filtrate turbidity significantly, turbidity values were $\leq$ 0.3 NTU (Fig. 5).

Increasing the filtration rate reduced the volume of water produced per run. However, the rate at which it was reduced was not the same for all bed combinations. Filter beds which were fluidized during backwashing (single anthracite, dual and mixed media) had a more significant decrease in water produced than those which were not fluidized (basalt and tuff) (Fig. 4). In fluidized filter beds, backwashing resulted in deposition of the finer grains on the top of the bed, causing higher head loss development at higher filtration rates. In filter beds which were not fluidized during backwashing, filter bed grains were almost uniformly distributed throughout the depth of the filter, thus resulting in a more gradual head loss when the filtration rate was raised.

Table 3. Summary of results from pilot plant experiments

Medium	Filt. rate m/h	Run length h	Produced water m^3/m^2	Area [1] ×1000 m^2	Bed cost $\$/m^2$
Monomedium					
A1-Anthracite	12.5	45.7	571.2	3.430	5.145
	15.0	34.2	513.0	2.874	4.312
	17.5	22.5	393.7	2.497	3.746
A2-Anthracite	12.5	59.8	747.5	3.407	5.110
	15.0	41.6	624.0	2.857	4.285
	17.5	27.6	483.0	2.475	3.713
A3-Anthracite	15.0	59.0	885.0	2.855	4.283
B1-Basalt	12.5	46.0	575.0	3.419	2.057
	15.0	37.2	558.0	2.858	1.720
	17.5	30.0	525.0	2.460	1.481
B2-Basalt	15.0	44.0	660.0	2.845	1.707
T1-Tuff	15.0	45.8	687.0	2.843	1.706
Dual media					
D1-Anthracite	12.5	41.5	518.7	3.440	4.541
Sand	15.0	30.5	457.5	2.886	3.810
	17.5	19.5	341.2	2.516	3.321
D2-Anthracite	12.5	53.6	670.0	3.415	4.508
Sand	15.0	36.7	551.2	2.868	3.785
	17.5	24.5	428.7	2.488	3.284
D3-Basalt	12.5	39.0	487.5	3.435	2.061
Sand	15.0	30.5	457.5	2.876	1.725
	17.5	24.8	434.0	2.476	1.486
Mixed media					
M1-GAC	12.5	45.7	567.5	3.430	6.329
Anthracite	15.0	33.4	501.0	2.877	5.308
Sand	17.5	21.3	372.7	2.504	4.620
M2-GAC	12.5	58.7	733.7	3.408	6.288
Anthracite	15.0	40.4	606.0	2.859	5.276
Sand	12.5	37.5	469.0	2.478	4.572
M3-Tuff	12.5	43.1	538.7	3.425	2.055
Basalt	15.0	33.0	495.0	2.868	1.721
Sand	17.5	27.5	481.2	2.467	1.480

[1] Calculated area for design flow rate of 1.0 m^3/day

Depth of Filter Bed. Filter bed depths were either 120 cm or 150 cm. About 80–90 percent of turbidity removal, and resulting head loss development, occurred at the top 40–60 cm of the filter beds (Fig. 6). Greater penetrations of flocs were obtained in filter beds with the coarser grains and at the higher filtration rates. Filtrate turbidities of $\leq$ 0.3 NTU were achieved with filter depths of 100 and 150 cm, respectively, for the finer and the coarser filter media tested.

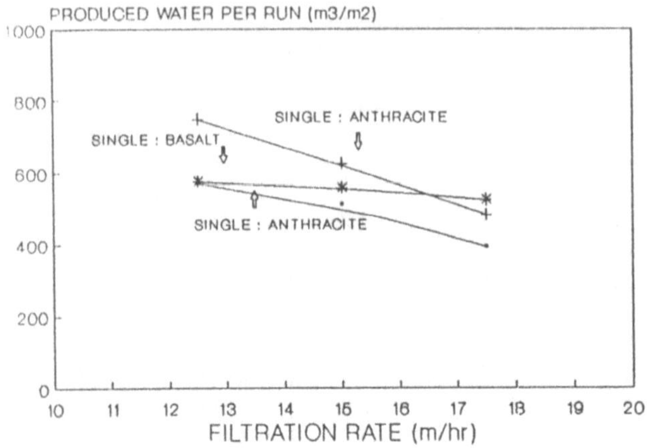

Fig. 4. Water produced per run for various media types as a function of filtration rate

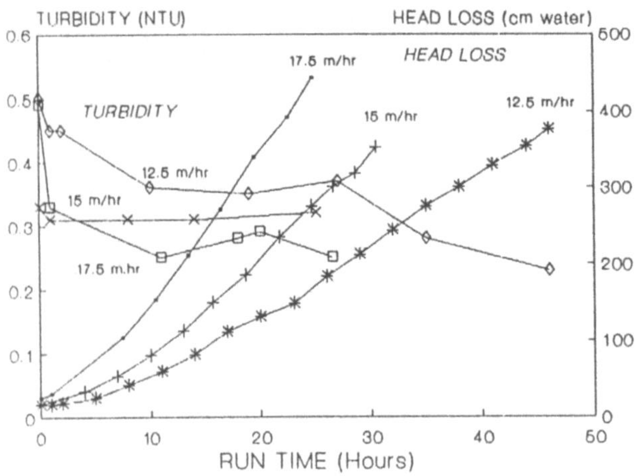

Fig. 5. Effect of filtration rate on turbidity removal and head loss (single anthracite media)

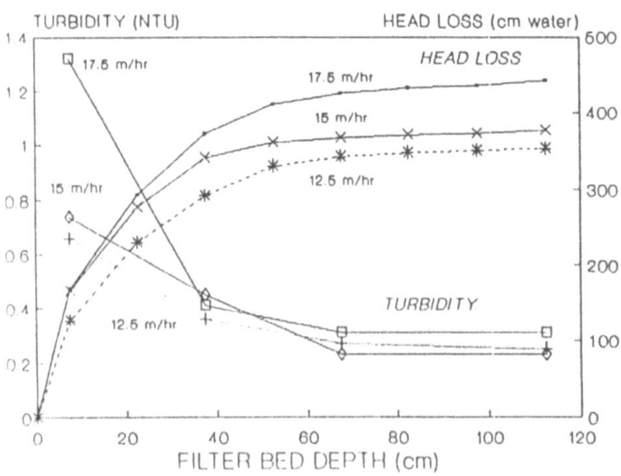

Fig. 6. Turbidity removal and head loss development vs. filter bed depth at various filtration rates

Backwashing. Various backwashing regimes consisting of water and air scour were tested. Filter beds containing anthracite and/or GAC were cleaned using air scour followed by water, while filter beds containing basalt or tuff were cleaned effectively only by using water and air simultaneously. Generally, filter beds could not be cleaned effectively at the water rates used (25–50 m/hr) without applying air scour. Air scour broke flocs to smaller sizes, allowing them to be removed by lower water rates. Optimum rate and duration of backwash was established for each filter bed. Two successive backwash regimes, 3 minutes air scour followed by 10 minutes water each, gave better resuls than a regime of air scour for 6 minutes followed by water for 20 minutes (Fig. 7). Air scour rate was 50 m/hr, followed by water at rates ranging from 25 to 50 m/hr depending on media grain size. Backwash rates used caused 15–17 percent bed expansion. A Water rate of 25 m/hr, when applied simultanously with air, was not enough to fluidize the filter bed.

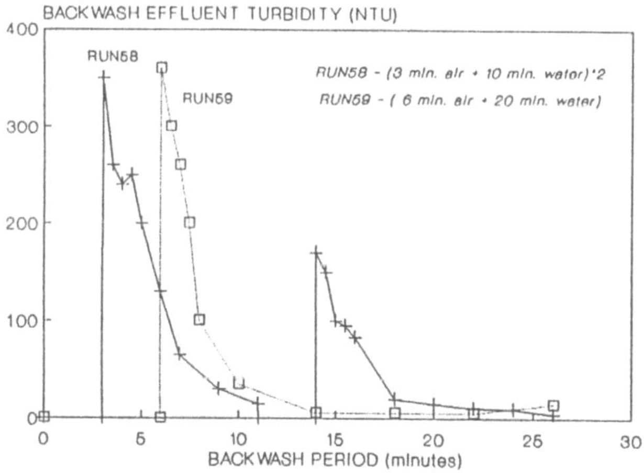

Fig. 7. Comparison of two different backwashing regimes

Chemical Pretreatment. Initial alum dose of 15 mg/l and a pH of 7.0 were selected by conventional jar tests. The first 50 percent of the initial dose (i.e. 7.5 mg/l) was applied to the filters and later was changed to help select the optimum dose for filtration. As Figure 8 shows, higher alum dose causes higher head loss by generating coarser flocs. Reducing alum dose to 5 mg/l caused significant reduction in head loss development rate without much affecting the filtrate turbidity. Filtrate turbidity started to deteriorate from 5 mg/l of alum, thus 6 mg/l was determined as the optimum dose.

Economic Evaluation

Parameters of filter beds capable of producing the target filtrate (turbidity ≤ 0.5 NTU) were provided by the process optimization model from data obtained by pilot plant runs. An area was calculated, based on the experimental results, for an imaginary plant supplying a daily flow rate of 1.0 m^3 of water (column 5,

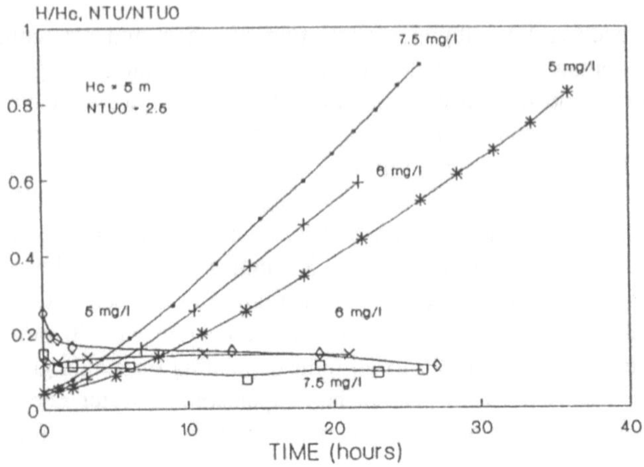

Fig. 8. Effect of alum dose on turbidity removal and head loss development (single anthracite media)

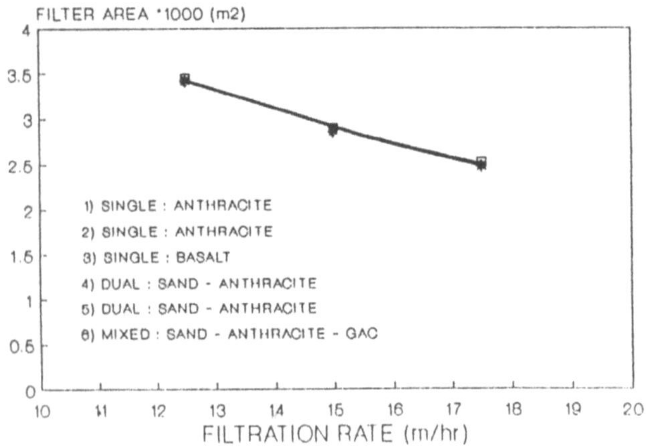

Fig. 9. Effect of filtration rate on filter area for various media types

Table 3). Increasing filtration rate from 12.5 to 17.5 m/hr decreases filter area by 27 percent (Fig. 9).

For a given filtration rate the calculated filter area is the same for all media types and materials. Based on the aforementioned fact that the costs of filter house, air and water scour facilities, entirely depend on filter area, their costs will be the same for all media types for a given rate.

The capital costs of filter media for different media types, for the calculated area, are presented in Figure 10. The costs of filter beds composed of basalt or tuff are significantly lower than those composed of anthracite or granular activated carbon.

Data provided by Eq. 3 show that the capital cost of the three filter system components (a) filter house (b) backwashing system and (c) filter bed, decreases with increasing filtration rate, however, each component with a different rate (Fig. 11). The fraction of each component in the total capital cost depends on filtration rate, thus justifying the assumptions on which the economic model is based.

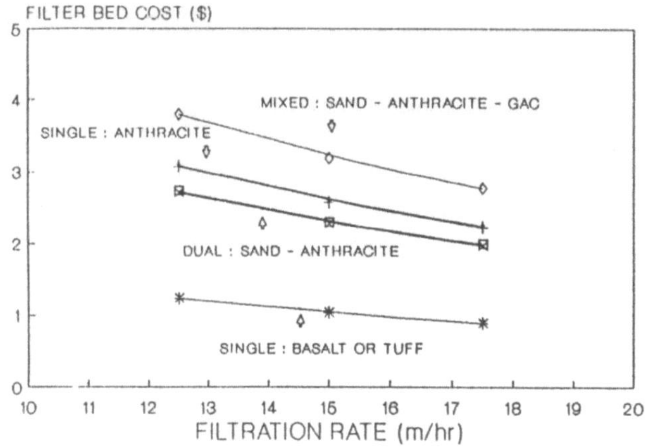

Fig. 10. Effect of media material on filter bed cost

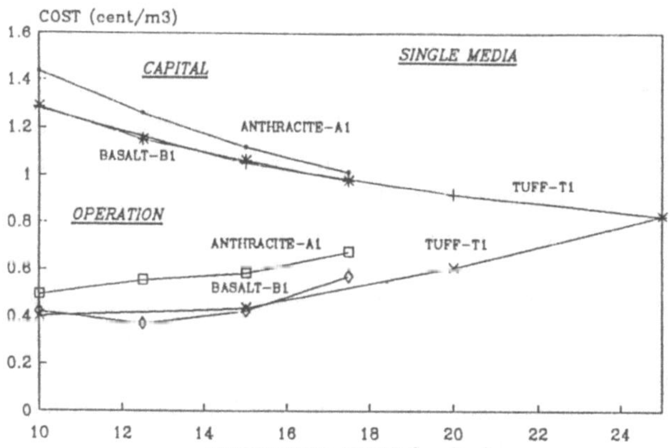

Fig. 11. Dependence of capital cost on filter system components

The effect of the media parameters is as follows:

Media Grain Size and Depth. Coarser media grains are normally accompanied by deeper filter beds in order to provide the desired filtrate quality. Media effective size of 1.3 mm provided the lowest treated water cost. Larger grain sizes required deeper media (bigger filter house and more media) and higher backwash rates (bigger hydraulic systems).

Filtering Material. The lower cost materials tuff and basalt exhibited lower capital costs than the higher cost material anthracite. However, cost differences decrease when filtration rate increases (Fig. 12). Tuff and basalt contribute less to capital cost than anthracite because they are cheaper materials and they can be backwashed simultaneously with air and water using lower backwash rates. Increasing filtration rate demands increasing filter depths for both tuff and basalt more than for anthracite (as a result of differences in filtration efficiencies of each material), thus differences between their costs become less.

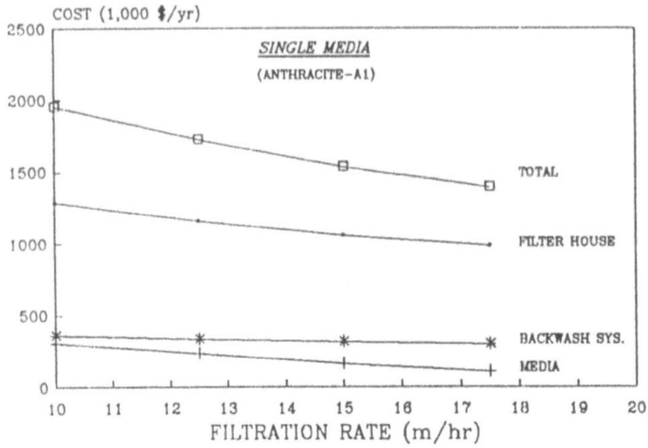

Fig. 12. Dependence of treatment costs on media material

Media Type. Greater filter bed depths and shorter filter runs are required more with single layer beds than with dual-layer or mixed beds to achieve desired turbidities. Backwash rates with single layer beds are lower (smaller hydraulic systems) and wasted water is insignificant (long filter runs), therefore filter systems with single layer beds demonstrated lower treatment costs.

Filter Type and Size. Two known Phenomena are demonstrated by the economic analysis utilizing pilot plant filter runs. One is that increasing filtration plant capacity decreases treatment costs. The second is that pressure filter plants are more economical than gravitational filters in small units and less economical in large units.

Conclusions

1. The filter optimization process should include two successive stages; process and economic optimization. The optimization model developed here was verified successfully by filtration of Lake Kinneret water through different filter media configurations.
2. Filtration rate is the main filter variable for reducing filter costs.
3. Tuff and basalt proved to be efficient and cost effective as single layer media in removing turbidity. The capital cost of a backwash pumping facility is a function of backwash flow rate, therefore the lower capital costs are associated with filter beds composed of basalt and tuff, since they utilire the lowest backwash rates.
4. Determination of optimal media parameters should be carried out at the optimal mode of backwashing and chemical treatment.
5. Process efficiency indicators, such as volume of water produced per filter run, are poor economic indicators and should be used carefully.

Acknowledgement. This paper is based on the doctoral thesis of Samir Hatukai. The work was supported by the Ministry of Science and Technology of Germany in conjunction with the Israel National Council for Research and Development.

References

[1] Adin, A., Baumann, E.R., Cleasby, J.L.: J. AWWA, 71:01:17 (1979)
[2] Mintz, D.M.: Special subject No. 10, IWSA Congr., Barcelona 1966
[3] Camp, T.R.: J. Sanit. Engrg. Div.-ASCE, 90:SA4:3 (1964)
[4] Adin, A., Hatukai, S.: Filtr. and Separ., 28:01:33–36 (1991)
[5] Huang, J.Y.C., Baumann, E.R.: J. Sanit. Engrg. Div.-ASCE, 97:SA2:171 (1971)
[6] Letterman, R.D.: J. Envir. Engrg. Div.-ASCE, 106:EE2:279 (1980)
[7] Ives, K.J.: First Pacific Chem. Engrg. Congr., Part I, Section 2, Separ. Techniques, Soc. of Chem. Engineers, Japan and AICHE, (1972) 99–107
[8] Logsdon, G.S., Clark, R.M., Tate, C.T.: J. AWWA, 72:03:134 (1980)
[9] Gumerman, R.C., Culp, R.L., Hansen, S.P.: Vol. 2, EPA Rep. EPA-600/2/79–162d.MERL, USEPA, Cincinnati, Ohio, 1979
[10] Wiesner, M.R., O'Melia, C.R., Cohon, J.L.: J. Envir. Engrg. Div.-ASCE, 113:03:567 (1987)

Avner Adin
Environmental Sciences Division
The Hebrew University of Jerusalem
Jerusalem 91904
Israel

Samir Hatukai
Amiad Filtration Systems
Kibutz Amiad 12335
Israel

Wastewater Treatment

Chemical Wastewater Treatment – Value for Money

H. Ødegaard and I. Karlsson

Introduction

The British often use the expression "value for money". In this paper we will try to demonstrate that this expression is very valid for chemical wastewater treatment. Different aspects of wastewater treatment will be examined, such as:

- Treatment efficiency
- Area requirement
- Sludge production
- Cost
- Ecological impact.

Chemical treatment is used in different ways in wastewater treatment plants, for instance as the only treatment process (primary precipitation or direct precipitation) or in combination with biological treatment processes in terms of pre-, simultaneous- and post-precipitation (precipitation before, simultaneous to or after the biological treatment).

Chemical treatment alone, as conventional biological treatment, should be categorized as secondary treatment. A combination of chemical and biological treatment is required for tertiary treatment with the goal of protecting sensitive receiving waters. In this paper we shall address both chemical and chemical/biological treatment, but we will focus on comparisons between:

a) Direct precipitation (primary precipitation) and conventional biological treatment on the level of secondary treatment
b) Pre-precipitation and simultaneous precipitation on the level of tertiary treatment with phosphate removal
c) Post-denitrification/pre-precipitation and pre-denitrification/simultaneous precipitation on the level of tertiary treatment with both phosphate and nitrogen removal

R. Klute / H.H. Hahn (Eds.)
Chemical Water and Wastewater Treatment III

Removal Efficiency

For a long time, chemical precipitation has primarily been considered to be a method for phosphorus removal. Indeed, very low values of phosphate in the effluent may be reached (< 0.1 mg P/l). There is no dispute among experts that chemical precipitation is by far the most efficient method for phosphate removal.

To a lesser extent, the focus has been on the fact that the addition of a coagulant (aluminium, iron or lime), results in simultaneous coagulation of colloids. During the past few years, however, the ability of chemical treatment to remove organic matter has been appreciated as a consequence of the need to upgrade wastewater treatment plants for nutrient removal.

Influence of Wastewater Characteristics

When chemical treatment is carried out on a raw or settled wastewater, the purification efficiency will obviously be influenced by the characteristics of the wastewater. This may vary considerably from one place to another. Let us consider organic matter.

If the landscape is hilly, the wastewater flows rapidly in the pipelines and the turbulence causes oxygen to be absorbed in the wastewater. Under these landscape circumstances considerable leaking-in of oxygen-rich surface water is often experienced as well. Because of the availability of oxygen, the microorganisms attached to the pipe wall and suspended in the wastewater will have the optimal possibility to degrade the most easily biodegradable organic matter (the organic acids) and convert some of this to cells. The total COD in the water does not change much on its way through the pipeline, while the fraction of soluble COD is decreased and the fraction of particulate/colloidal COD is increased.

Contrary to this is a situation where the landscape is flat and the movement of the wastewater through the pipeline is gentle. Long pumping pipes are often needed and the wastewater is normally anaerobic. In this situation, the colloidal and particulate organic matter is hydrolysed by anaerobic bacteria in the pipeline, resulting in an increase in soluble COD (organic acids) and a decrease in particulate COD.

It is obvious that the characteristics of the wastewater in these two situations are quite different with respect to what can be achieved in the treatment plant. In the first case, a large fraction of the COD is very easily removed by coagulation while in the second case a large fraction of the COD is easily consumed in biological degradation. The COD in the first case is a poorer carbon source for denitrification than the COD in the second case.

This example demonstrates that proper characterization of the wastewater may be used directly for selection of the optimal treatment process in a given case.

Treatment Efficiency in Direct Chemical Treatment

It has already been demonstrated by many [1] that a major part (65–90 %, depending on the situation as discussed above) of the organic matter in wastewater is colloidal or particulate. Consequently, a reduction in the same removal range can be obtained with chemical treatment of raw wastewater.

Ødegaard [2] published data from 87 chemical treatment plants in Norway. The Norwegian circumstances are best described by a situation with a hilly landscape and with a lot of surface water reaching the sewer, resulting in a high oxygen content in the wastewater when it arrives at the treatment plant. Considering the fact that these plants do not have biological treatment, the removal of organic matter is very good, as the characteristic wastewater situation would also indicate.

Table 1. Average treatment results from 87 Norwegian treatment plants in 1990 [2]

Parameter	N[1)]	n[2)]	Influent mg/l	Effluent mg/l	Removal[3)] %
BOD_7	23	183	167 ± 95	27.2 ± 12.7	83.7
COD	87	1193	463 ± 251	104 ± 38	77.5
TOC	58	820	76.7 ± 45.8	23.8 ± 11.2	69.0
Tot. P	87	1270	5.24 ± 2.53	0.26 ± 0.16	95.0
SS	78	931	233 ± 186	16.6 ± 9.6	92.9
Tot. N	37	246	33.7 ± 21.3	26.6 ± 9.4	21.1

1) Number of plants 2) Number of monthly averages 3) Based on average infl. and effl. concent.

In the future, the focus will be more on three groups of pollutants; heavy metals, organic micropollutants and bacteria/viruses, especially when discharging to the marine environment. As pointed out by several authors [1], these pollutants are strongly associated with colloidal matter. They are removed to a significant extent, therefore, by chemical treatment of raw wastewater (primary precipitation). This treatment method should consequently be considered appropriate for discharge to the marine environment in non-sensitive areas.

Chemical treatment may also be used favourably as pre-treatment before biological treatment for further removal of organic matter or nitrogen. As demonstrated above, the removal of organic matter in the primary precipitation step is on the order of 65–85 % depending on the characteristics of the wastewater. In a traditional pre-denitrification plant for N-removal this is unfortunate since part of the carbon source has been removed. Pre-precipitation should not be used in a predenitrification plant if the raw water COD alone is to act as carbon source unless the wastewater contains a very high fraction of its COD in soluble form.

The organic matter remaining after chemical coagulation may, however, be expected to be that of lowest molecular weight and this organic matter, therefore, has a higher fraction of readily biodegradable organic matter per kg COD_{total} than has the raw water. The COD left after coagulation of the particles is consequently a more valuable carbon source per kg COD than that of the uncoagulated wastewater (especially when rapidly biodegradable COD is asked for).

Primary precipitation should primarily be used in N-removal plants when the raw water does not contain sufficient easily biodegradable carbon to sustain the required denitrification or when high rates (compact plants) are asked for. Typically, therefore, primary precipitation should be used together with postdenitrification or more preferably in a plant based upon a combination of pre- and post-denitrification (Fig. 1). In the designs shown in Figure 1, the soluble organic matter left after primary precipitation is used as a carbon source for pre-denitrification of a certain amount of nitrate that is being recycled from the nitrification step. The major part of the denitrification, however, takes place in the post-denitrification step after carbon source addition.

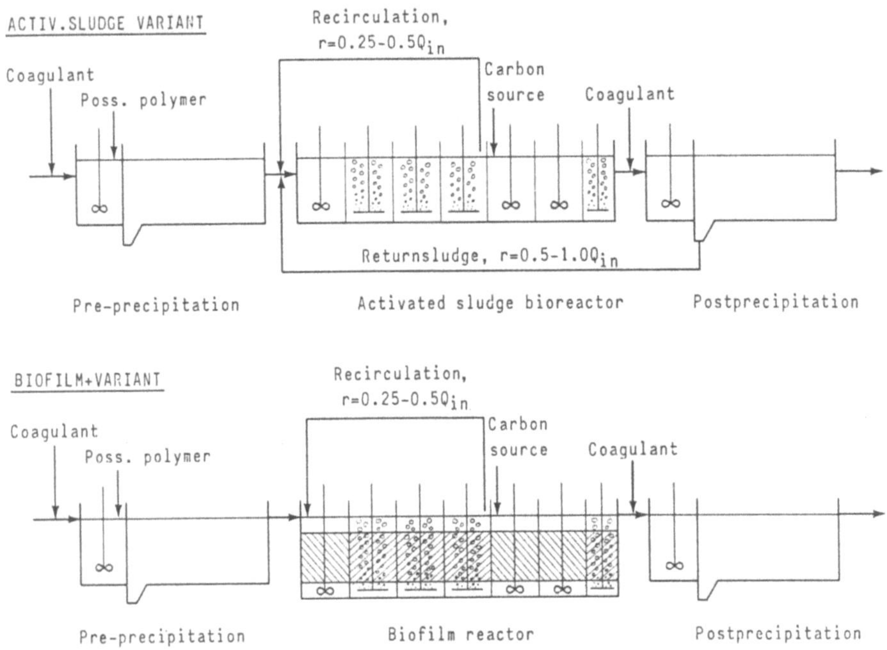

Fig. 1. Useful designs for nutrient removal plants based on primary precipitation as pre-treatment

It has been shown [3] that degradation in biofilms is negatively influenced by particles in the water with which the biofilm is in contact. Since organic particles have to be hydrolysed before they may diffuse into the biofilm, we think that pre-precipitation is especially favourable when the biological process downstream is based upon a biofilm reactor. In such a case the chemical/biological plant will be very compact.

Area Requirement

In wastewater treatment plants the separation reactors, most often settling tanks, represent a major fraction of the area required for the plant. It is well known that a major part of the pollutants leaving a treatment plant with the effluent, is in particulate form. Notably is this true in activated sludge plants, where the secondary settling tank is the most crucial step with respect to the effluent quality, since the process itself is totally dependent upon the return of the sludge. For this reason, the maximum hydraulic surface load is normally quite low ($< 1\ m^3/m^2\,h$). Similarly, the separation of biofilm sludge is also difficult in many instances since flocculation is poor and a relatively high fraction of the suspended solids leaving a biofilm reactor appears as very small particles with low settleability.

While there is a limited possibility to improve settleability of activated sludge by chemical addition, great improvement can be achieved by chemical addition after a biofilm reactor. In the latter case, the fine particles will be flocculated and removed together with the precipitated chemical sludge.

When considering separability of the chemical flocs themselves, settling tanks in Scandinavia were originally designed for a quite low overflow rate, typically in the range of 1–1.5 $m^3/m^2\,h$. This was based more on experiences from coagulation of drinking water than on real experimental verification.

Especially in Sweden, but also to some extent in Norway, post-precipitation plants were built with settling as the last separation step. At the low overflow rates used, excellent separation results were obtained. In many plants, the post-precipitation step acted, in fact, as the guarantee for a good treatment result, since sludge loss from the biological step was caught in the chemical step. A separate post-precipitation step is, however, a costly "safety valve" and with the introduction of nitrogen removal, many plants have switched to pre-precipitation in order to achieve N-removal without volumetric expansion of the plant.

The settling tanks in the Norwegian primary precipitation plants were designed for the same overflow rate as those of the post-precipitation step. This has turned out to be more than adequate. In some cases, however, when the hydraulic loading is increased over the hydraulic design load, improved settleability has been required. In other plants the primary settling step has been transformed into a primary precipitation step in order to be able to accomodate better N-removal. In both these cases anionic polymers have been used as flocculants in order to improve settleability.

Ødegaard et al. [4] demonstrated on a pilot scale the potential of reduced settling area by using anionic polymers as flocculants. The floc blanket settling rate (i.e. overflow rate) was on the order of 2 m/h without polymer and it did not vary much with the dosage of the coagulant. With addition of the anionic polymer, however, the overflow rate could be doubled or tripled without losing effluent quality. The possible overflow rate, however, was found to be strongly dependent upon the polymer dosage, that typically would be in the range of 0.2–0.5 mg/l.

The use of polymers is now being introduced in many Scandinavian plants. Figure 2 shows the effluent tot P-concentration versus overflow rate at the new deep

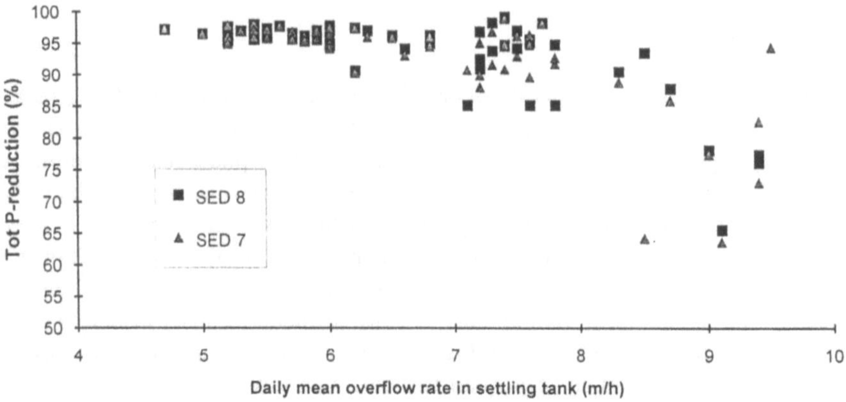

Fig. 2. Effluent tot P versus overflow rate in the new primary precipitation step at VEAS [5]

floc blanket clarifiers at the VEAS treatment plant in Oslo [5]. In the situation from where the data originates, JKL (ferric chloride/sulfate) at a dosage in the range of 100–150 mg/l, depending on raw water quality, and an anionic polymer at a dosage of 0.15–0.20 mg/l were used. It is demonstrated that tot P-removals of more than 90 % may be achieved up to an overflow rate of 7.5 m/h, well in accordance with the laboratory results mentioned above.

It is important to stress that such high overflow rates can only be obtained after an optimization of chemical mixing, flocculation and clarifier design. The results certainly show, however, that the potential for high floc settling rates and, consequently, for the use of high overflow rates, is substantial in chemical wastewater treatment.

It may be argued that this would lead to higher operation costs caused by the polymers that have to be added. When calculated on a yearly cost basis, taking both investment and operation cost into consideration, Ødegaard et al. [4] showed, however, that polymer addition would lead to a considerable overall cost saving.

With the possibility of influencing the floc separation characteristics and the design of mixer/flocculator/clarifier, settling tanks in primary precipitation plants may, without problems, be designed with more than the double (maybe triple) overflow rate that is used for the design of activated sludge settling tanks.

Area Savings by Using Chemical Treatment Ahead of Biological Treatment (Pre-precipitation)

Not only may a primary precipitation step be very compact, but the biological step may also be, in a nitrogen removal plant, when pre-precipitation is used. This shall be demonstrated by reference to two different investigations, the first one at the Klagshamn treatment plant in Malmø, Sweden, based on activated sludge, and the second one based on the biofilm experiments in the Norwegian nutrient removal research programme (FAN). These examples were chosen because they include operations both with and without pre-precipitation.

There are three reasons for the fact that pre-precipitation leads to a much smaller bioreactor:

1. Since a major part of the suspended and organic matter that adds to sludge production in the bioreactor is removed in the primary treatment step, the activated sludge bioreactor volume needed is correspondingly less for maintaining a given sludge age. Similarly the biofilm reactor may be smaller since the major part of the organics is already removed during pre-precipitation.
2. The nitrification rates increase as a result of an increased sludge age in the activated sludge process and as a result of a lower organic loading on the biofilm reactor.
3. The denitrification rate in the post-denitrification reactor is higher as a result of the higher biodegradability of the added carbon source.

These points were clearly demonstrated at the Klagshamn works [6, 7] and at the pilot plant at Nordre Follo treatment works [8] as shown in Tables 2 and 3. Both tables clearly demonstrate the positive effects in terms of area requirement that may be achieved by using pre-precipitation in nitrogen removal plants.

Table 2. Operational data from the Klagsham wastewater treatment plant [6, 7]

Parameter	Pre-DN/presettl.[1)]	Post-DN/pre-prec.[2)]
Equivalent bioreactor volume (m^3) to keep sludge age at 20 d at 3.5 kg SS/m^3	6000	4300
Total sludge production [g SS/m^3_{flow}]	230	240
Sludge production in biological step [g SS/m^3_{flow}]	70	50 [3)]
Nitrification rate [g NH_4-N/(kg VSS · h)]	4.0–4.5	4.5–5.0
Denitrification rate [g NO_x-N/(kg VSS · h)]	1.0–2.0	3.0–4.0 [4)] 8.0–9.0 [5)]
Tot. N in effluent at given volume [mg N/l]	16	5

1) Internal carbon source only
2) Added carbon source (methanol/ethanol)
3) Carbon source dosage of about 90 g COD/m^3
4) Methanol as carbon source
5) Ethanol as carbon source

Table 3. Operational data from Nordre Follo moving bed biofilm pilot plant [8]

Parameter	Pre-DN/pre-siev.[1)]	Post-DN/pre-prec.[2)]
Necessary residence time in biofilm reactor [hrs]	6 [3)]	3 [4)]
Nitrification rate [g NH_4-N/(m^3 · d)]	0.2–0.3	0.3–0.4
Denitrification rate [g NO_x-N/(m^3 · d)]	0.1–0.2	0.5–0.6
Spec. sludge prod. in bioreactor [g SS/(g COD_{rem})]	0.75	0.35
Total sludge production	185 [5)]	210

1) Internal carbon source only
2) External carbon source added (acetic acid)
3) At N-removal efficiency of 60–70 %
4) At N-removal efficiency of 80–90 %
5) Incl. calculated value for post-precipitation

Sludge Production

The primary negative factor associated with chemical treatment is sludge production. In a simplified way, sludge production (SP in terms of g SS/m^3_{flow}) may generally be expressed as:

$$SP = SS_{in} - SS_{out} + K_c \cdot D .$$

This expression demonstrates that sludge production is mainly influenced by two factors: the removal of suspended solids and the dosage of the given coagulant. The coefficient K_c is coagulant dependent. Based on laboratory analysis of sludge production (see Table 4), the coagulant coefficient, when given as g SS/g metal, seems to be lower for a ferric iron coagulant than for an aluminium-based coagulant. When calculated as g SS/mol metal, however, K_c is about the same for both metals. K_c seems to increase with the basicity of the aluminium-based coagulant.

When investigating 87 Norwegian chemical treatment plants, Ødegaard [2] found the mean sludge production data shown in Table 5. It seems that the plants based on $iron_{III}$-coagulation have considerably less sludge production than those based on aluminium coagulation. One should be careful, however, when comparing the data, partly because the data base for aluminium plants is much bigger than that for iron-plants and partly because most of the plants that use iron are the bigger ones where the wastewater is most diluted. In Table 6 the mean K_c-value is calculated for plants with the same type of coagulant.

Table 4. Sludge production coefficients found in laboratory tests with constructed wastewater

Coagulant	K_c g SS/g Me[1)]	Basicity OH/Me ratio
JKL (Fe)	3–5	0
AVR (Al + Fe)	6–7	0
ALG (Al)	8–9	0
PAX 14 (Al)	8–10	1.0
PAX XL-1 (Al)	10–12	2.0

[1)] Me – metal as Fe or Al

Table 5. Measured sludge production on Norwegian chemical wastewater treatment plants, 1990 [2]

Coagulant	Number of plants	Sludge production	
		kg DS/m^3	kg DS/pe · year
Alum (AVR or ALG)	45	0.283 ± 0.20	35.7 ± 14.9
$Iron_{III}$ (JKL)	8	0.167 ± 0.07	21.3 ± 13.5
All plants (Al and Fe)	53	0.267 ± 1.19	21.8 ± 15.6

Table 6. Calculated mean sludge production coefficients from the study of the Norwegian plants

Coagulant name	Formula	Number of plants	Mean K_c g SS/g Me
ALG	$Al_2(SO_4)_3$	13	7.0
AVR	$Al_2(SO_4)_3 + Fe_2(SO_4)_3$	32	5.0
JKL	$FeClSO_4$	1 [1)]	3.6

[1)] The VEAS plant

There is a considerable spread among the data but it is interesting to see that the K_c-values from the field study show acceptable agreement with the ones from the laboratory study. Generally the laboratory study was carried out with a relatively concentrated wastewater and high dosages while the wastewaters in the field study were far more dilute and lower dosages were used.

One should not use the above data for design purposes since the water will change from one plant to the next. Obviously several factors, not included here, such as alkalinity, hardness, etc. will have a bearing on sludge production. The data are only shown here in order to demonstrate that the important factor to control if one wants to reduce sludge production is the dosage of the coagulant.

Methods to Reduce Sludge Production

Quite often, and especially where the water is soft, a much higher dosage is used compared to what is necessary in order to reach the treatment goal. Especially in Scandinavia, the dosage is determined by the standard of phosphate in the effluent. As demonstrated in Figure 3, one needs to get down below a certain pH level in order to achieve the very low phosphate values that are normally required. Since a certain amount of the phosphate appears as particulate phospate, the optimum pH is often quite low, for instance around 6.0 for alum coagulation and around 5.5–6.0 for ferric coagulation.

In order to get down to these pH levels, many operators choose to use the coagulant as an acid. They overdose (compared to the stoichiometric need) in order to reach the desired pH. By doing so, however, the sludge production is increased proportional to the overdosing. This is probably the main reason for the excessive sludge production that we see at many plants.

As demonstrated in Figure 3, one might accept a higher precipitation pH without losing efficiency, if the floc separation was improved. In the example in Figure 3, for instance, the soluble phosphate concentration is below 0.1 mg P/l at pH levels as high as 6.7 for both aluminium coagulation and ferric coagulation. Improved floc separation can be obtained by using polymers as shown earlier. At the VEAS plant in Oslo, for instance, a low total phosphate concentration in the effluent is reached at a coagulation pH as high as 6.8–7.0 since the effluent SS concentration is kept below 10 mg/l by using an anionic polymer flocculant [9].

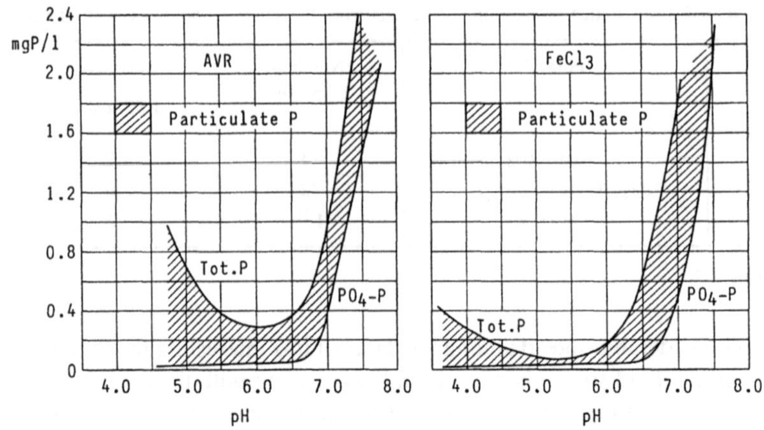

Fig. 3. Demonstration of the influence of pH on the concentration of total P and soluble PO_4-P after coagulation of a typical Norwegian wastewater

In pre-precipitation plants which include a biological step for N-removal, removal of phosphate is of less importance. In fact, one has to be careful not to remove too much phosphate or else phosphate limitation in the biological step may occur. Nordeide et al. [10] demonstrated that a phosphate concentration of at least 0.5 mg P/l should be left after the pre-precipitation step in order not to limit the nitrification in a downstream biofilm process.

The water leaving the primary precipitation step should be as well composed for the subsequent nitrification step as possible. This means that both the pH and alkalinity should be as high as possible (the use of metal coagulant taken into consideration). This has lead to the development of the prepolymerized coagulants. As demonstrated by many authors [11] these coagulants may be used at a higher pH and they consume less alkalinity than do the unprepolymerized ones. That does not necessarily mean that the metal dosage and hence the sludge production is lowered. In fact, the specific sludge production coefficient seems to be higher the higher the basicity, as demonstrated above.

An alternative route to lower sludge production in these cases is to reduce the metal dosages by replacing some of them with a cationic polymer. This effect is demonstrated in Figure 4. The data are taken from an investigation on constructed wastewater carried out at the Norwegian Institute of Technology in order to study how sludge production may be reduced by the use of polymers. It is not easy to demonstrate this effect, since both the effluent particle concentration and the sludge production are dependent upon the coagulant dosage. In Figure 4 some data with PAX XL-60 (prepolymerized Al-chloride with high basicity) as the main coagulant and Fennofix 40 as the substituting cationic polymer is shown. The raw water has a suspended solids concentration of 250–300 mg SS/l.

The upper line shows the measured sludge production ($SS_{separated} + SS_{effl}$) versus SS in the effluent on the y-axis for different dosages of the main coagulant

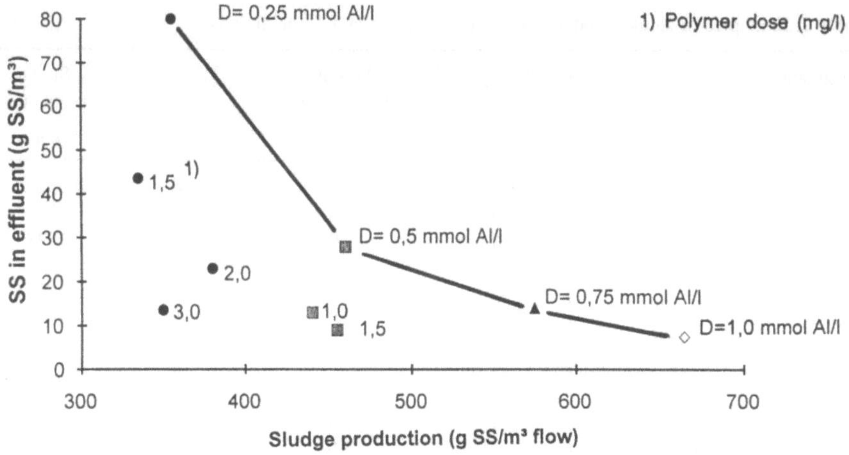

Fig. 4. Example of reduced sludge production as a consequence of the use of a cationic polymer as replacement for the main coagulant. Coagulant: PAX XL-60. Polymer: Fennofix 40

(PAX XL-60). As dosage goes up, sludge production goes up and effluent SS concentration goes down. In order to have less than 15 mg SS/l in the effluent, the dosage had to be 0.75 mmol Al/l giving a sludge production of 575 mg/l (560 mg/l separated as sludge). The group of closed square points represents a main coagulant dosage of 0.5 mmol Al/l, but now with the addition of various amounts of a cationic polymer (Fennofix 40) in the dosage range of 1.0–1.5 mg/l. The dosage of polymer (in mg/l) is given as the numbers at each point. At the given Al dosage the SS concentration in the effluent was about 28 mg SS/l without polymer. At 1 mg polymer/l added, however, the effluent SS concentration was brought down to the same level as was needed with the higher coagulant dosage without polymer (13 mg SS/l) while the sludge production was reduced to about 425 mg/l.

The group of closed circular points represents a main coagulant dosage of only 0.25 mmol Al/l. At this dosage the SS concentration was around 80 mg SS/l. With a dosage of polymer of 3.0 mg/l the SS concentration was again brought down to the low level (12 mg SS/l) while the sludge production was down to 360 mg/l.

Table 7 shows several more examples. They all demonstrate that it is possible to reduce the sludge production substantially by using a substitute coagulant in the form of a cationic polymer.

In some cases the lowering of sludge production in this way is difficult since the effluent concentration may become too high. This may be remedied by using an anionic polymer as flocculant aid as well. Table 8 shows some examples from the laboratory investigation on constructed wastewater.

In the first of the examples in Table 8, a considerable reduction of metal dosage is obtained at a cationic polymer addition of 2.5 mg/l. The effluent SS concentration may, however, be higher (19 mg/l) than required (for instance 15 mg/l). By the addition of an anionic polymer as well (1 mg/l), the dosage of cationic polymer could be reduced and the SS concentration came down to the acceptable limit

Table 7. Examples of sludge production at different coagulant combinations

Coagulant combination	Dosage Al mmol/l	Dosage cation. polym. mg/l	$SS_{effl.}$ mg/l	Sludge production $mg/l_{treat.}$
ALG	0.75	0	36	474
ALG + Cat-Ff 40VHV[1)]	0.5	1.0	30	380
ALG + Cat-Ff 40VHV	0.5	2.0	12	348
PAX 14	0.75	0	11	515
PAX 14 + Cat-Ff 40VHV	0.50	2.0	10	400
PAX 14 + Cat-Ff 40VHV	0.25	3.0	19	300
PAX XL-1	0.5	0	14	464
PAX XL-1 + Cat-P 650Bc[2)]	0.25	2.0	14	362
PAX XL-1 + Cat-Ff 40VHV	0.135	3.0	17	333
PAX XL-60	0.75	0	14	575
PAX XL-60 + Cat-Ff 40[3)]	0.50	1.0	13	445
PAX XL-60 + Cat-Ff 40	0.25	3.0	12	365

[1)] Cationic polymer: Fennofix 40VHV (Kemira) [2)] Cationic polymer: Praestol 655 BC (Stockhausen)
[3)] Cationic polymer: Fennofix 40 (Kemira)

Table 8. Examples from the use of both cationic and anionic polymer

Coagulant combination	Dosage Al mmol/l	Dosage cation. polym. mg/l	Dosage anion. polym. mg/l	$SS_{effl.}$ mg/l	Sludge prod. mg/l
PAX XL-1	0.5	0	0	14	464
PAX XL-1 + Ff 40VHV	0.135	2.5	0	19	336
PAX XL-1 + Cat-Ff 40VHV + An-A302[1)]	0.135	2.0	1.0	15	335
ALG	0.75	0	0	36	474
ALG + Cat-Ff 50[2)]	0.5	2.0	0	15	360
ALG + Cat-Ff 50 + An-A305[3)]	0.5	0.5	1.0	12	360

[1)] Anionic polymer: Fennopol A302 (Kemira) [2)] Cationic polymer: Fennofix 50 (Kemira)
[3)] Anionic polymer: Fennopol A305 (Kemira)

(15 mg/l) with a considerable savings in sludge production. The same is shown in the second example in Table 8 with aluminium sulfate as the main coagulant. In this case a significant improvement of results (both with respect to effluent quality and sludge production) was obtained with the introduction of the polymers.

It should be remembered that the influent SS concentration in these examples was 250–300 mg/l and that sludge production includes the suspended solids in the effluent. This means that the sludge production caused by the chemical treatment itself (the precipitated sludge) is brought down to under 100 mg SS/l by these measures.

Cost Considerations

It is always very difficult to make cost comparisons, especially since the local circumstances have such an important influence on the cost. Nevertheless some evaluations were carried out below. Figure 5 shows the assumptions upon which all the evaluations are based.

Table 9. Assumptions in the cost and energy evaluations for a 100 000 pe plant

Plant size: 100 000 pe   Specific flow: 400 l/pe · d

Specific loads: 70 g BOD_7/(pe · d), 100 g SS/(pe · d), 2.25 g P/(pe · d), 12 g Tot N/(pe · d)

Treatment alternative	Removal [%]				Sludge production
	BOD	SS	P_{tot}	N_{tot}	g SS/m^3
Primary precipitation	75	90	90	10[3]	290
Biological[1]	90	90	30	10[3]	200
Biol./chem. (simult. precip.)[1]	90	90	85	10[3]	240
Biol./chem. (pre-precip.)	90	90	90	15[3]	310
Biol./chem. with N-removal[1] (pre-DN/simult. precip.)	95	90	90	70[3]	240
Biol./chem. with N-removal[2] (post-DN/pre-precip.)	95	90	90	85[3]	330

[1] Activated sludge with presettling   [2] Based on biofilm process and external carbon source
[3] No nitrogen removal in sludge water from digestion is assumed

Table 10 shows the specific costs (NOK/m^3_{flow}) calculated for the processes evaluated. The data are based on cost estimates carried out by Henze and Ødegaard [12]. It is clearly demonstrated that the total cost for primary precipitation is lower than that for biological treatment, that pre-precipitation is cheaper than simultaneous precipitation and that post-denitrification with an external carbon source may be cheaper than pre-denitrification based on an internal carbon source. The differences between the process alternatives in the different categories (secondary, tertiary, etc.) are not very substantial. In these calculations, however, no attention has been given to the biogas that may be produced from the sludges.

The cost should rather be evaluated in relation to the benefits that are gained by the treatment. In order to be able to evaluate cost relative to benefit, we must, however, be able to describe the benefits of the different processes. One way of doing this is to express the reduction in oxygen consumption potential (OCP). Oxygen consumption may be caused by both primary oxygen consumption (degradation of BOD and NH_4-N) and secondary oxygen consumption (degradation of algae caused by the nutrients, phosphorus and nitrogen). The calculations that follow are based on the assumptions that:

1 kg BOD gives maximum 1 kg in primary oxygen consumption (OC)
1 kg Tot N gives maximum 4 kg in primary oxygen consumption (OC)
1 kg Tot P gives maximum 100 kg in secondary oxygen consumption (OC)
1 kg Tot N gives maximum 14 kg in secondary oxygen consumption (OC).

Table 10. Cost comparisons between different treatment methods for a 100 000 pe plant (sludge treatment = anaerobic digestion and dewatering for all alternatives), 20 years life time of equipment, 7 % interest

Treatment alternative	Capital cost NOK/m^3_{flow}	O & M cost NOK/m^3_{flow}	Total cost NOK/m^3_{flow}
Primary precipitation			
Wastewater treatment	0.64	0.58	1.22
Sludge treatment	0.22	0.21	0.43
Total	0.86	0.79	1.65
Biological treatment[1] (incl. primary settling)			
Wastewater treatment	0.89	0.58	1.47
Sludge treatment	0.17	0.16	0.33
Total	1.06	0.74	1.80
Biol./chem. treatment[1] (simultaneous precip.)			
Wastewater treatment	0.92	0.76	1.68
Sludge treatment	0.19	0.18	0.37
Total	1.11	0.94	2.05
Biol./chem. treatment[1] (pre-precipitation)			
Wastewater treatment	0.82	0.70	1.52
Sludge treatment	0.25	0.23	0.48
Total	1.06	0.94	2.00
Biol./chem. treatm.[1] (pre-DN/simult. prec.)[3]			
Wastewater treatment	1.40	1.11	2.51
Sludge treatment	0.20	0.19	0.39
Total	1.60	1.30	2.90
Biol./chem. treatm.[2] w/N-rem. (post-DN/pre-prec.)[4]			
Wastewater treatment	1.11	1.21	2.32
Sludge treatment	0.25	0.23	0.48
Total	1.36	1.44	2.80

[1] Based on activated sludge
[2] Based on biofilm process
[3] Without external carbon source
[4] With external carbon source

These are all values close to what one may expect based on theoretical considerations [13]. In a given case, the depth of the discharge in the receiving water will determine how much of the nutrients will lead to algal growth and thereby oxygen consumption. We will analyse two situations:

a) A situation where all the nutrients lead to algal growth and corresponding oxygen consumption.
b) A situation where 50 % of the nutrients lead to algal growth.

The algal impact is also dependent upon the type of receiving water that we are considering. We will evaluate two different alternatives:

a) A fresh-water situation (a lake or a brackish water inner fjord) where only phosphorus is considered to be limiting for algal growth.
b) A marine situation where phosphorus limits algal growth 20 % of the time and nitrogen limits algal growth 80 % of the time.

Based on these assumptions, the oxygen consumption reduction equivalents for each process is calculated for freshwater and marine and for the two situations regarding the yield of nutrients. Based on this, the specific cost per OC-equivalent removed (NOK/kg OC) is calculated, see Table 11.

Table 11. Cost/benefit-analysis for different treatment alternatives

Treatment alternative	Spec. cost NOK/m^3_{flow}	Spec. cost per OC-equiv. [NOK/(kg OC)] Freshwater[1]	Marine[1]
Primary precipitation	1.65	2.58 (4.25)	6.11 (8.25)
Biological	1.80	5.14 (6.84)	7.35 (8.78)
Biol./chem. (simult.-precip.)	2.05	3.20 (5.02)	6.83 (8.72)
Biol./chem. (pre-precip.)	2.00	2.94 (4.65)	6.45 (8.51)
Biol./chem. w/N-removal (pre-DN/sim.-precip.)	2.90	3.89 (5.86)	5.32 (8.17)
Biol./chem. w/N-removal (post-DN/pre-precip.)	2.80	3.66 (5.44)	4.56 (7.18)

[1] In parentheses: a situation where only 50 % of the nutrients discharged lead to algal growth

It can be seen from Table 11 that in the case of freshwater discharge, primary precipitation is the method that gives most value for money followed by pre-precipitation and simultaneous precipitation. In the marine discharge alternative the chemical/biological process based on pre-precipitation and the biofilm process with external carbon source gives most value for money.

Ecological Considerations

Cost analyses do not always give an adequate picture of the consumption of resources. Many people say that chemical treatment may be cheap but that it is less "ecological" than biological treatment. We think that such a statement is made without due consideration of the aspects involved and we will discuss this in the following.

It is important when ecological aspects are considered, for instance in terms of energy consumption, that all elements that contribute to the total consumption must be included. For instance, a compact treatment plant, like a chemical one, may require less energy for concrete, while it requires more energy for chemicals. In the following we shall evaluate the problem in two different ways:

a) An evaluation of energy consumption based on the major energy-associated elements: chemicals, air and biogas.

b) An analysis of all ecological aspects based on a model for evaluating the ecological value of the products.

In the first comparative analysis we will evaluate the different processes from our cost comparison example with respect to the energy required for chemicals and air and the excess energy resulting from the biogas that can be produced from the sludge. Other elements are also important. For instance will the energy needed for heating and ventilation play a major role in covered plants. If this energy consumption had been taken into account, this element would be in favour of compact plants. Since we want to be as general as possible here, we exclude, however, all energy elements other than the chemicals, the air and the biogas.

Coagulants. Table 12 shows the energy required to produce different coagulants ready for use at the plant.

Table 12. Total energy (kWh/ton product) for different coagulants

Energy requiring element	Coagulants		
	JKL	AVR	PAX
Raw material (incl. transport)	50–100	200	150
Production	35–40	50	100
Transport final product	60	70	60
Total	< 200	320	310

The table demonstrates that the energy consumption for using JKL (iron-chloride/sulphate) at a dosage of 150 g/m^3 is about 0.03 kWh/m^3. Table 13 shows the energy requirement for chemicals depending on the dosage of iron-chloride. In the case of simultaneous precipitation, a dosage of 100 mg/l is assumed.

Air. In the biological processes, more or less air is required depending on the loading of the plant. In the following we will use 1.0 kWh/kg $BOD_{rem.}$ for the normally loaded plants and, in addition, 4 kWh/kg NH_4-$N_{rem.}$ in the plants where nitrification is required. This leads to the energy consumption per m^3 of flow in our water as shown in Table 13.

Biogas. Energy may be produced from the sludge. The raw water organic matter found in the sludge varies depending on removal of organic matter and transformation of organic matter to CO_2. When calculating with a fraction of organic matter (VSS/SS) in the sludge of 60–75 % and a degradation of organic matter of 40–50 % depending on the type of sludge, and with a gas production of 0.9 l/g $VSS_{degraded}$ and an energy value of 6 kWh/m^3_{gas}, the energy production calculated from gas is shown in Table 13. The energy needed for heating the digester is calculated to be 40 % of the energy produced, appearing as negative energy in the biogas column of Table 13.

Table 13. Specific energy usage (Wh/m^3_{flow}) for the different treatment methods used in the example

Treatment alternative	Energy consumpt./product. [Wh/m^3_{flow}]			
	Chemicals[1]	Air	Biogas	Net
Primary precipitation	−30	0	+450 – 180	+240
Biological	0	−105	+350 – 140	+105
Biol./chem. (simult. precip.)	−20	−105	+350 – 140	+85
Biol./chem. (pre-precip.)	−30	−35	+475 – 190	+220
Biol./chem. w/N-removal (pre-DN/simult. precip.)	−20	−185	+325 – 130	−10
Biol./chem. w/N-removal (post-DN/pre-precip.)	−30 – 60[2]	−110	+475 – 190	+85

[1] Coagulants and carbon source (methanol) in the case of post-DN [2] Methanol

From Table 13 it is evident that the methods based on primary precipitation as the only step or as pre-treatment step come out very favourably in a comparison like this. The potential for energy production from sludge is significant, but even if this is not considered, the methods based on chemical pre-treatment consume less energy than those based on pre-settling and biological treatment thereafter. Table 14 shows the energy consumption per oxygen consumption equivalent removed.

Table 14. Energy/benefit-analysis for different treatment alternatives

Treatment alternative	Specific enery requir. Wh/m^3_{flow}		Specific energy/OC-equivalent removed $kWh/kg\ OC_{equiv.}$			
			Freshwater[1]		Marine[1]	
	NBG[2]	IBG[2]	NBG[2]	(IBG)[2]	NBG[2]	(IBG)[2]
Primary precipitation	−30	+240	−50	(+375)	−110	(+890)
Biological (AS)	−105	+105	−300	(+300)	−430	(+430)
Biol./chem. (simult.-prec.)	−125	+85	−195	(+130)	−415	(+285)
Biol./chem. (pre-precip.)	−65	+220	−95	(+325)	−210	(+710)
Biol./chem. with N-rem. (pre-DN/simult.-precip.)	−205	−10	−275	(−15)	−375	(−20)
Biol./chem. with N-rem. (post-DN/pre-precip.)	−200	+85	−260	(+110)	−325	(+140)

[1] All nutrient discharge lead to algal growth [2] NBG – No biogas included; IBG – Biogas included

Table 14 demonstrates that in terms of energy consumption, the treatment alternatives that include primary precipitation all consume less energy per oxygen-equivalent removed than the corresponding process of the same category that uses pre-settling and activated sludge. It is also demonstrated that the treatment method that requires the least energy per oxygen consumption-equivalent is, in fact, primary precipitation. This method should therefore be considered to be more "ecological".

There are models which can be used to estimate the environmental friendliness of products or processes in terms of life-cycle analysis. In these analyses all aspects of effects tied to the products from raw material to waste products are considered. In the following we shall use such a programme (SimaPro 2.0) [14] that takes into account raw materials, transport, production and waste. The process/product is evaluated relative to a response model that includes eutrophication, acidification, green-house effect, smog formation, toxicity, energy consumption, waste production, etc. The outcome of the calculations are Ecopoints, the lower the better from an ecological viewpoint.

We have chosen to compare primary precipitation, conventional biological treatment and pre-precipitation (see Table 15) without taking the effect of biogas into consideration, since the higher energy saving caused by biogas production in the case of primary precipitation will have an overshadowing effect.

Table 15. Ecopoints after the SimaPro 2.0-model [14]

Element considered	Treatment alternative		
	Chemical (primary precip.)	Biological (conv. activ. sludge)	Biol./chem. (pre-precip.)
Construction	20 000	55 000	45 000
Chemicals	110 000	–	110 000
Air	–	630 000	225 000
Total	130 000	685 000	380 000

Even though one should be careful in looking for quick conclusions, it is quite interesting that the Eco-model gives the same results as the calculations in Table 14.

Summary

This paper has documented that chemical wastewater treatment gives "value for money" in the sense that such treatment gives a low cost per removed unit of total oxygen consumption in the receiving water and a low energy consumption per unit of total oxygen consumption removed. The total oxygen consumption evaluation takes into consideration both the primary oxygen consumption caused by organic matter and ammonium as well as the secondary oxygen consumption from degradation of algae resulting from the discharge of nutrients.

It has been demonstrated that in comparison to conventional activated sludge treatment (with pre-settling), chemical treatment plants (primary precipitation) are cheaper in terms of cost per unit of total oxygen consumption removed and that they require less energy per unit of total oxygen consumption removed. Chemical treatment is considered to be the most suitable method when discharging municipal sewage to non-sensitive marine receiving waters and should be categorized as secondary treatment.

When comparing tertiary treatment methods that include phosphate but not nitrogen removal, it has been demonstrated that pre-precipitation/activated sludge plants are cheaper in terms of cost per unit of total oxygen comsumption removed than simultaneous precipitation/activated sludge plants and that they also require less energy per unit of total oxygen consumption removed. When comparing tertiary treatment methods that include both phosphate and nitrogen removal, it has been demonstrated that pre-precipitation/post-denitrification plants with external carbon source are cheaper in terms of cost per unit of total oxygen consumption removed than plants based on pre-denitrification and simultaneous precipitation. They also require less energy per unit of total oxygen consumption removed.

References

[1] Ødegaard, H.: Particle Separation in Wastewater Treatment. Documentation 7'th EWPCA-Symposium, Munich, May 1987. ATV, St. Augustin, Germany

[2] Ødegaard, H.: Norwegian Experiences with Chemical Treatment of Raw Wastewater. Wat. Sci. Tech. *25* (12) (1992) 255–264

[3] Boller, M., Gujer, W., Nyhuis, G.: Tertiary Rotating Biological Contactors for Nitrification. Wat. Sci.Tech. 22 (1/2) (1990) 89

[4] Ødegaard, H., Grutle, S., Ratnaweera, H.: An Analysis of Floc Separation Characteristics in Chemical Wastewater Treatment. In: Chemical Water and Wastewater Treatment II, R. Klute and H.H. Hahn (Eds.). Springer, Berlin Heidelberg New York 1992, pp. 97–114

[5] Ryrfors, P.: Personal communication, 1994

[6] Andersson, B., Aspegren, H., Nyberg, U., la Cour Jansen, J., Ødegaard, H.: Evaluation of Pre-precipitation in a Wastewater Treatment System for Extended Nutrient Removal. In: Chemical Water and Wastewater Treatment II, R. Klute and H.H. Hahn (Eds.). Springer, Berlin Heidelberg New York 1992, pp. 341–355

[7] Nyberg, U., Andersson, B., Aspegren, H., Ødegaard, H.: The Use of Polymer in the Pre-precipitation Step of a Wastewater Treatment System for Extended Nutrient Removal. Accepted for publication at the 6th Gothenburg Symposium, Gothenburg, June 1994

[8] Ødegaard, H., Rusten, B.: Norwegian Experiences with Nitrogen Removal in a Moving Bed Biofilm Reactor. Dokum. 9th EWPCA-Symp., Munich, May 1993. ATV, St. Augustin, Germany

[9] Sagberg, P., Sæther, R., Baggerud Berge, R.: Increasing the Surface Load at a Direct-Precipitation Plant, VEAS, Norway. In: Chemical Water and Wastewater Treatment, H.H. Hahn and R. Klute (Eds.). Springer, Berlin Heidelberg New York 1990, pp. 271–282

[10] Nordeidet, B., Rusten, B., Ødegaard, H.: Phosphorous Requirement for Tertiary Nitrification in a biofilm. 2nd International Specialized Conference on Biofilm Reactors. Paris Sept./Oct. 1993

[11] Ødegaard, H., Fettig, J., Ratnaweera, H.: Coagulation with Prepolymerized Metal Salts. In: Chemical Water and Wastewater Treatment, H.H. Hahn and R. Klute (Eds.). Springer, Berlin Heidelberg New York 1990, pp. 189–220

[12] Henze, M., Ødegaard, H.: An Analysis of Wastewater Treatment Strategies for Eastern and Central Europe. Accepted for publication in Wat. Sci. Tech, 1994

[13] Baalsrud, K., Bjerkeng, B.: Tiltaksanalyse for Indre Oslofjord. Norwegian Institute for Water Research, December 1991 (in Norwegian)

[14] Anon: Introduct. to life cycle analysis. SimaPro Manual. PRé Consultants, Amersfoort, Holland 1993

Hallvard Ødegaard
Division of Hydraulic and Environmental Engineering
The Norwegian Institute of Technology
The University of Trondheim
N-7034 Trondheim – NTH
Norway

Ingemar Karlsson
Kemira Kemwater
Kungstensgatan 38
S-113 59 Stockholm
Sweden

The Use of Polymer in the Pre-precipitation Step of a Wastewater Treatment System for Extended Nutrient Removal

U. Nyberg, B. Andersson, H. Aspegren, and H. Ødegaard

Abstract

The Klagshamn wastewater treatment plant is operated with extended nutrient removal. The process configuration is based on pre-precipitation and post-denitrification in a single sludge activated sludge system. As the nitrogen removal at the plant is strongly affected by the sludge age, the capacity of the system should be increased if the load of particulate material on the biological treatment is decreased.

In order to improve the suspended solids removal in the pre-precipitation process, a full-scale test with polymer addition was performed at the plant. The test clearly showed that the concentrations of suspended solids in the primary effluent decreased when a polymer was added to the pre-precipitation step. The decreased load of particulate material on the biological treatment resulted in an increased sludge age and an improved nitrification capacity. Model simulations showed that this improvement allowed the flow to the biological treatment to be increased by about 15 %.

Introduction

The Klagshamn wastewater treatment plant in Malmö has recently been upgraded to include nitrogen and phosphorus removal. The upgraded plant is based on pre-precipitation and post-denitrification in a single sludge activated sludge system with addition of an external carbon source (Fig. 1). The flow is divided equally between two identical parallel flow trains. The annual average wastewater flow to the plant is about 15 000 m^3/d, and the corresponding hydraulic detention time in the bioreactor basins is about 7 hours.

Most of the particulate organic material, as well as phosphorus in the influent, is removed in the pre-precipitation step. By adding of methanol or ethanol to the bioreactor, the particulate organic material in the raw wastewater is replaced by a more readily biodegradable carbon source. This results in a lower sludge production and higher utilization rates in comparison to a process where the particulate organic material in the raw wastewater is to be utilized for denitrification. Despite the comparatively short detention time in the bioreactor, concentrations around 6–8 mg N/l are achieved in the secondary effluent, as an annual average.

R. Klute / H.H. Hahn (Eds.)
Chemical Water and Wastewater Treatment III

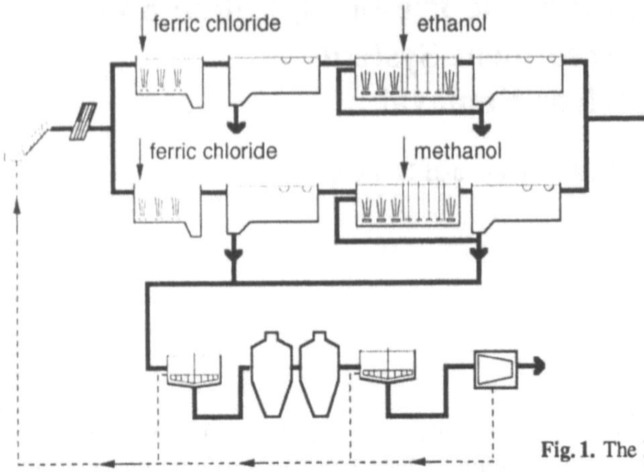

Fig. 1. The Klagshamn plant

Correspondingly, phosphorus concentrations around 0.5 mg P/l and BOD_7 concentrations of well below 10 mg/l are reached. The process has been more thoroughly described by Andersson et al. (1991), Andersson et al. (1992), Andersson et al. (1993), Nyberg et al. (1991) and Nyberg et al. (1992).

As most of the slowly degradable and inert particulate material in the raw influent wastewater is removed by pre-precipitation, a high sludge age can be maintained in the bioreactor. This favors the growth of nitrifiers and improves the nitrification process. Furthermore, as the nitrification process as well as the overall nitrogen removal are strongly affected by the biological sludge production, the capacity of the system should increase if the concentration of suspended solids in the primary effluent is decreased. This may be accomplished by the use of a polymer as a flocculant, added after the main coagulant (iron chloride).

The effects on both the primary and secondary treatment, as a result of polymer addition to the pre-precipitation step, was evaluated at the plant during a 6 month full-scale test. In addition, the effect on the secondary treatment was studied with the aid of model simulations. The observed effects will be described and discussed below.

Experimental Set-Up

As the plant is divided into two identical flow trains, parallel tests could easily be performed. The polymer addition tests were divided into two phases. In the first phase, the polymer was added to one of the flow trains, whereas no polymer was added to the parallel one. During the second phase the dosage was shifted to the other flow train.

The polymer used in the full-scale tests was an anionic one (Fennopol A392), chosen from batch tests conducted in cooperation with Kemira Kemwater. Related to the average flow, the polymer dosage was set at about 0.3 ppm and the dosage of ferric chloride corresponded to an iron dosage of around 10 g Fe/m^3. The dosages of ferric chloride and polymer were kept constant throughout the test.

Samples from the two flow trains were collected within the normal daily sampling program at the plant. Samples were taken five days a week, Sunday to Thursday, and were analyzed for suspended solids, organic material (BOD_7) and phosphorus and nitrogen fractions. Samples for the determination of mixed liquor suspended solids were also collected five days a week from the bioreactor basin and the return sludge. Ammonia utilization rates with sludge from the full-scale system were measured twice a week during the last test period by means of a laboratory procedure described by Nyberg et al. (1991).

Results

The quality of the primary effluent using polymer addition differed significantly from one flow train to the other, which was probably due to differences in mixing conditions with respect to the polymer addition. This indicates how important mixing is for the final result in a coagulation and flocculation process. The data presented in, and used as a basis for, the ensuing discussion were obtained from the second flow train during the second phase of the test. This flow train showed the best improvement in the removal of suspended solids. These results can be regarded as obtainable in the other flow train as well following an optimization of the coagulation and flocculation process including polymer addition.

The impact of polymer addition on the secondary treatment is not as direct as on the primary treatment. The sludge retention time is long, which means that the sludge is formed in the course of weeks and during that time it is constantly changing due to the composition of the influent wastewater. Consequently, the full impact on the secondary treatment will not be seen until at least one sludge age later than the starting time for the polymer addition. This was illustrated in advance by means of model simulations using the simulation package EFOR. The result is shown in Figure 2, where a 50% increase in the removal of suspended solids in the pre-precipitation process was assumed when polymer was added. The

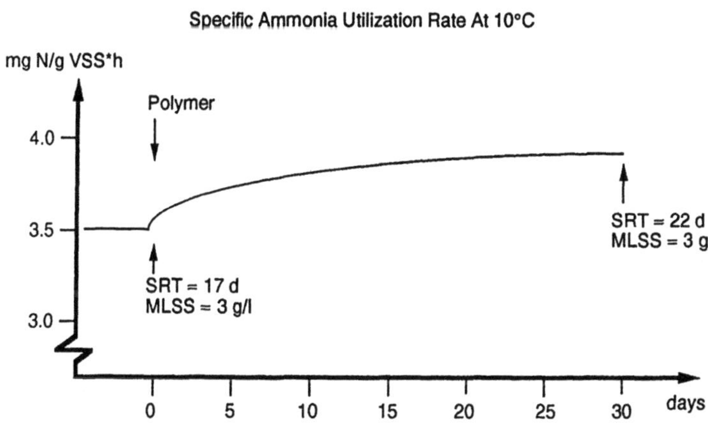

Fig. 2. Specific ammonia utilization rate versus time after start-up of polymer addition. Results from mathematical simulations with EFOR

improvement of the specific ammonia utilization rate, as a result of the increased sludge age, should not be expected until more than 30 days after the start-up of the polymer addition.

Primary Treatment

The effect of the polymer addition on the removal of particulate material in the pre-precipitation step can be seen in Figure 3. The average concentration of suspended solids in the primary effluent decreased from 59 to 28 mg/l immediately after the start of the polymer addition. The total phosphorus concentrations in the primary effluent decreased from 2.7 to 1.8 mg P/l, and the concentrations of BOD_7 decreased from 55 to 38 mg/l on the average. The average values in the primary

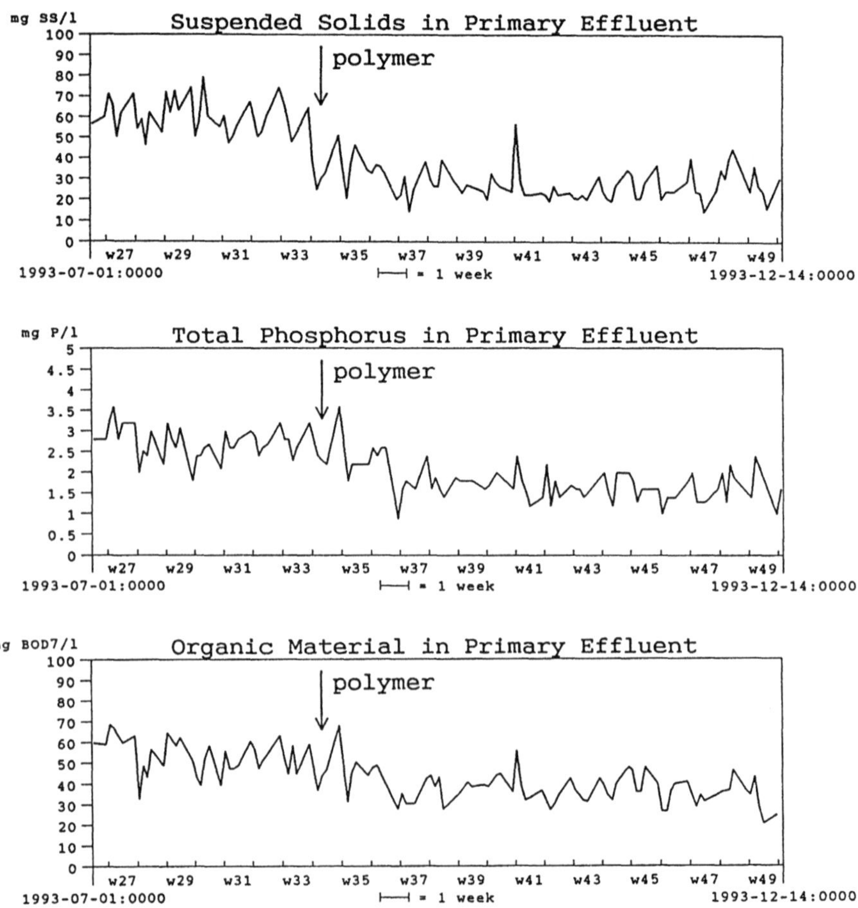

Fig. 3. Concentrations of suspended solids, total phosphorus and organic material BOD_7 in the primary effluent

effluent during the periods without and with polymer addition are summarized in Table 1.

Table 1. Average values in the primary effluent without and with polymer addition

Average	Without polymer addition	With polymer addition
Flow [m^3/d]	7570	8900
Surface load primary clarif. [m/h]	0.63	0.74
Suspended solids [mg/l]	59	28
Total phosphorus [mg/l]	2.7	1.8
Organic material BOD_7 [mg/l]	55	38
Total nitrogen [mg/l]	32	29

Although the hydraulic load was higher during the period with polymer addition, the load of suspended solids on the biological treatment step decreased by about 40 % as a result of the polymer addition. The removal of particulate organic material and phosphorus increased simultaneously by about 20 %.

Secondary Treatment

As a result of the lower load of particulate material on the biological treatment step, the biological sludge production also decreased. Consequently, the sludge age could be increased even if the concentrations of mixed liquor suspended solids in the bioreactor were maintained at a constant level. The MLSS concentration and the sludge age in the biological reactor before and after start of polymer dosage are shown in Figure 4. The sludge age is presented as a moving average over 20 days. The sludge age increased from about 17 days to 21 days as a result of the lower concentrations of suspended solids in the primary effluent. The MLSS concentration was 3.1 g SS/l on the average. The sudden drops in sludge age at the end of the polymer addition period are due to loss of sludge from the secondary clarifiers during heavy stormwater flow conditions.

It has previously been shown by Andersson et al. (1992), as well as by the EFOR simulations, that a decrease in the load of particulate material on the biological treatment step entails increased specific ammonia utilization rates. These rates reflect the fraction of nitrifiers in the sludge. During the test with polymer addition, the ammonia utilization rates were measured in both flow trains. The measured values are presented in Figure 5 along with rates measured at the beginning of 1993, the latter ones serving as reference values. The ammonia utilization rates in the figure are expressed in terms of NO_3-N equivalents. This means that nitrite formed during the tests is recalculated into nitrate on the basis of electron equivalents. The rate increased in the flow train to which polymer was added. After about 2 months, the ammonia utilization rate in the flow train with polymer addition had reached a value which was about 15 percent higher than in the one without polymer addition.

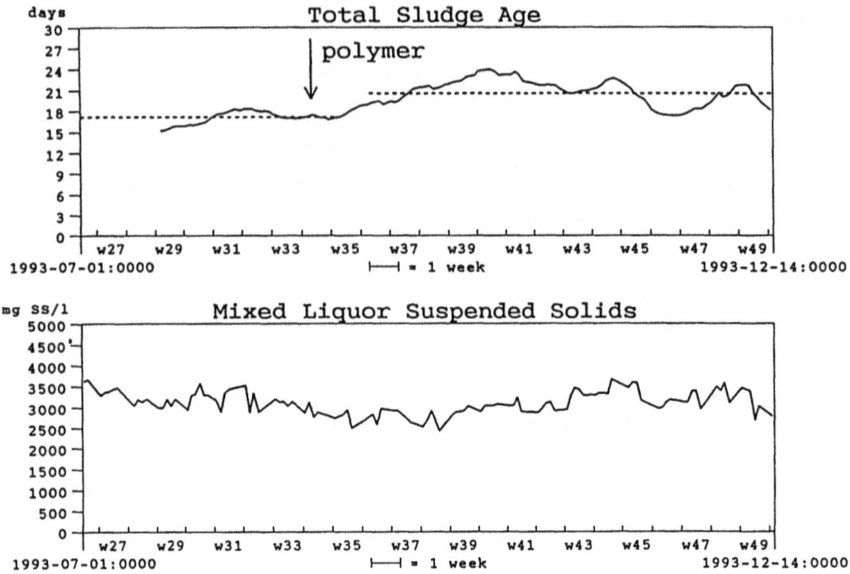

Fig. 4. Total sludge age and MLSS concentration in the biological reactor

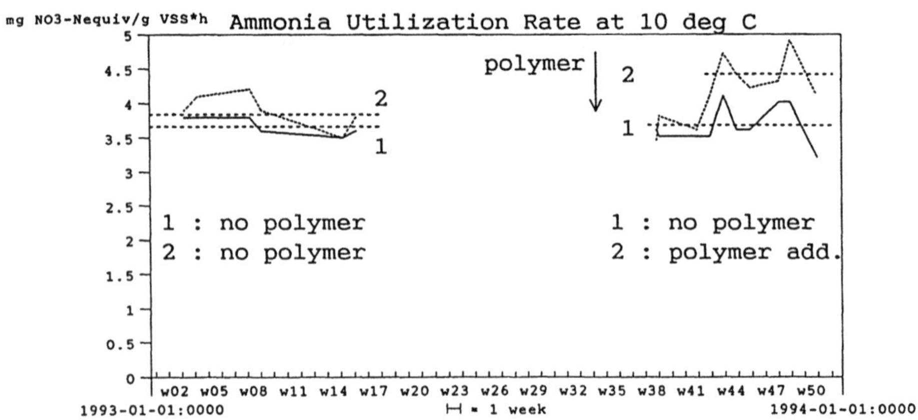

Fig. 5. Specific ammonia utilization rates without and with the addition of polymer as flocculant

Discussion

The advantages resulting from an increase in sludge age and ammonia utilization rate as a consequence of increased removal of suspended solids in the primary treatment step may not be obvious. To assess the improvements in the biological treatment, simulation studies were performed using the simulation package EFOR. The results from the full-scale tests formed the basis for the simulations. The characteristics of the primary effluent without and with polymer addition were used as influent wastewater in the simulations. The mixed liquor suspended solid level in the bioreactor was set to about 3.1 g/l in all the calculations.

The simulation results showed that the sludge age could be increased from about 17 days to about 20 days with the measured improvement of the suspended solids removal in the primary effluent by the polymer addition. The nitrification activity increased by about 12 %, which was attributed to an increase in the autotrophic biomass. Whether the improved removal of suspended solids also results in a higher removal of autotrophic biomass from the influent wastewater can be disputed. If this is the case, there would be some loss of nitrification capacity attributable to seeding from the raw wastewater. However, the concentrations of autotrophic biomass in the influent wastewater were set to zero in the simulations. Consequently, the results show that regardless of the possible effect of seeding of autotrophic biomass from the raw wastewater, the fraction of nitrifiers in the sludge will increase due to improved removal of suspended solids in the primary treatment step. The benefit of the improvements in the biological treatment can be illustrated by the following two examples.

When the biological treatment step is exposed to considerable variations in nitrogen load, the nitrification process is overloaded during some periods of the day and underloaded during others. Consequently, there will be peaks of ammonia in the secondary effluent during some hours of the day. However, an increase in the fraction of nitrifiers will improve the nitrification capacity of the system. With such an increase in the fraction of nitrifiers, the system will thus be less sensitive to variations in nitrogen load.

Furthermore, an increase in nitrification activity leads to an increased capacity of the system. Consequently, the flow to the system could be increased without a concomitant increase in the nitrogen removal effiency. This implies, however, that the carbon dosage is increased with the flow and that the denitrification process is not rate limited.

The two situations outlined above were more closely studied by means of simulation calculations. The outcome is illustrated by the three cases in Figure 6. The first case represents the initial conditions before the start of polymer addition. The second case represents a steady state condition with polymer addition to the pre-precipitation process. The third case represents a system with polymer addition, where the flow is increased to a level where the sludge age is the same as in the system without polymer addition. The carbon dosage in the last case was increased proportionally to the increase in flow. A peak load of nitrogen was added to the influent wastewater of the three different systems. The peak load was dimensioned so as to imitate the recycling of nitrogen-rich supernatant from the sludge treatment for a period of 6 hours a day.

The figure demonstrates two things. First, it is clear that the peak of the ammonia concentration in the secondary effluent decreases as expected when the sludge age is increased. Second, it is obvious that the flow to the system with polymer addition could be increased by about 13 % in comparison to the system without polymer addition.

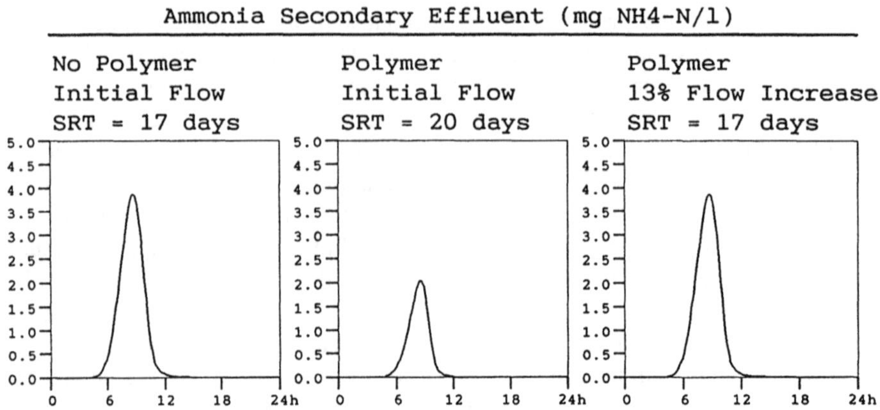

Fig. 6. Response of a peak load of ammonia in systems with 17 days SRT and in a system with 20 days SRT

It was also of interest to analyze whether or not the capacity of the system could be increased even further if the concentration of particulate material in the primary effluent could be decreased even more. A survey of pre-precipitation processes in Norway has shown that the concentration of suspended solids in the primary effluent was below 20 mg SS/l as an annual average in most of the plants (Ødegaard, 1992). Generally speaking, the characteristics of Norwegian wastewater are in many ways similar to those of the wastewater at the Klagshamn plant. With reference to Norwegian results with pre-precipitation, it seems perfectly possible to reach suspended solids concentrations of well below 30 mg SS/l in the primary effluent at the Klagshamn plant.

Further simulations using the calibrated model were performed in order to estimate the increase in capacity due to such an optimization of the pre-precipitation process. The calculation showed that the flow to the biological treatment step could be increased by 15 % in comparison to a situation without polymer addition if the concentration of suspended solids in the primary effluent was reduced to 20 mg/l.

Conclusions

Full-scale tests at the Klagshamn plant have clearly shown that the removal of particulate material in a pre-precipitation process could be increased by means of the addition of a polymer. The positive effects of the decreased load of particulate material on the biological treatment were an increase in sludge age and nitrification capacity.

The increase in nitrification capacity improves the stability of the nitrification process and, consequently, the sensitivity to variations in nitrogen load decreases. Furthermore, the increase in capacity makes it possible to increase the flow to the biological treatment step without impairing the quality of the secondary effluent. Simulations based on mathematical modelling (EFOR) indicate that the capacity of the plant could be augmented by about 15 % as a result of an optimized pre-precipitation process including addition of a polymer.

Even if an exact determination of the biological sludge production and sludge age is difficult in a full-scale system, the trends of the calculated parameters from the full-scale system as well as the results based on model simulations are well defined by the thorough sampling and measuring program used throughout the test.

The tests have also demonstrated significant differences with regard to the effect of polymer addition depending on where and how the polymer is added to the water flow.

Literature

[1] Andersson, B., Nyberg, U., Aspegren, H., la Cour Jansen, J.: Extended Nutrient Removal at the Klagshamn Wastewater Treatment Plant in Malmö. EWPCA Conf. Advanced Sewage Purification to the Protection of the North and Baltic Sea in Lübeck/Travemünde, Germany, November 7–8, 1991

[2] Andersson B., Nyberg U., Aspegren H., la Cour Jansen, J., Ødegaard, H.: Evaluation of Pre-precipitation in a Wastewater Treatment System for Extended Nutrient Removal. In: Chemical Water and Wastewater Treatment II, R. Klute and H.H. Hahn (Eds.). Springer, Berlin Heidelberg New York 1992, pp. 341–355

[3] Andersson, B., Nyberg, U., Aspegren, H., la Cour Jansen, J., Ødegaard, H.: Nutrient Removal at a Wastewater Treatment Plant in Malmö. IFAT 93, Munich, Germany, 1993

[4] Nyberg, U., Andersson, B., Aspegren, H.: Nitrogen Removal in an Activated Sludge System Without Primary Clarification. Vatten/Journal of Water Management and Research *47* (4) (1991) 278–285

[5] Nyberg, U., Aspegren, H., Andersson, B., la Cour Jansen, Villadsen, I.: Full Scale Application of Nitrogen Removal with Methanol as Carbon Source. Wat. Sci. Techn. *26* (5–6) (1992) 1077–1086

[6] Ødegaard, H.: Norwegian Experiences with Chemical Treatment of Raw Wastewater. Wat. Sci. Techn. *25* (12) (1992) 255–264

Ulf Nyberg, Bengt Andersson,
and Henrik Aspegren
Malmö Water & Sewage Works
S-205 80 Malmö
Sweden

Hallvard Ødegaard
Department of Hydraulic and Environmental Engineering
The University of Trondheim
N-7034 Trondheim
Norway

The Use of Reagents in Up-flow Submerged Biofilters

B. Pujol, P. Sagberg, H. Lemmel, and M. Hamon

Summary

This paper presents results concerning three wastewater treatment plants combining biological treatment (by upflow biofiltration) and chemical treatment in order to achieve:

- Suspended solids and phosphorus removal in a lamella settler before biofiltration enhancing dissolved organic matter removal and nitrification (Perroy – Switzerland).
- Simultaneous nitrification and phosphorus removal in biofilters, after adding Al salts and polymers at the inlet of the biofilters (Cologne – Germany).
- Denitrification in upflow biofilters, (Oslo – Norway) after chemical preprecipitation

The results show advantages of such an association which makes it possible to reach high effluent quality and reliability with respect to stringent standards (C, N, P).

Introduction

In order to have compact plants and to achieve high removal efficiency and high reliability, primary lamella settlers combined with submerged biofilters can successfully be used for wastewater treatment (Boller et al. 1993, Canler and Perret 1993, Smith et al. 1991, Jepsen et al. 1992). The submerged biofilters allow simultaneous suspended solids retention (physical filtration) and biological transformation of organic matter (C, N) by aerobic bacteria attached to a granular support. Both functions (physical and biological) can be boosted by the addition of reagents just before the reactor.

This paper concerns the use of reagents in combination with the Biofor®, an upflow biofiltration system developed by Degremont. More than fifty facilities ensuring the treatment of 2 000 000 m^3/day (corresponding to several million inhabitants) are equipped with the Biofor®, some of which are in large cities: Oslo, Cologne, Grenoble, Chateaugay (suburb of Montreal), and soon Colombes (suburb of Paris).

R. Klute / H.H. Hahn (Eds.)
Chemical Water and Wastewater Treatment III

The use of reagents will be illustrated by results carried out in pilot and full scale plants.

In this paper, three reagent applications will be presented:

- chemical treatment upstream from a biofilter (Perroy plant – Switzerland – 7500 p.e)
- using reagents for phosphorus removal in a nitrifying Biofor® (Cologne plant: 1 450 000 p.e)
- addition of an extra carbon source (methanol) for denitrification (Oslo plant: 450 000 p.e).

After a presentation of the Biofor characteristics, each of these applications will be detailed and the results discussed, showing the advantages of reagent addition which make the treatment easier. However, as reagent addition increases operational costs, the control of the doses to be injected on-line, and the monitoring of excess sludge production must be made carefully.

The BIOFOR® Process

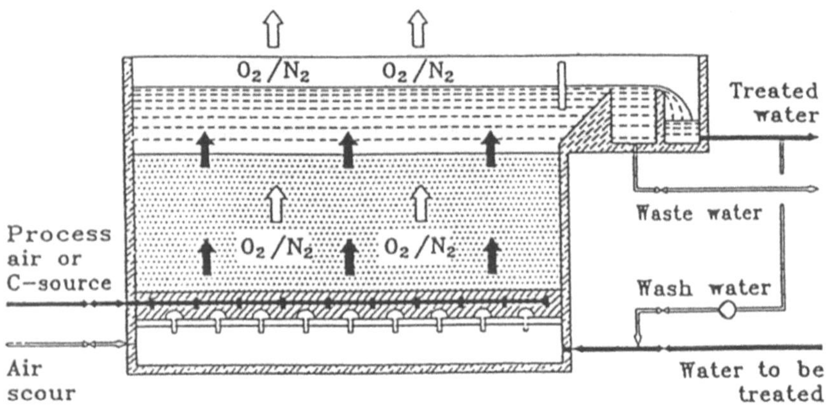

Fig. 1. The BIOFOR® Diagram

This process (Fig. 1) is based on four principles:

- a single layer of a carrier material (Biolite®)
- a specific aeration system
- an upflow distribution of air and water
- adapted washing sequences.

Depending on the application, appropriate systems provide either:

- process air (for conversion of the carbonaceous matter and/or nitrification)
- an easily assimilated substrate (methanol) for denitrification.

Finally, phosphorus removal, where needed, is performed chemically with specific equipment (chemical feed: coagulants, flocculation aids).

In general, the density of the filter medium Biolite® is greater than 1.

The clay medium generally belongs to the phyllosilicate family, but sand or other silicates can be used as well. Particle size is usually between 3 and 6 mm, with smaller sizes being reserved for tertiary treatment applications (nitrification, phosphorus removal).

Bed depth is 3 meters, but can be set at 2 m or 4 m for specific applications. The filter contact time ranges from 1/4 hr to 1 hour. It is important to note that process reaction times are significantly lower than those required by conventional processes. This is an additional advantage offered by biofiltration in cases where unit compactness is a priority.

Normal operation produces sludge that must be regularly withdrawn in order to maintain the filter in good working order. The filter run ranges from 24 to 48 hours. This feature is an advantage over down-flow systems, which have shorter run times due to surface clogging (Jepsen et al. 1993).

The system is aerated with fine bubbles distributed at the bottom of the media by flexible membranes (Oxazur®). The upflow distribution of air and water makes it possible to retain homogeneous suspended solids without the creation of short-circuits nor the trapping of gas bubbles (embolism).

Materials and Methods

On-site measurements and investigations were carried out on three plants. Duration of monitoring varied in the range of three weeks (Perroy – Cologne) to half a year (Oslo).

Samples (24 hour samples or two-hour samples) were made proportionally to the flow through the filter. As far as the sampling points are concerned, the inlet and outlet of every step was taken into account for PERROY in order to analyse simultaneously the efficiency of chemical treatment in lamella settlers and the efficiency of the biofilters.

At Cologne and Oslo, only the inlets and outlets of the Biofor® were investigated. Depending on the objectives of the treatment, different parameters were analyzed (COD, BOD, SS, N, P, etc.) in conformity with the standards applied in the respective countries.

Reagent dosages were controlled by on-line measurements, depending on flow (Perroy), or PO_4 values (Cologne) and NO_3 values (Oslo).

The water temperature was also noted, in particular, to examine nitrification conditions.

Results and Discussion

Upstream Chemical Treatment: PERROY Plant (Switzerland)

The plant was implemented in 1989 and it is located on the banks of Lake Geneva. High effluent quality levels must be reached:

Suspended solids	≤ 20 mg/l
BOD_5	≤ 20 mg/l
Total phosphorus	≤ 1 mg/l
$N\text{-}NH_4$	≤ 5 mg/l (as low as possible).

The guarantees must be met throughout the year, especially during the grape harvest period which is accompanied by strong load variations.

This industrial wastewater input is important, because urban wastewater represents only one fourth of the pollution load during the grape harvest period.

The total capacity of the plant is 7500 p.e (difficult to express in terms of equivalent person due to winery effluents). The total flow taken into account for the design is 100 m^3/d, with an applied load equal to 560 kg BOD/d.

The total surface area of the completely covered compact plant, built on four levels, is 360 m^2 for a flow of 1000 m^3/d. The treatment line is mainly composed of:

- a chemical treatment in a lamellar settling tank (7.5 m^2)
- a first Biofor® stage for C removal (2×17.5 m^2)
- a second Biofor® stage for N removal (17.5 m^2).

Three different periods were considered (Table 1), periods one and two were continuous, but measurements were stopped on some days between period 2 and 3 (because of excessive rain flows).

Effluent standards were widely respected during all the periods with very low ammonia (< 1 mg/l) and total phosphorus (< 0.2 mg/l) concentrations.

The second biofilter step makes a polishing treatment possible and gives very reliable efficiency during the grape harvest period.

Expressed in percentage for each step, the results are as follows (Table 2):

Table 2. Total efficiency (PERROY)

	Average values [%]				
	COD	BOD	SS	NH_4	P
Chemical treatment + lamellar settling	50	53	60	25	82
BIOFOR I + II	88	93	90	97	72
Total efficiency	94	97	96	98	95

These results are obtained from a chemical treatment step having very low dosages: 20 to 30 g Fe/m^3 and < 1 g polymer/m^3.

Table 1. PERROY treatment plant – main results (1990) (results obtained during CEMAGREF evaluation)

Parameters	Raw water g/m³	Inlet first stage g/m³	Inlet second stage g/m³	Outled second stage g/m³
Period 1: Dry weather (558 m^3/d)				
Suspended solids	272	52	10	5
COD	535	196	43	25
BOD_5	228	73	12	6
N-NH_4	17.2	12.6	0.7	0.2
P_{total}	5.4	0.9	0.5	0.2
Period 2: Rainy weather (1224 m^3/d)				
Suspended solids	164	53	10	4
COD	190	104	19	16
BOD_5	70	36	4	3
N-NH_4	9.8	8.4	0.1	0.1
P_{total}	3.1	0.9	0.5	0.1
Period 3: Grape harvest (846 m^3/d)				
Suspended solids	185	116	28	8
COD	738	476	129	34
BOD_5	348	208	53	11
N-NH_4	12.1	8.0	3.0	0.2
P_{total}	4.6	1.0	0.5	0.1

Total sludge production represents about 0.7 kg SS/kg COD eliminated (except during rainy days), biological sludges 0.35 kg SS/kg COD. These values are in accordance with published values by Canler and Perret 1993.

These first results illustrate the advantages of such an association between chemical and biological treatments, making the plants highly compact, efficient and reliable.

Simultaneous Treatment: Nitrification and Phosphorus Removal: Cologne/Stammheim – Germany

The plant in Cologne is the first and largest application of biofiltration in the world for tertiary nitrification with simultaneous phosphorus removal.

The plant (Fig. 2) has been in operation since the end of 1992. It consists of an activated stage (A–B process) completed by a biofiltration step: 48 Biofors® (73 m^2 per Biofor®).

The total treatment capacity represents 1 450 000 p.e and the hydraulic loads are summarized as follows:

- Dry weather: 4.7 m^3/s
- Rainy weather: 9.2 m^3/s.

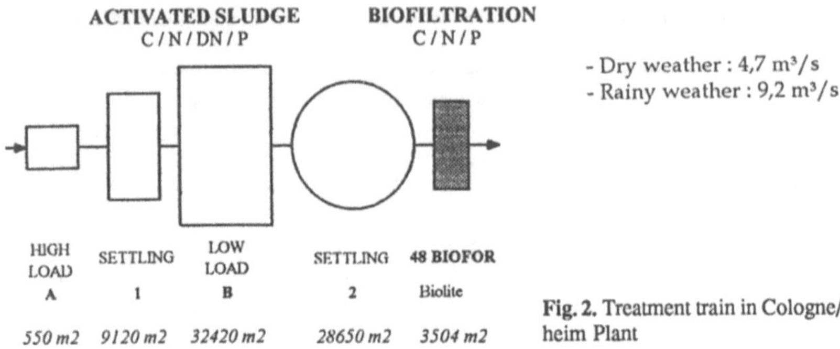

Fig. 2. Treatment train in Cologne/Stammheim Plant

Tests and measurements were conducted in February and March 1993 in order to confirm that performance levels complied with specifications.

Tests were accomplished over a 15-day period (dry weather simulation) and a 5-day period (rainy weather simulation). The water temperature inside the filter varied between 13 °C and 16 °C.

The feedwater velocity ranged from 2 to 11 m/h, with an average value close to 5 m/h. The air flow rate remained constant at 4 Nm/h throughout the duration of testing. Chemicals (Kemazur 48 + polymer AS37) were injected upstream from the filters, to promote chemical precipitation of phosphorus. The average dose for the coagulant was 3 $g/(m^3 W \cdot W)$ and less than 0.1 $g/(m^3 W \cdot W)$ for the polymer.

Backwashing was triggered automatically every 24 hours. The number of filters in operation was adjusted and based on the feedwater flow rate. The analytical results collected pertain to 2-hour average samples.

Specified performance levels were met in:

- 100 % of cases for COD, BOD, SS, $N\text{-}NH_4$ (Fig. 3)
- 99 % of cases for total phosphorus.

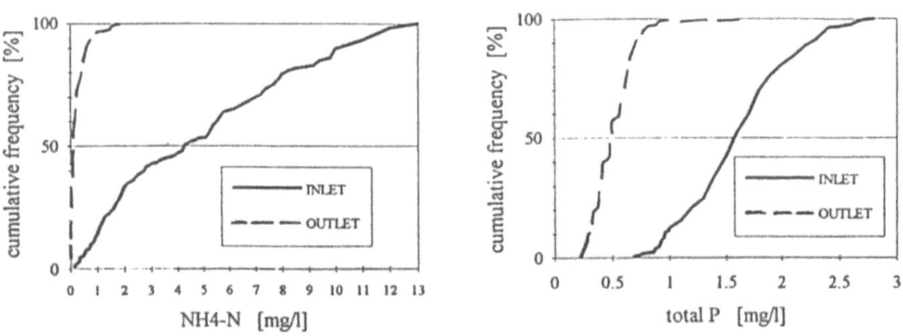

Fig. 3. Cologne: NH_4 and phosphorus removal

Table 3 summarizes the average values observed during the compliance tests; the values indicated in parentheses are the average concentrations stipulated in the specifications.

Table 3. Cologne/Stammheim compliance tests

	BOD	COD	SS	P_{total}	$N\text{-}NH_4$
Inlet [mg/l]	6.5 (10)	45 (50)	10 (15)	1.65 (1.5)	4.4 (3)
Outlet [mg/l]	< 3 (10)	35 (50)	3.7 (15)	0.5 (0.8)	0.2 (3)
Efficiency [%]	> 54	22	63	70	95.5

The remarkable performance achieved in $N\text{-}NH_4$ and P_{total} removal for brief contact times (20 minutes to 1 hour) is particularly noteworthy. Such results could not have been reached using a conventional biological process (activated sludge) with such low inlet concentrations.

External Carbon Source Addition for Denitrification: Oslo

The preceding examples demonstrate the merits of up-flow biofiltration when a high degree of nitrification is required. Obviously, these "multi-purpose" reactors can also provide a high level of denitrification. In this application, the denitrifying filter is placed downstream from the nitrification. The lack of assimilable carbon in this stage of the treatment dictates the addition of an exogenous carbon source. In wastewater treatment, that source is methanol, a carbonaceous organic substrate that is easily assimilable by the heterotrophic denitrifying bacteria.

Denitrification by biofiltration with an exogenous carbon source has frequently been used in drinking water purification (Richard 1988), but its use in wastewater applications is recent. The first up-flow denitrifying filter was commissioned in 1994 for the city of Oslo, Norway (450 000 p.e).

In this case, the complete treatment line consists of three steps:

- removal of suspended particles in a settling tank, with the addition of chemicals
- nitrification in 32 up-flow reactors, each with a unit surface area of 87 m^2
- denitrification in 32 up-flow reactors, each with a unit surface area of 65 m^2.

Although process results are not yet available due to plant construction, the following data were obtained from a half-year pilot study involving units with surface areas of 0.2 m^2 (bed depth = 3 m).

The pilot unit was used to fine-tune the parameters for the full-scale plant under construction. The graph (Fig. 4) illustrates process reliability in terms of removal rates (there is a little dispersion of points around the straight section of the curve). The $N\text{-}NO_3$ content at the pilot unit outlet was lower than 2 mg/l for 70 % of all samples, despite high feedwater velocity (> 10 m/h). The higher values for $N\text{-}NO_3$ ($\leq$ 5 mg/l) correspond to sharp increases in the feedwater velocity (in experiments conducted on the pilot unit).

Sludge production was close to 1 kg SS/kg $N\text{-}NO_3$ eliminated. Methanol/nitrate ratio was about 3.

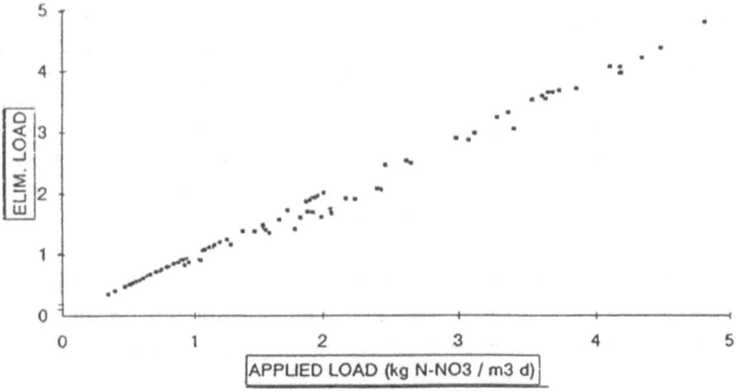

Fig. 4. Oslo: Denitrification results (pilot plant)

Conclusion

Upflow biofiltration associated with chemical reagents is a very attractive wastewater treatment train. High quality effluents can be obtained with respect to stringent standards (C, N, P).

The results presented for the three plants give the following outlet concentrations:

	DCO mg/l	SS mg/l	$N\text{-}NH_4$ mg/l	$N\text{-}NO_3$ mg/l	P_{tot} mg/l
Perroy	< 40	< 10	< 0.2	–	< 0.2
Cologne	< 45	< 10	< 0.2	–	< 0.5
Oslo	< 40	< 10	< 1	< 2	–

These plants are compact and automated with flexible operational conditions. A regulation system is necessary in order to reduce operational costs and sludge production by correctly correlating inlet concentrations and reagent needs.

References

Boller, M., Gujer, W., Tschui, M.: Parameters Affecting Nitrifying Biofilms Reactors. Proceedings of 2nd specialized conference on Biofilms Reactors, Paris Sept. 1993, pp. 15–25

Canler, JP., Perret, J.M.: Biological Aerated Filters. Assessment of the process based on 12 sewage treatment plants. Proceedings of 2nd specialized conference on Biofilms Reactors, Paris Sept. 1993, pp. 39–49

Jepsen, S.E., Laursen, K., Lacour Jansen, J., Harremoës, P.: Denitrification in Submerged Filters of Nitrified Wastewater and Chemical Pre-precipitated Wastewater. In: Chemical Water and Wastewater Treatment II, R. Klute and H.H. Hahn (Eds.). Springer, Berlin Heidelberg New York 1992, pp. 373–388

Richard, Y.: Operating Experiences of Full Scale Biological Denitrification Plants France. JWEM *3* (1988) 154–167
Smith, A.T., Hardy, J.P.: High Rate Sewage Treatment Using Biological Aerated Filters. JIWEM - 6 April 1992, 179–183

R. Pujol, H. Lemmel, and M. Hamon
Degremont
Département Recherche
38 rue du Président Wilson
F-78230 Le Pecq
France

P. Sagberg
VEAS Oslo
Postboks 245
N-3470 Slemmestad
Norway

The SIROFLOC Sewage Treatment Process: A High Rate Process for Sewage Clarification

N. A. Booker, E. Cooney, G. Öcal, and A. J. Priestley

Abstract

A process for rapid clarification of sewage has been developed, in Australia, by the CSIRO and extensively tested in pilot scale studies in Melbourne, Sydney and the UK. The results of the pilot studies were used to design a 5 Ml/day prototype of the SIROFLOC™ sewage treatment process (STP). This was constructed at the Malabar sewage treatment works (STW), in Sydney, and has been successfully operated since October 1992.

The process utilises finely divided particles of magnetite combined with an inorganic coagulant to aid the rapid separation of colloidal and suspended solids from sewage. The magnetite and inorganic coagulant are recovered and reused within the process, thus minimising the operating costs. The use of magnetite allows for the rapid coagulation and clarification of sewage; raw sewage can be treated to a high quality within 15 minutes.

Since the process is a coagulation process, it is ideally suited to the removal of insoluble matter from sewage. The 5 Ml/day prototype plant at Malabar has operated continuously to produce treated sewage containing 30 mg/l TSS and 6 mg/l oil and grease, equivalent to reductions of up to 90 %. Phosphate removal was better than 90 %, consistent with the use of an inorganic coagulant and total coliforms were reduced by up to 3 logarithms. Significant reductions in heavy metals were achieved by the process, with removal efficiency dependent on the particular metal. Soluble contaminants present in the sewage, such as ammonia, were not significantly removed by the process. Biochemical and chemical oxygen demand (BOD and COD, respectively) were generally reduced by at least 50 %, consistent with the removal of their insoluble fraction from the sewage.

Introduction

In the development of our urban environment the pressures on civic infrastructure become more and more intense. The modern pace of life expresses itself in the need for faster, higher capacity service systems that must have a minimal impact on our environment. Nowhere has this been more evident than in the wastewater handling and treatment field. Urban growth and increased environmental discharge standards

R. Klute / H.H. Hahn (Eds.)
Chemical Water and Wastewater Treatment III

have meant that increasing pressures are being placed on existing sewerage systems and treatment plants. The traditional approach of providing time and "nature" to solve our wastewater problems is a luxury not readily available in most cities. In order to meet this challenge, wastewater engineers and scientists have had to broaden their horizons to encompass a wide range of different technologies. This work has led to a better understanding of biological treatment processes, has allowed tighter control and increased treatment rates to be achieved. It has also led to the development of a number of physico-chemical treatment processes that are characterised by high treatment rates and high solids removal efficiencies. The challenge now is to combine the best qualities of the biological processes with the high rate physico-chemical processes. The research discussed in this paper addresses the development of a high rate physico-chemical process for the treatment of municipal wastewater. An earlier paper [1] on this subject, delivered to the Gothenburg Symposium, in 1992 by Dr. Priestley, discussed the development of the process from the bench scale through to pilot scale trials. Development has successfully proceeded from this scale to the operation of a full scale treatment plant at Malabar in Sydney, the performance of which is discussed here.

Process Development

The SIROFLOC™ sewage treatment process (STP) was developed from the successful SIROFLOC™ water treatment process [2]. Development began in 1986 when a series of exploratory jar tests was performed to confirm the potential of the process for sewage treatment. These jar tests were followed up by large scale batch tests performed in a 90 litre reactor using primary settled sewage. The first continuous pilot scale testing of the process concept was performed on a converted SIROFLOC™ water treatment pilot plant. The trials [3] were performed at the CSIRO field test station at Lower Plenty in Melbourne; the plant treated raw sewage that had been macerated and passed through a high rate primary settler. The raw sewage treated was approximately two hours old and was pumped from a sewer which received mainly domestic sewage. The pilot plant had a throughput of about 2.4 m^3/h and was operated for eight hours a day, 5 days a week. The long term operability of the process was determined over a period of a month, where the plant was run continuously 24 hours a day, seven days a week.

In Sydney, three large treatment plants operate at North Head, Bondi and Malabar; based on primary sedimentation, they treat all the sewage emanating from the Sydney metropolitan area and discharge effluent into the Pacific Ocean. Extensive public pressure forced the Sydney Water Board to reduce the environmental impact of the discharge of sewage into the near coastal region. Initially, three deep ocean outfall pipes were constructed, one at each of these three plants, to take the primary treated sewage about 5 kilometres offshore. At the same time as construction of the outfalls began, the Water Board began evaluation of techniques to improve the degree of treatment achieved by their treatment plants. Their requirement was for compact high rate processes that could reduce sewage suspended solids and grease to concentrations similar to those achieved by conventional, biological, secondary treatment. The need for compact, and hence high rate, processes was driven by the degree of urbanisation that had been allowed to occur in close proximity to these three plants. Development of conventional biological treatment processes at these plants was considered to be extremely expensive. The main processes considered were those based on chemical coagulation and sedimentation, namely dissolved air flotation (DAF) and chemically assisted sedimentation (CAS). Other processes evaluated were cross flow microfiltration and Biocarbone. The Water Board became interested in the results achieved by the SIROFLOC™ STP pilot plant at Lower Plenty, Melbourne, and commissioned the process licensees (Davy John Brown Pty. Ltd.) to construct another pilot SIROFLOC™ STP plant for operation at Malabar, Sydney [4]. This plant treated 8.4 m^3/h of raw sewage and ran for about eighteen months from December 1989 to mid-1991. Another pilot plant was constructed by Davy Energy and Environmental Limited (UK) for Yorkshire Water and was operated at both Blackburn Meadows STWs (Sheffield, UK) and at Seamer, UK.

The performance of the Malabar pilot plant encouraged the Water Board to commission Davy John Brown Pty. Ltd. to design and construct a full-scale prototype of the SIROFLOC™ STP at Malabar, treating 5 Ml/day of raw sewage. The purpose of the prototype plant was to test the process on a scale that would enable full engineering evaluation of each unit within the process. Operation of the plant began in November 1992 and has continued to the present. The plant is manned by two Water Board operators, covering two shifts, and is operated generally for twelve hours a day. For a period of about a month the plant ran unmanned overnight to allow continuous 24 hour a day operation. It is equipped with two process control computers, one continuously on line, the other accumulating historical trend data whilst acting in a standby capacity. The controller automatically starts the process from cold, on operator request, and will go through a controlled shut-down upon detection of equipment failure or major set-point deviation. The installed process equipment, including the control room, covers a total area of only 280 m^2, a truly compact plant.

A schematic flowsheet of the process is shown in Figure 1. Raw sewage is screened through a disk strainer with a screen aperture of 300 μm; this removes the bulk of fibrous material from the sewage prior to it contacting the magnetite. The screened sewage is mixed with a stream of clean magnetite and an inorganic coagulant (alum or ferric chloride). The magnetite particles used are less than 20 μm in diameter and present a large surface area to the coagulating suspended solids. As a result, adsorption of the incipient floc particles is very rapid and effective. After only two minutes of mixing, the magnetite and its adsorbed solids are passed across the poles of a pair of permanent magnets in order to magnetically flocculate the magnetite. The flocculated magnetite slurry is then passed into a conventional 6m diameter clarifier and allowed to settle under gravity; the rise rate in the clarifier is 10 m/h. The thickened magnetite slurry is then collected and pumped to a tank where it is mixed with sodium hydroxide to raise the slurry pH and release the solids and coagulant from the magnetite particles. The magnetite

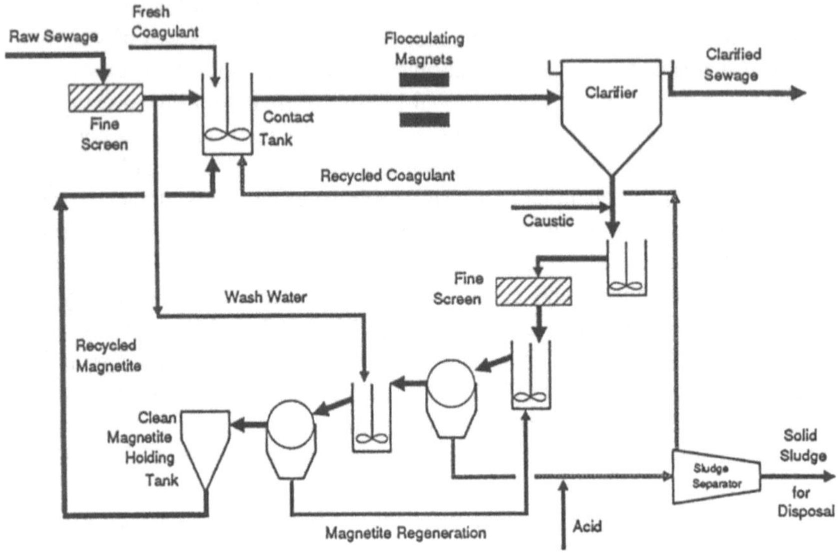

Fig. 1. SIROFLOC™ STP schematic layout

is then passed through a fine screen, to separate fine fibrous matter from the magnetite, and then washed, in a counter-current manner, with a stream of treated sewage equivalent to about 3 % by volume of the plant sewage flow. The cleaned magnetite is recycled to the contact stage of the process. The magnetite wash water contains all the material removed from the sewage by the magnetite and also the inorganic coagulant, as a sewage concentrate stream. This concentrate is collected and acidified to allow the organic material present to flocculate and separate. The clarified acid stream is collected and recycled to the contact stage of the process where any value from the recovered coagulant can be realised.

The separation of solids from the acidified coagulant stream is achieved in a decanter centrifuge, however, a DAF unit was also trialed for this duty. Lime neutralisation of the sludge produced by the process was tested in laboratory bench scale tests. Anaerobic digestion of the sludge has been evaluated in pilot, single-stage anaerobic digesters at Lower Plenty, Melbourne, using sludge shipped from the Malabar prototype plant.

Crossflow microfiltration and ultra-violet irradiation were evaluated in pilot scale tests for disinfection of the sewage treated by the process at Lower Plenty, Melbourne. A 3 m^2 crossflow microfiltration unit, having a polypropylene membrane of 0.2 μm pore size was borrowed from Memtec Pty. Ltd., in order to evaluate this process for product disinfection. A 100 W, quartz-tube, ultra-violet disinfection unit was borrowed from Ultra-Violet Technologies, Australia and operated at a range of flow rates to determine the UV dose response of SIROFLOC™ STP effluent.

Process Performance

The SIROFLOC™ STP is effectively a chemical coagulation process and as such is effective for the removal of suspended solids and colloidal material from sewage. The process is controlled by monitoring the turbidity of both raw and treated sewage and relating that to the TSS of each stream. A target turbidity of between 10 and 20 NTU was sought since this related to a product TSS of generally less than 30 mg/l. The plant operators attempted to control the coagulant dose manually depending on the raw sewage strength and the turbidity of treated effluent. Figure 2 shows the coagulant dose required to maintain the treated sewage turbidity at 20 NTU over a range of raw sewage strengths, from 100 to 200 NTU (approximately 150 to 350 mg/l TSS). However, due to diurnal variation and other rapid fluctuations in sewage strength, it was difficult to maintain the coagulant dose at the optimum, as evidenced by the wide envelope of points shown in Figure 2. Since the coagulant dose was difficult to maintain manually at the optimum, the plant was generally operated at a constant coagulant dose.

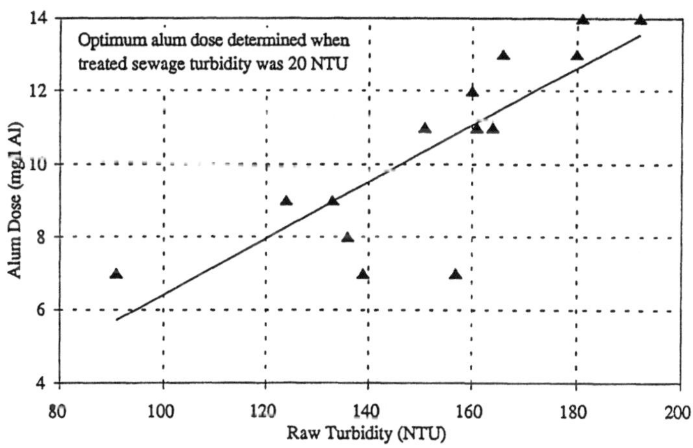

Fig. 2. Raw turbidity vs. optimum alum dose

Solids Removal

Figure 3 shows a plot of the raw and treated sewage total suspended solids (TSS) versus plant operating time for a period when coagulant (alum) dose was constant at 12 mg/l Al. TSS was determined in both streams from 12 hour composites of hourly stream samples. As can be seen from the graph, the treated sewage TSS was maintained in the order of 30 mg/l despite wide fluctuations in the raw sewage TSS; the removal efficiency averaged better than 85 % for this period. Table 1 shows the performance of the process over a range of coagulant doses.

The results in Table 1 indicate that the process can be operated to achieve the required effluent quality and maintain TSS reductions in excess of 85 % simply by

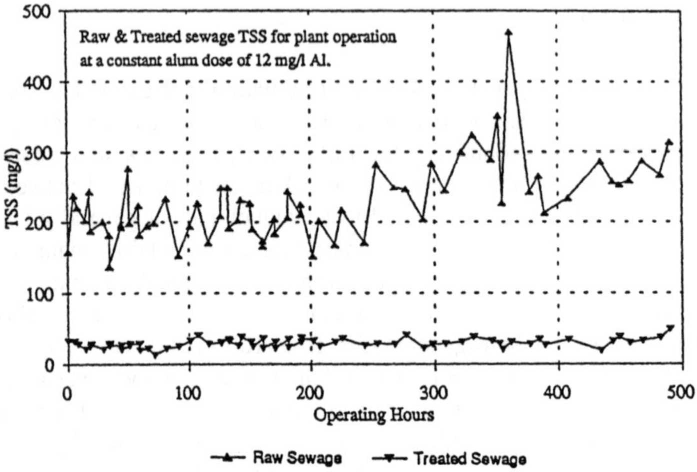

Fig. 3. TSS versus operating hours

Table 1. Suspended solids removal related to coagulant dose

Coagulant Dose mg/l Al	Hours operated	Raw sewage TSS		Treated sewage TSS		Removal
		Average mg/l	Std. dev'n. mg/l	Average mg/l	Std. Dev'n. mg/l	%
0	21	222	14	117	9	47
6	177	190	40	52	15	72
12	289	259	60	32	9	88
14	201	206	30	29	10	86

adjusting the coagulant dose. The results also show that the presence of magnetite in the coagulation stage gives a base TSS reduction of about 50% before any coagulant is added. This occurs because the surface of the magnetite particle has some positively charged sites at the pH at which coagulation occurs and consequently can neutralise and adsorb some of the negatively charged species present in the sewage. Once the surface charge on the magnetite is neutralised, further reductions in TSS can only result from the addition of positively charged coagulant. However, sufficient magnetite particles need to be present in order that magnetic flocculation can be effectively achieved. If insufficient magnetite is present, the surface loading on the magnetite can restrict the magnetite particles from coming close enough together for the attractive magnetic forces to provide effective flocculation and resist floc shear. Excessive shear can result in floc carry-over into the treated sewage and high losses of magnetite.

Oil and Grease Removal

The Sydney Water Board was particularly interested in the removal of oil and grease (O & G) from the raw sewage since grease can accumulate within the outfall pipes. The fear is that grease balls can be produced and that these can float back onto the suburban bathing beaches, carrying faecal contamination. Table 2 shows the relationship between oil and grease removal and coagulant dose achieved on the plant.

Table 2. Oil ans grease removal related to coagulant dose

Coagulant Dose mg/l Al	Hours operated	Raw sewage O & G		Treated sewage O & G		Removal
		Average mg/l	Std. dev'n. mg/l	Average mg/l	Std. Dev'n. mg/l	%
0	21	55	4	30	3	45
6	177	44	10	11	5	75
12	289	59	14	6	2	90
14	201	48	6	4	2	91

The process exhibits a very good ability to cope with oils and greases present in the sewage. The results obtained showed a higher degree of O & G reduction compared to conventional coagulation processes achieving similar TSS reductions; CAS achieved about a 65 % reduction in O & G where the SIROFLOC™ STP was achieving a 90 % reduction. The enhanced performance is due, in part, to the large surface area of magnetite, presented to the sewage, acting as an adsorbent for the grease. The O & G reduction was fully sustainable since the O & G were effectively stripped from the magnetite at the pH used for magnetite regeneration.

Other Results

A number of other parameters were measured in both the raw and treated sewage, these included COD, BOD_5, nutrients, bacteria and metals; Table 3 shows some of the average results obtained by the process.

The COD and BOD_5 reduction seen by the process is consistent with removal of their insoluble fractions from the sewage. The Malabar plant achieved about a 50 % reduction in these parameters, compared with about a 70 % reduction achieved at Lower Plenty in Melbourne. The Malabar plant receives sewage that has a mix of industrial and domestic effluents which takes about eighteen hours to reach the treatment plant, during dry weather. These factors combine to increase the soluble fraction of both the COD and BOD_5 either by direct input or anaerobic hydrolysis within the sewer. The sewage at Lower Plenty, by contrast, is principally of domestic origin and has a short life in the sewer before reaching our treatment plant. The pilot SIROFLOC™ STP plant operated near Sheffield, in the UK, also achieved reductions in BOD_5 on the order of 50 %, at similar TSS reductions..

The sewage received by the Sheffield plant had a similar history to that treated at Malabar.

The removal of phosphate by the process is consistent with that achieved by chemical coagulation processes. The phosphate remaining in the treated sewage was mainly present as insoluble phosphates; filtered samples of treated sewage had soluble phosphorus contents generally less than 0.1 ppm. Removal of ammonia by the process was negligible since the SIROFLOC™ STP is ineffective at removing soluble matter from sewage. The reductions in total nitrogen seen are consistent with the removal of insoluble proteinaceous material from sewage.

Table 3. Other SIROFLOC™ STP performance results

Parameter	Raw sewage		Treated sewage		Removal
	Average	Std. dev'n.	Average	Std. dev'n.	%
Chemical oxygen demand [mg/l]	627	65	286	50	54
Biochemical oxygen demand [mg/l]	266	21	136	37	49
Total phosphorus [mg/l P]	7.3	0.8	0.8	0.3	89
Total nitrogen [mg/l N]	33	4	27	1	18
Ammonia [mg/l NH_3]	23	5	25	2	0
Total coliforms [counts/100 ml]	8×10^7	3×10^7	9×10^5	1×10^6	99
Copper [μg/l]	286	32	39	23	86
Zinc [μg/l]	289	74	42	15	85

Plate counts of raw and treated sewage showed considerable reductions in faecal coliforms and total coliforms from the sewage. This is not surprising since bacteria tend to colonise suspended matter in sewage and thus will be removed with TSS.

Heavy metals were reduced considerably as evidenced by the reductions in both copper and zinc, shown in Table 3. These two metals were used as indicators for other heavy metals, since their concentrations were high enough to measure easily with accuracy. Other metals such as mercury, lead and cadmium were present in the raw sewage but at concentrations near the limit of detection of the inductively coupled plasma spectrophotometric technique used (< 5 μg/l). These metals were all removed from the sewage to some extent since they all appeared at measurable concentrations in the centrifuged sludge.

Operating Costs

The major operating costs associated with the SIROFLOC™ STP result from the chemicals dosed into the process. These costs break down to the following proportions of the total operating cost:

Sulphuric acid = 25 %
Coagulant (alum) = 25 %
Caustic soda = 20 %
Polymer = 10 %
Magnetite losses = 10 %
Power = 10 %

Sulphuric acid is used in the process to maintain the correct coagulation pH in the contact stage and to aid the recovery of coagulant. Pilot plant trials at Lower Plenty were able to balance the acid demands of coagulation against those required for coagulant recovery such that the acid demand of the process did not change when coagulant was recovered. This was not the case at Malabar where the acid demand of the process increased by between 40 and 50 % when coagulant recovery was practised. This difference in acid consumption is attributed to the higher alkalinity of Malabar raw sewage (220 mg/l $CaCO_3$) compared to Lower Plenty where the sewage is relatively soft.

The prototype plant was initially operated without the coagulant recovery and recycle section operating. This section was commissioned once the centrifuge performance had been mechanically optimised and a suitable polymeric flocculation aid found. A high molecular weight, branched, high charge density, cationic polymer gave best clarification results in the decanter centrifuge, even at the low operating pH of 2.0. The coagulant recovery system yielded the highest effective aluminium recovery at a pH close to 2.0; however, separation of the flocculated organic material from the acidic coagulant proved difficult to control at this pH. The increased cost of acid required to reduce the coagulant stream pH offset the gains made by the increased coagulant recovery. A process optimum was found at a higher pH where floc separation was appreciably easier and acid consumption reduced by about 30 % . The recycled coagulant stream still gave some coagulant benefit, reducing fresh coagulant demand in contact by about 20 % .

The real benefit of operating the coagulant recovery system was in reducing the volume of effluent from the process. Acidifying and separating the solids from the process wash water stream resulted in a centrifuge sludge as the only process by-product. The centrifuge produced a sludge averaging 21.8 % w/w solids which could be effectively neutralised with lime.

DAF was trialed for this separation duty. It achieved an effluent of superior clarity with about an eighth of the polymer dose used in the centrifuge. However, the sludge it produced averaged only 4.8 % w/w solids. The decision to use either a decanter centrifuge or DAF for this separation rests on the relative economics of the two systems. The DAF process, although occupying more area, has a lower installed capital cost than centrifugation. Provided the sludge solids do not need to be increased, the operating costs associated with DAF would be less than for centrifugation.

Disinfection

The Sydney Water Board plans to use the 5 Ml/day SIROFLOC™ STP prototype plant to supply process water to the whole of the 625 Ml/day Malabar sewage treatment plant. This is expected to save a large component of their current operating costs which is estimated to pay for the capital investment in the SIROFLOC™ STP plant within three years. In order to achieve this objective the effluent from the SIROFLOC™ STP plant must be disinfected to less than 200 total coliform counts/100 ml. Pilot scale trials of ultra-violet disinfection are planned to begin in March 1994, in an attempt to determine the UV dose required to achieve this quality.

Pilot scale tests of ultra-violet disinfection have been performed at Lower Plenty on SIROFLOC™ STP treated effluent. These trials followed laboratory UV absorption tests performed on treated effluent. The treated sewage transmitted 25 to 30 percent of the UV radiation at 254 nm despite having a turbidity of 10 NTU. The soluble COD remaining in the treated sewage was obviously responsible for most of the UV absorption. Pilot scale tests were performed over a range of UV dose rates and sewage samples were taken and analysed for coliforms. The pilot scale tests revealed that a UV dose rate of 24 000 $\mu Ws/cm^2$ was sufficient to reduce total coliforms in the SIROFLOC™ STP effluent to less than 200/100 ml.

Table 4. Disinfection of SIROFLOC™ STP pilot plant effluent

	Raw sewage	SIROFLOC treated	UV* disinfected	CMF disinfected
TSS [mg/l]	180	15	15	0.1
Total coliforms [per 100 ml]	1.7×10^8	2.4×10^6	110	< 2
E-coli [per 100 ml]	1.3×10^7	2.8×10^5	22	< 2

* UV dose was 24 000 $\mu Ws/cm^2$

Trials were also performed at Lower Plenty to evaluate the efficiency of cross-flow microfiltration (CMF) for disinfection of SIROFLOC™ STP treated sewage. A 3 m^2, 0.2 μm pore size polypropylene membrane was operated at a constant flux rate of 200 $l/(h \cdot m^2)$ and variable trans-membrane pressure (declining from 80 to 40 kPa). The backwash frequency required to sustain continuous filtration depended on the solids loading to the filter and was varied from 6 to 20 minutes. The unit proved to be very effective at reducing the coliform plate count in the sewage to below detection and reduced the TSS in treated sewage to below 1 mg/l. The clarification rate of the SIROFLOC™ STP pilot plant was increased from 9.4 to 14 m/h to determine the effect of increased clarifier carry-over on membrane performance. The increased clarifier rise rate increased the TSS in plant effluent from 15 to 47 mg/l. The CMF unit maintained good disinfection and TSS removal from the sewage over this range of flows; however, the flux rate could not be maintained without reducing the backwash interval to every 6 minutes. Some typical results from the disinfection trials using either UV or CMF are shown in Table 4.

Sludge Treatment

The sludge produced by centrifugation of the acidified sewage concentrate stream must be disposed of in some manner. Upon starting this work, Sydney Water Board had made no clear decision regarding the fate of sludge produced by their sewage treatment plants. The NSW government placed a ban on sludge incineration within the Sydney metropolitan area, so this avenue of disposal was unavailable. The disposal of sludge to the ocean off Sydney is also to cease by 1994. The Water Board is developing composting technology with the aim of using this as a route to land disposal of sludge.

Anaerobic digestion of sludge and the simultaneous production of methane has been the focus of our research into the disposal of SIROFLOC™ STP sludge. Large batches of centrifuge sludge are regularly shipped from Malabar to our laboratory at Lower Plenty, where pilot anaerobic digestion studies are carried out. Four 10 litre pilot scale digesters are operated various conditions in an attempt to maintain continuous single stage digestion of the sludge. Hydraulic retention times ranging from 10 to 20 days and COD feed rates between 2 and 4 g/(l · day) have been trialed. The results obtained to date indicate that the first stage of anaerobic digestion, acidogenesis, occurs rapidly and can produce volatile fatty acids (VFAs) in excess of 3 g/l (as acetate) within 21 days of start-up. This has presented a problem to subsequent methanogenesis since the elevated VFA levels can inhibit the growth of the methanogens. Sustainable methanogenesis has only been achieved if the concentration of VFAs in the digester is controlled to low levels by controlling the feed rate to the digester. This protocol allows the slower growing methanogens to catch up with the faster doubling acidogens. By careful control of VFAs, the feed rate to the digesters was gradually brought up to the target rate. Biogas production rates of between 200 and 300 ml/g COD digested have been achieved from COD conversions of about 40 %. At present this performance has only been achieved on a 70 : 30 blend of SIROFLOC™ STP sludge and primary settled sludge. The aim of our current research is to find the conditions required to allow sustainable methanogenesis of up to 100 % SIROFLOC™ STP sludge.

Current Status

The development of the SIROFLOC™ STP has successfully progressed from jar tests to a full scale plant in under 8 years. The results obtained on the full scale plant matched or were superior to those obtained in pilot scale trials. Suspended solids were reduced by up to 90 % from raw sewage; similar removals of oil and grease were obtained. The process coped well with wide variations in raw sewage quality and the effluent quality could be effectively controlled by the coagulant dose. The process produces a waste sludge of up to 20 % w/w solids that is amenable to partial disposal via anaerobic digestion. Present research on the process is linked more to process applications rather than further process development.

The high rate and rapid start-up that can be achieved in the SIROFLOC™ STP makes it attractive for situations where there may be rapid fluctuations in sewage flow. The infiltration of rainwater into sewer networks can result in massive increases in sewer flows that may lead to sewer flooding and overflow into stormwater systems. The feasibility of using the SIROFLOC™ STP to treat sewer overflows is currently being investigated. These same properties also make the process attractive for regions where seasonal population changes occur, for example coastal holiday towns. In this circumstance the process could be utilised to supplement any existing treatment process to reduce either the volumetric demand or solids load.

As mentioned previously, the Sydney Water Board is planning to use the 5 Ml/day prototype plant for the production of process water from sewage. The expected savings to the Water Board by limiting the potable water required for this duty amount to about $A1 million/year. Some mechanical changes will be required to push the throughput of the plant from 5 Ml/day to the required 6 Ml/day.

The concentration of ammonia remaining in the effluent after SIROFLOC™ STP treatment may, under certain circumstances, be too high for discharge to the environment. We are currently working on the development of adsorption processes for the removal of ammonia from physico-chemically treated sewage, the rate of which would be commensurate with that of the SIROFLOC™ STP.

Acknowledgments. The authors would like to thank the Sydney Water Board for their considerable support in the development of this process. The development of the SIROFLOC™ STP would not have been possible without the input of the process licensee, engineering contractor Davy John Brown Pty. Ltd., in particular, Aidan Smith, Dr. Ron Brooks and Damien Twining. Thanks are due to Chris Hallam for his technical support at Lower Plenty and to Ron Hood and the operators on the Malabar prototype plant.

References

[1] Booker, N.A., Priestley, A.J., Ritchie, C.B.: Rapid Sewage Clarification with Magnetite Particles. In: Chemical Water and Wastewater Treatment II, R. Klute and H.H. Hahn (Eds.), Springer, Berlin Heidelberg New York 1992, pp. 485–496

[2] Anderson, N.J., Priestley, A.J.: Colour and Turbidity Removal with Reusable Magnetite Particles – V Process Development. Wat. Res. *17* (1983) 1227

[3] Booker, N.A., Keir, D., Priestley, A.J., Ritchie, C.B., Sudarmana, D.L., Woods, M.A.: Sewage Clarification with Magnetite Particles. Wat. Sci. Techn. *23* Kyoto (1991) 1703

[4] Akyel, G., Booker, N., Cooney, E., Priestley, A.: Rapid Sewage Clarification Using Magnetite Particles. Proc. 15th Federal Convention, Australian Water and Wastewater Association, Gold Coast, 1993, pp. 137–144

Nicholas Booker
CSIRO, Division of Chemicals and Polymers
Private Bag 10, Rosebank MDC
Clayton, Victoria, 3169
Australia

Full-Scale Experiences from Biological Nitrogen Removal Combined with Different Phosphorus Removal Alternatives at Öresundsverket in Helsingborg

L.-E. Jönsson, K. Jönsson, and P. Magnusson

Abstract

Öresundsverket in Helsingborg is the first municipal wastewater treatment plant in Sweden, designed and revamped for biological nitrogen removal. Different alternatives of phosphorus removal were used during the first two years of operation. The main purpose of this paper is to illustrate how the biological nitrogen removal has been influenced by the different phosphorus removal alternatives.

Pre-precipitation had a positive effect on the nitrification rate, because of a lower carbon to nitrogen ratio. The other processes tested gave equivalent nitrification rates.

Denitrification rates determined with acetate as single carbon source gave equivalent rates to all tested processes. Denitrification with methanol had an adaptation time of at least 1.5–2 months.

The characteristics of the activated sludge varied substantially, diluted sludge volume index (DSVI) between 100 and 400 ml/g. The recirculation in the biological treatment is an important factor that influenced DSVI. A higher amount of easily biodegradable carbon had a positive effect on DSVI.

Pre-precipitation gave high DSVIs with large variations from week to week.

Öresundsverket – a Presentation

Introduction

By the end of the 1980s an intense debate in Sweden on the state of the seas surrounding our country had led to demands for the introduction of reductions in the nitrogen discharged from public sewage treatment plants near the coast. Öresundsverket, the municipality of Helsingborg's sewage treatment works, is the first to have been designed and modified to meet extensive nitrogen reduction requirements. Modifications to Öresundsverket were carried out from 1990 to 1991.

During the statutory inquiry under the Environmental Protection Act (Miljöskyddslagen) that preceded modification of the plant, the Franchise Board (Koncessionsnämnden) laid down targets for which the plant was to be designed as

R. Klute / H.H. Hahn (Eds.)
Chemical Water and Wastewater Treatment III

well as criteria applicable during a trial period stretching from 1 June 1992 to 1 April 1995. The following annual mean values for discharged wastewater were set:

Table 1. Discharge limits

Parameter	Target value mg/l	Trial period criteria mg/l
Total phosphorus	0.3	0.3
Total nitrogen	8	12
BOD_7	10	10

Discharged water should be 80% oxygenated

Specification of the Plant

Öresundsverket is designed to have the following capacity:

Table 2. Design data

Mean inflow	m^3/d	67000
Q_{dim}	m^3/h	3250
BOD_7	kg/d	15000
P_{tot}	kg/d	490
N_{tot}	kg/d	2700

Öresundsverket is an activated sludge plant. Its configuration can be seen in the flow diagram in Figure 1.

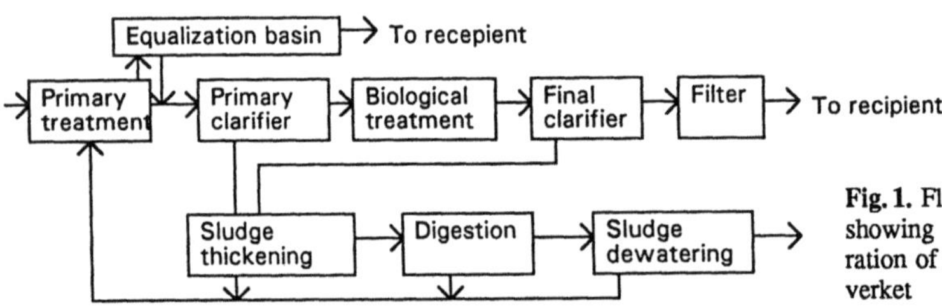

Fig. 1. Flow diagram showing the configuration of the Öresundsverket

What makes Öresundsverket special is its division into four trains running from primary treatment to the filters. Each train is parallel with a separate sludge flow, making it possible to test simultaneously several different process configurations in the plant.

Primary Treatment. Incoming water passes two 10 mm screens. The water then passes two parallel, aerated grit chambers. After the grit chambers the wastewater is divided into four parallel trains. Flow metering is carried out separately for each of these trains. The maximum inflow to biological treatment is controlled by the possibility to divert water to the equalization basins. It is possible to dose water diverted to the equalization basins with chemical coagulants.

Each train has two primary clarifiers with plastic chain scrapers. It is possible to dose the primary clarifiers inflow channel with chemical coagulants so using this for pre-precipitation.

Biological Treatment. The activated sludge plant is divided into six zones. The way in which these zones are operated can be varied. Figure 2 shows the different zone functions and sizes in one of the four trains:

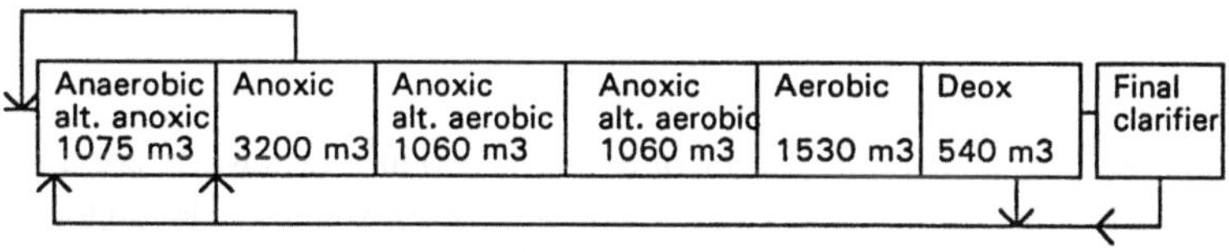

Fig. 2. Flow diagram – biological treatment

Each train has four final clarifiers with a combined area of 1320 m^2 and a volume of 5095 m^3. This configuration provides great flexibility. However the plant must be operated using pre-denitrification. Post-denitrification can only be employed if the filters are used.

The aeration unit consists of three HV-Turbo compressors with a capacity of 6600–14700 m^3/h. Air is transported to the aeration tanks via a system of galvanised or stainless steel pipes to 2336 rubber membrane aerators. Each train has four zones which can be aerated, each of which has an oxygen meter which controls the flow of oxygen to the zone. Oxygen content should be at least 2 mg/l.

Those zones which can operate anoxically have mixers. Recirculation from the anoxic to the biological plant inlet is 810 m^3/h, corresponding to 150 % of average dry weather flow.Recirculation from the deox zone to the anaerobic or anoxic zone can be regulated in three stages, 810 m^3/h, 1620 m^3/h or 2430 m^3/h, corresponding to 150, 300 and 450 % of average dry weather flow respectively. Throughout this period sludge return from final clarifiers was 700 m^3/h, corresponding to 130 % of average dry weather flow.

The biological treatment during this period was driven with two different configurations:

- a pre-denitification process with the RAS and recirculation from deox led to the first (anoxic) zone.
- the UCT-process where the RAS and recirculation from deox are led to the second (anoxic) zone, and with a recirculation from the second (anoxic) zone to the first (anaerobic) zone.

In order to permit simultaneous precipitation it is possible to dose chemical coagulants (primarily iron sulphate) into the inlet to the biological stage.

Filter. The final stage consists of twelve 50 m^2 downstream dual-media filters. The filter material is 0.7 m expanded clay on top of 0.3 m filter sand. The filters can be flushed with air or water. To allow direct precipitation it is possible to dose chemical coagulants into the filter inlet channel.

The filters are flushed somewhat differently to conventional downstream filters. A volume above the filter beds is filled with water and only when this has passed through the filter, and the filter material has been allowed to settle, is an outlet to the flush water run-off opened. Flush water does not therefore flow through continuously.

So as to fulfil the 80% oxygenation requirement, water is aerated in the final stage before the outlet tube.

Wastewater Inflow

Öresundsverket receives an average inflow of about 50 000 m^3 wastewater per day, of which about a fifth is industrial in origin. Industrial wastewater in Helsingborg is primarily from the chemical, food and printing industries and from a landfill plant.

The temperature of wastewater varies from 8 °C in winter to 19 °C insummer. The pH is about 7.25, although very high and very low values do occur.

The composition of the wastewater is shown in Table 1; the figures are the averages of diurnal flow-proportional samples in the period 1992–1993. Overflow from the sludge dewatering operations is not included. Wastewater arriving at Öresundsverket can be viewed as normal Swedish wastewater apart from its phosphorus content which is slightly lower than average. It has also been found that wastewater arriving at the plant sometimes has a strong inhibiting effect on nitrification.

Table 3. The characteristics of wastewater arriving at Öresundsverket

Parameter	Unit	Average	σ_n
Inflow	m^3/h	2220	±503
N_{tot}	mg/l	34	±7
NH_4-N	mg/l	22	±5
P_{tot}	mg/l	5.6	±1.7
COD	mg/l	420	±117
COD_{filt}	mg/l	160	±53
Alkalinity	mg HCO_3^-/l	250	±32
Conductivity	$\mu S/cm$	1600	±710
Suspendedsolids	mg/l	196	±50
COD/N_{tot}	–	12.6	±2.8

Investigated Processes

The following combinations of different processes and pre-treatments have undergone full-scale trials over a period of almost two years:

- Pre-precipitation with normal primary clarification surface area, with and without the addition of methanol, using UCT-process configuration.
- Simultaneous precipitation with half the normal primary clarification surface area.
- Biological phosphorus removal with primary sludge hydrolysis and normal primary clarification surface area,using UCT-process configuration.
- Biological phosphorus removal without primary sludge hydrolysis and normal primary clarification surface area, using UCT-process configuration.

Pre-denitrification was carried out in all cases.

Pre-precipitation

Pre-precipitation was tested in one train for six months. The inlet channels of the primary clarifier basins, which did not have flocculation chambers, were dosed with a quantity of PIX-C quality iron chloride from Kemira Kemwater equivalent to 10–12 g Fe/m^3. The COD_{tot} and P_{tot} contents in the effluent from the basins were 70 and 35 %, respectively, compared with a control train which was operated without the addition of iron but with a doubled surface area load. Filtered and total COD were separated by the same amount so that the COD_{filt}/COD_{tot}-ratio remained unchanged. No absolute separation results can be determined because the effect of the grit chamber is unknown. The average values recorded during the trial period are contained in Table 4.

Table 4. Average analyses of diurnal samples during pre-precipitation period

Sample point	P_{tot} mg/l	COD_{tot} mg/l	COD_{filt} mg/l	SS mg/l	Surface overflow rate m/h
Pre-precipitation effluent	1.4	150	100	38	0.8
Control effluent	4.1	210	140	61	1.6
Grit chamber influent	5.8	430	170	214	

In the three last months of the trial period the inlet channel tothe first anoxic zone of the biological stage was dosed with a constant 25 g/m^3 of methanol.

The carbon to nitrogen ratio COD_{tot}/N_{tot} in the effluent from the pre-precipitation basins is shown in Figure 3. The ratio is calculated as a 30 day moving average. The figure also shows the effect of the methanol dosage on the ratio. The proportion of methanol in the total COD was 0.14–0.24.

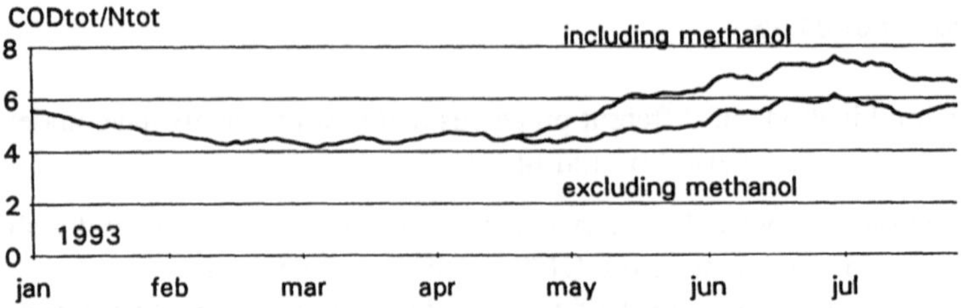

Fig. 3. Carbon to nitrogen ratio during pre-precipitation period

Simultaneous Precipitation

Simultaneous precipitation was tested in one train for fifteen months. For the first two months, PIX-C quality iron chloride from Kemira Kemwater was added. After this, a saturated solution of divalent iron sulphate of two different qualities from Kemira Kemwater and Kronos Titan A/S were used.

The iron sulphate was supplied in solid form as $FeSO_4 \cdot 7\,H_2O$ containing about 17.5 % Fe, and was the by-product of titanium dioxide manufacture. The inlet channel to the first anoxic zone of the biological stage was dosed with an amount equivalent to 12–21 g Fe/m^3, an average of 15 g Fe/m^3.

Throughout the trial period, the train was operated with only one primary clarifier basin – double the normal surface area load – so as to raise the COD_{tot} to the biological stage. The carbon to nitrogen ratio COD_{tot}/N_{tot} in the effluent from the primary clarifier basin is shown in Figure 4. The ratio is calculated as a 30 day moving average.

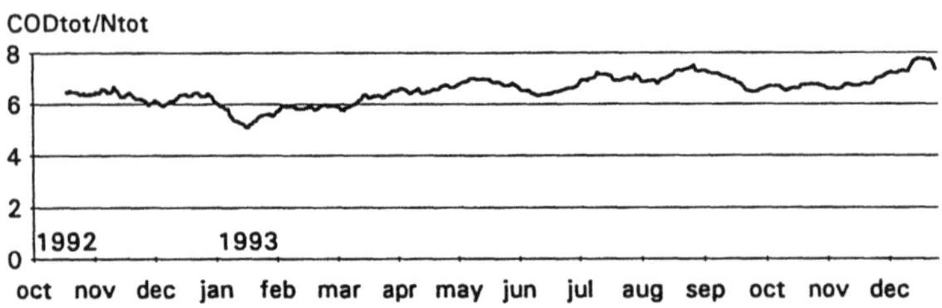

Fig. 4. Carbon to nitrogen ratio during simultaneous precipitation period

Biological Phosphorus Removal with Primary Sludge Hydrolysis

Biological phosphorus removal using the UCT-process in combination with primary sludge hydrolysis was tested in one train for just over a year. Hydrolysis was achieved by increasing the residence time of the primary sludge in the two primary clarifier basins to several days. Soluble products of hydrolysis were washed out of the sludge by pumping it around the basins. Hydrolysis meant that N_{tot}, P_{tot} and above all COD_{tot} in the effluent from the primary clarifier basins increased, partly because of the soluble products of hydrolysis but mainly because of the increased

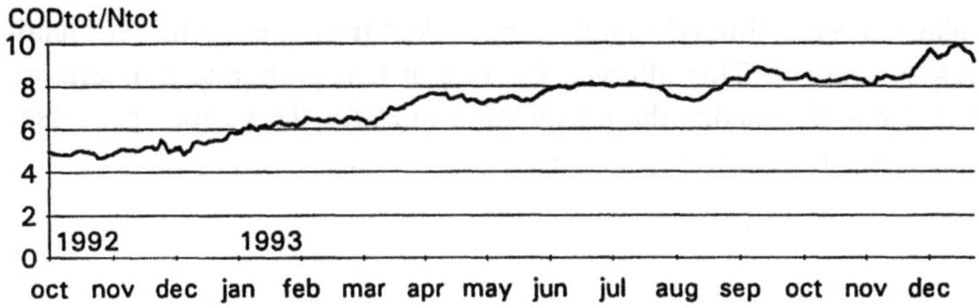

Fig. 5. Carbon to nitrogen ratio during period of biological phosphorus removal combined with primary sludge hydrolysis

amount of suspended solids due to inadequate control of the washing-out procedure. The carbon to nitrogen ratio COD_{tot}/N_{tot} in the effluent from the primary clarifier basins is shown in Figure 5. The ratio is calculated as a 30 day moving average.

Biological Phosphorus Removal Without Primary Sludge Hydrolysis

Biological phosphorus removal using the UCT-process was tested in one train for nine months. The carbon to nitrogen ratio CODtot/Ntot in the effluent from the primary clarifier basins is shown in Figure 6. The ratio is calculated as a 30 day moving average.

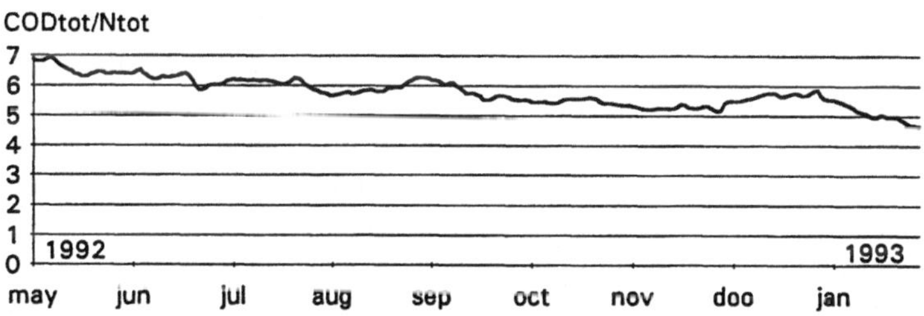

Fig. 6. Carbon to nitrogen ratio during period of biological phosphorus removal without primary sludge hydrolysis

Nitrification Rate

The nitrification rate of the activated sludge was determined a number of times during the investigation periods. This was carried out in the laboratory using the following method [1]: Mixing activated sludge from an aerated zone in the biological stage with a synthetic nutrient solution containing $(NH_4)_2SO_4$ and $NaHCO_3$. The sludge and nutrient solution was tempered to the current temperature in the biological stage. The SS and VSS contents of the mixture were analysed. The temperature of the mixture was maintained in a water bath and the mixture was aerated. At regular intervals, filtered samples were taken from the mixture for analysis of their NO_3-N content. This allowed the rate at that particular temperature to be determined. Table 5 contains the nitrification rates for the various processes adjusted to 15 °C using the following equation [2]:

$$r\ (15\,^\circ\mathrm{C}) = r\ (T\,^\circ\mathrm{C}) \cdot e^{0.12(15-T)}\,.$$

Table 5. Nitrification rates at 15 °C for investigated processes

Process	Organic part of sludge VSS/SS	Carbon to nitrogen ratio COD_{tot}/N_{tot}	Nitrification rate (15 °C) mg N/g VSS · h	Nitrification rate (15 °C) mg N/g SS · h
Pre-precipitation + methanol	0.79–0.80	6–7.5	2.8–3.1	2.3–2.4
Pre-precipitation	0.73–0.75	4.5	3.0–3.3	2.0–2.4
Simultaneous precipitation	0.61–0.72	5.5–7.5	2.0–2.9	1.3–2.0
Biological phosphorus + hydrolysis	0.72–0.78	6–9.5	1.0–2.8	0.7–2.1
Biological phosphorus	0.81–0.83	5.5–7	1.8–2.5	1.5–2.0

It can be seen that because of the lower carbon to nitrogen ratio pre-precipitation has had a positive effect on nitrification rates. Pre-precipitation was also beneficial in the one case that nitrification was totally inhibited in the plant. This pre-precipitation train recovered a great deal faster than the other trains.

The rates of the three other process configurations were fairly similar. The rates found in the process of biological phosphorus removal with primary sludge hydrolysis varied greatly. The low rates were due to inadequate control of the primary sludge hydrolysis which caused very high carbon to nitrogen ratios and incomplete nitrification, and cannot be seen as representative.

The addition of methanol in the pre-precipitation process increases the carbon to nitrogen ratio significantly while the nitrification rate decreases negligibly. This indicates that methanol gives a low sludge yield.

Figure 7 illustrates the relationship between the nitrification rate and the carbon to nitrogen ratio in the influent to the biological stage. All the tested processes are included in theFigure. For the pre-precipitation process the ratio does not include the methanol dosage.

The scattering of the points in the figure can be due to several reasons, including:

- sporadic inhibiting effects. These inhibiting effects have been verified in the influent to the plant.
- sporadic incomplete nitrification because of a low sludge age.
- varying carbon quality in the influent to the biological stage because of a high degree of industrial wastewater load to the plant and different pre-treatments.

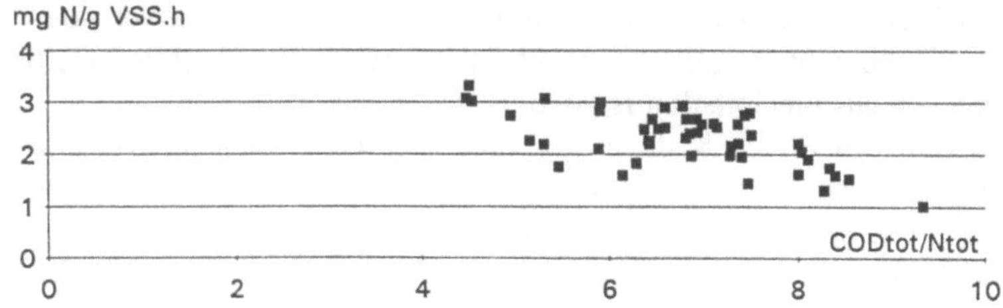

Fig. 7. Nitrification rate at 15 °C for all investigated processes as function of C/N ratio of influent to biological stage

Denitrification Rate

The denitrification rate of the activated sludge was determined a number of times during the investigation periods. This was carried out in the laboratory using the following method [1]: Mixing activated sludge from a non-aerated zone in the biological stage with a synthetic nutrient solution containing KNO_3 and KH_2PO_4 and a specified external carbon source, in our case, sodium acetate or methanol. The sludge and nutrient solution were tempered to the current temperature in the biological stage. Prior to the start of the test we ensured that the oxygen content of the sludge was close to zero. The SS and VSS contents of the mixture were analysed. The temperature of the mixture was maintained in a water bath. At regular intervals, filtered samples were taken from the mixture for analysis of their NO_3-N content. This allowed the rate at that particular temperature to be determined.

Table 6 contains the denitrification rates for the various processes adjusted to 15 °C using the following equation [3]:

$$r\ (15\,°C) = r\ (T\,°C) \cdot e^{0.08(15-T)}\,.$$

Table 6. Denitrification rates at 15 °C for investigated processes

Process	Organic part of sludge	Denitrification rate (15 °C)			
		Acetate	Acetate	Methanol	Methanol
	VSS/SS	mg N/g VSS · h	mg N/g SS · h	mg N/g VSS · h	mg N/g SS · h
Pre-precipitation + methanol	0.79–0.80	2.7–3.0	2.1–2.5	0.7–1.1	0.5–0.9
Pre-precipitation	0.73–0.75	2.9–3.7	2.1–2.6		
Simultaneous precipitation	0.61–0.72	1.7–4.5	1.7–3.1		
Biological phosphorus + hydrolysis	0.72–0.78	2.5–3.9	1.9–3.0	0–0.4	

The denitrification rates measured when acetate was the only carbon source are on the same order of magnitude for all the processes tested. However, the range for simultaneous precipitation is large.

The time series in Figure 8 shows the rates measured when methanol was the only carbon source for activated sludge from the pre-precipitation process configuration. It can be noted that an adaptation time of at least 1.5–2 months is required. The decrease in the rate towards the end of the trial period can at least in part be attributed to a lower proportion of methanol in the COD_{tot} of the influent to the biological stage.Too low an anoxic residence time caused by a very high recirculation may alsohave decreased the rate.

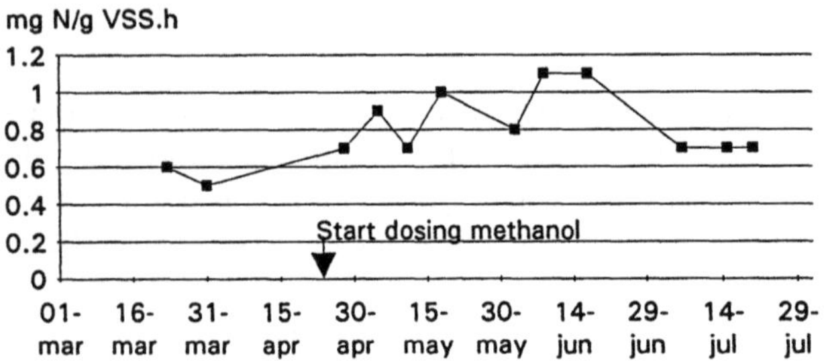

Fig. 8. Denitrification rate at 15 °C for methanol illustrating adaptation period

Sludge Properties

To characterize the activated sludge's sedimentation properties, the Diluted Sludge Volume Index (DSVI) is measured. A test sample is taken from the deox zone on each train twice a week. In the period 1 June 1992 to 15 January 1994 the DSVI varied between 100 and 480.

It should be noted that it may sometimes be hard to find any direct correlation between the DSVI and the sedimentation properties of the sludge in the final clarifiers. Sludges with a high DSVI have in some cases settled better than sludges with lower DSVIs.

The variations in DSVI on the four trains are presented below. The aerobic sludge age calculated as 30 days moving average.

Train 1. Throughout the period, train 1 has been operated with simultaneous precipitation and without an anaerobic zone. The changes made during the period are recorded below (Fig. 9).

When the recirculation was shut down on 8 March 1993, the DSVI decreased only to rise again when recirculation was restarted.

Train 2. Just as train 1, train 2 was operated with simultaneous precipitation. The changes made during the period are recorded below (Fig. 10).

Shutting down the recirculation on 8 March 1993 seems to have had a beneficial effect on the DSVI. When the recirculation is restarted on 15 June 1993 the DSVI rises, to fall again when the recirculation is shut down on 13 Aug. 1993. The reaction is very rapid both on start up and shut down.

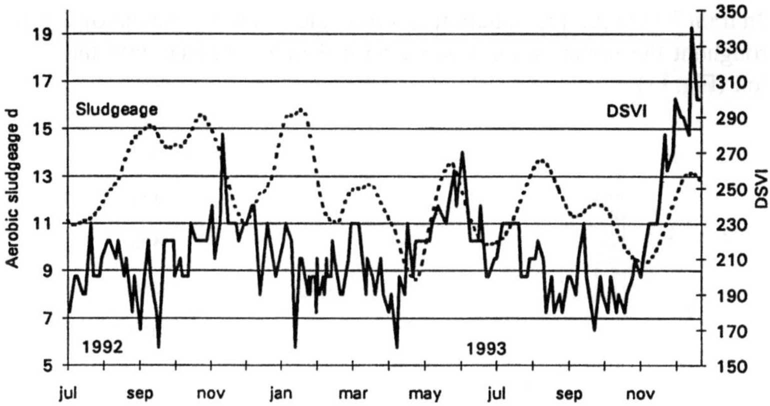

Fig. 9. DSVI and sludge age on train 1.

09.10.1992 Recirculation from deox to anox changed from 150 to 300 %.
14.12.1992 Change of chemical coagulant from PIX-C to iron sulphate.
08.03.1993 Recirculation from deox to anox shut down because nitrification was totally inhibited.
23.03.1993 Recirculation from deox to anox started with a flow of 150 %.
05.04.1993 Recirculation from deox to anox changed from 150 to 300 %.
12.11.1993 Recirculation from deox to anox changed from 300 to 150 %.
21.12.1993 Recirculation from deox to anox shut down.

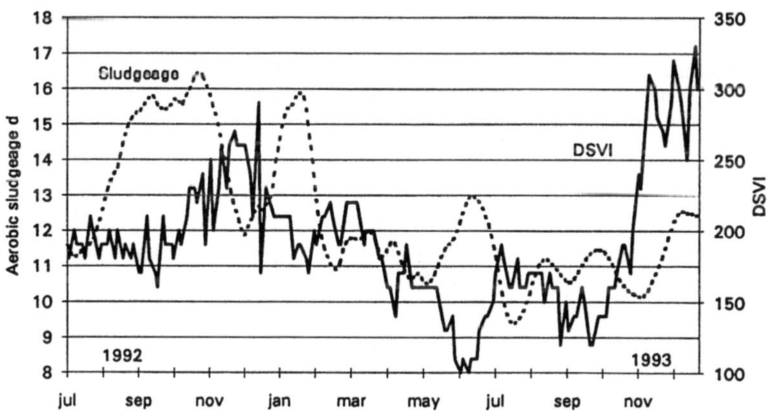

Fig. 10. DSVI and sludge age on train 2.

09.10.1992 Recirculation from deox to anox changed from 150 to 300 %.
13.10.1992 One of the pre-precipitation tanks shut down to increase the load on the activated sludge plant.
14.12.1993 Change of chemical coagulant from PIX-C to iron sulphate.
08.03.1993 Recirculation from deox to anox shut down because nitrification was totally inhibited.
15.06.1993 Recirculation from deox to anox started with a flow of 150 %.
13.08.1993 Recirculation from deox to anox shut down.
12.11.1993 Recirculation from deox to anox started with a flow of 150 %.
21.12.1993 Recirculation from deox to anox shut down.

Train 3. On train 3 both pre-precipitation and biological phosphorus removal were tested. Throughout the period, train 3 was operated in accordance with the UCT configuration (Fig. 11).

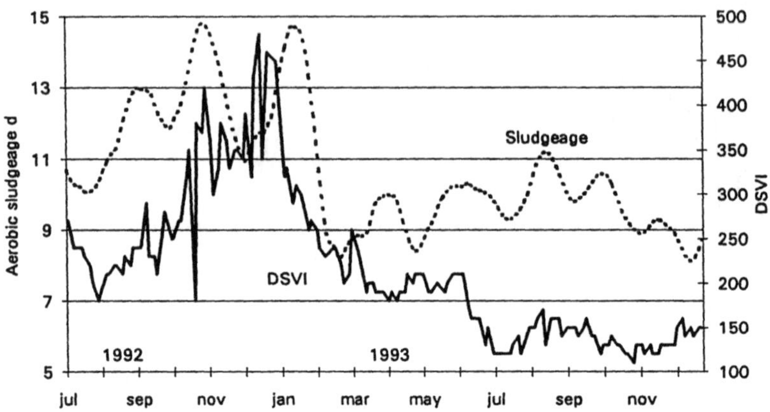

Fig. 11. DSVI and sludge age on train 3.
14.06.1992 Dosing of anaerobic zone with 25 mg/l acetic acid.
24.07.1992 Dosing with acetic acid stopped. Anaerobic zone dosed with PIX-C.
08.09.1992 Dosing of PIX-C moved to pre-precipitation.
19.12.1992 Dosing of PIX-C stopped. Trial of primary sludge hydrolysis started in one primary clarifier.
08.03.1993 Recirculation from deox to anox shut down becausenitrification was totally inhibited.
23.03.1993 Recirculation from deox to anox started with a flow of 150%.
24.03.1993 Trial of primary sludge hydrolysis started in both primary clarifiers.
21.09.1993 Dosing of anaerobic zone with 15 mg/l acetic acid.
06.12.1993 Dosing of anaerobic zone with acetic acid stopped.

A clear drop in the DSVI can be seen at the beginning of theperiod when acetic acid was being added. Soon after the addition of acetic acid was stopped on 24 July 1992 the DSVI starts to rise. This rise does not stop until pre-precipitation ceases on 19 December 1992, when it starts to drop dramatically. This could be due to an increased amount of volatile fatty acids from the hydrolysis in the primary clarifiers.

No effect of the shut down on 8 March 1993 can be noticed on this train.

Train 4. Just as in train 3, both pre-precipitation and biological phosphorus removal were tested in train 4. Throughout the period, train 4 was operated in accordance with the UCT configuration (Fig. 12).

During the period 21 January 1993 to 12 July 1993, when pre-precipitation was carried out on train 4, this train had the highest DSVI except for a period at the end when methanol was added. In this period there are large variations from weekto week in the DSVI. When pre-precipitation ceases there is a further decrease in DSVI. It is hard to trace any effect of the increases in recirculation on 6 May 1993 and 1 July 1993.

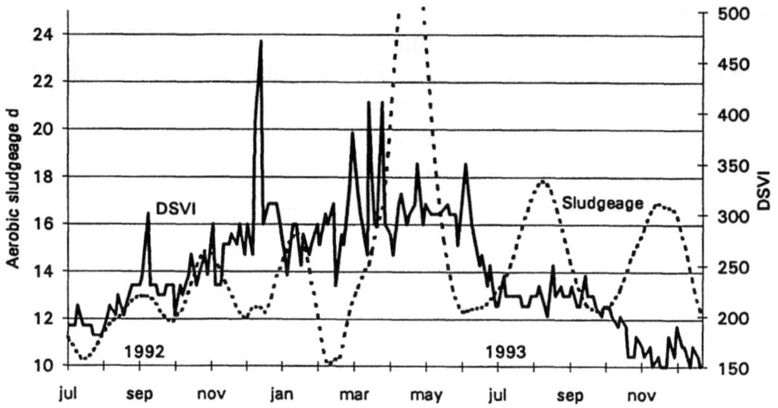

Fig. 12. DSVI and sludge age on train 4.
21.11.1992 Simultaneous precipitation with PIX-C shut down.
21.01.1993 Pre-precipitation with PIX-C started.
08.03.1993 Recirculation from deox to anox shut down because nitrification was totally inhibited.
23.03.1993 Recirculation from deox to anox started with a flow of 150 %.
22.04.1993 Dosing of the anaerobic zone with methanol (25 mg/l) started.
06.05.1993 Recirculation from deox to anox increased from 150 to 300 %.
19.06.1993 One of the primary clarifiers shut down to increase the load on the activated sludge plant.
01.07.1993 Recirculation from deox to anox increased from 300 to 450 %.
12.07.1993 Dosing of methanol and PIX-C shut down. Recirculationfrom deox to anox decreased to 150 %.
01.10.1993 Trial of primary sludge hydrolysis started in one primary clarifier.
20.10.1993 Start of seeding of WAS from train 3 to train 4.
28.12.1993 Shut down of seeding of WAS from train 3 to train 4.

The dosing of methanol seems to have a positive effect on DSVIs, an effect that also can be noticed when hydrolysis in the primary clarifiers and seeding with WAS from train 3 is started.

Conclusions

- Pre-precipitation with 10–12 g Fe/m^3 iron chloride to primary clarifiers resulted in COD_{tot}, P_{tot} and SS contents of 70, 35, and 60 %, respectively, compared to a control train that was operated without the addition of iron.The COD_{filt}/COD_{tot}-ratio remained unchanged from the pre-precipitation.
- Pre-precipitation had a positive effect on the nitrification rate, because of a lower carbon to nitrogen ratio. The other processes tested gave equivalent nitrification rates.
- Denitrification rates determined with acetate as single carbon source gave equivalent rates to all tested processes.
- Denitrification rates determined with methanol as single carbon source was a function of the proportion of methanol in the COD_{tot} in influent water to the biological treatment.
- Denitrification with methanol had an adaptation time of at least 1.5–2 months.

- Methanol gave a low sludge yield.
- The recirculation in the biological treatment is an important factor that influences DSVI. An increased recirculation gave a higher DSVI, and when the recirculation was shut off, the DSVI decreased.
- A higher amount of easily biodegradable carbon, for instance, acetate, methanol or volatile fatty acids, had a positive effect on DSVI.
- Pre-precipitation gave high DSVIs with large variations from weekto week.

Acknowledgement. The results reported in this paper are from a project financed by Statens Naturverdsverk (the Swedish EPA) and the city of Helsingborg.

References

[1] Naturvårdsverket rapport 3861. Kvävereduktion vid kommunala avloppsreningsverk. Karakteriseringsmetoder för vatten och slam. Statens Naturvårdsverk 1990

[2] Henze, M., Harremoës, P., la Cour Jansen, J., Arvin, E.: Spildevandsrensning Biologisk och kemisk. Polyteknisk Forlag, Lyngby 1992

[3] Ødegaard, H.: Fjerning av næringsstoffer ved RENSING AV AVLØPSVANN. Tapir Forlag Universitetet i Trondheim 1992

Lars-Erik Jönsson and Peter Magnusson
Helsingborg Water & Sewage Works
Gåsebäcksv 4
S-252 27 Helsingborg
Sweden

Karin Jönsson
Department of Environmental Engineering
Lund Institute of Technology
S-221 00 Lund
Sweden

Upgrading Point Loma Treatment Plant in San Diego with Improved Advanced Primary Treatment

B. Hansson and A. Langworthy

Abstract

The Point Loma Waste Water Treatment plant in San Diego operates as an advanced primary treatment plant at a mean capacity of 190 MGD (720 000 m^3/day). The discharge from the plant is through a four mile long outfall tunnel into the Pacific Ocean. Many scientists have documented that the BOD discharge does no harm to the marine environment, which is why most efforts have been focused on increasing the solids removal.

Through an intensive testing program, ordered by a federal court, the chemical treatment capacity of the Point Loma plant was investigated. The solids removal and floc formation at the plant can be increased by combining the existing ferric chloride dose with the use of poly-aluminium chloride (PAC). This will remove more colloidal material through better coagulation but will also increase the sludge blanket levels. Increased sludge removal capacity must be installed to maintain a consistent good treatment result.

If the inorganic salts are combined with a cationic polymer, the dosage of ferric chloride and PAC can be reduced by approximately 25 %. If an anionic polymer is dosed prior to the sedimentation tanks the sludge volume will also decrease.

The treatment system with ferric chloride, PAC and cationic polymer at the influent of the plant, and an anionic polymer prior to the sedimentation basins, will improve the solids removal, if the produce sludge is removed reality. This system may lead to removal rates acceptable to the new federal legislation for deep ocean discharge in the United States.

1. Background

The existing San Diego Metropolitan Sewerage System was designed and built in the early 1960s to solve the water quality problem within the San Diego Bay. Ten separate communities were connected and the wastewater was led to the Point Loma peninsula where a treatment plant was built in 1963. The discharge was into the Pacific Ocean through a 2.5 mile long tunnel.

Until 1985 this plant operated as a primary treatment facility that nominally removed 40–60 % of the incoming suspended solids. The plant was then converted

R. Klute / H.H. Hahn (Eds.)
Chemical Water and Wastewater Treatment III

into a chemically-assisted primary plant to produce a higher suspended solids removal rate. The use of ferric chloride, followed by a proprietary organic polymer resulted in a denser, more rapidly settling floc. This increased the removal rate to 70–80 % of the total suspended solids. No major capital improvements or physical alterations were done.

It was established early on that the iron salt, ferric chloride, is the most effective for San Diego's influent wastewater. Additionally, use of ferric material has resulted in significant lowering of hydrogen sulfide in the wastewater plant and digester gas. The digester gas does not require scrubbing prior to combustion in internal combustion engines that drive generators providing electrical power to the facility. Ferric chloride is dosed both at a pumping station at the raising main prior to the plant and at the plant before the aerated grit chambers.

Influent flow to the Point Loma plant is dependent on the number of pumps in operation at the last pumping station. These pumps are operated as discreet units in either an off or on mode. The flow varies by discreet changes and does not reflect the diurnal cycle one normally associates with wastewater flows. Flow pacing of chemicals is adjusted to the number of pumps in operation at the pumping station.

This was the status when the Point Loma plant entered a period of intensive testing of chemicals to upgrade the plant to meet more stringent effluent controls.

2. Present Situation

The present situation at Point Loma is an average treatment of 190 million gallons per day (MGD). The plant is designed for a much smaller flow. Normal usage of chemicals is 40 ppm dry base ferric chloride (corresponds to 100 ppm of 40 % liquid ferric chloride) and 0.2 ppm of anionic polymer. The sludge is being digested anaerobically and dried on sludge drying beds on Fiesta Island, a heavily-used recreational area.

Since the flow has increased without increasing the size of the plant, the operation is carried out at a surface overflow rate much higher than previously thought. Normal surface load is 2 m/h but during peak periods with 240 MGD, the surface load increases to 3 m/h. The present treatment at Point Loma seems to handle these high loads well. The use of chemical addition allows the sedimentation tanks to operate well beyond their original design parameters.

The retention time at the plants is presented in Figure 1.

The total retention time is only about 2 hours at the plant. This makes it possible to treat large amounts of wastewater in a very limited area. Additional space for new construction is scarce at the Point Loma plant.

The sedimentation tanks are all designed to thicken the sludge within the basin. No sludge thickening equipment is installed at the plant. Average sludge density is approximately 3 % dry solids being pumped into the digesters. The retention time in the digesters is 15 days, which is why the sludge removing capacity is limited at Point Loma. An improved result may be obtained by achieving a higher sludge density. The digesters operate at their maximum hydraulic capacity.

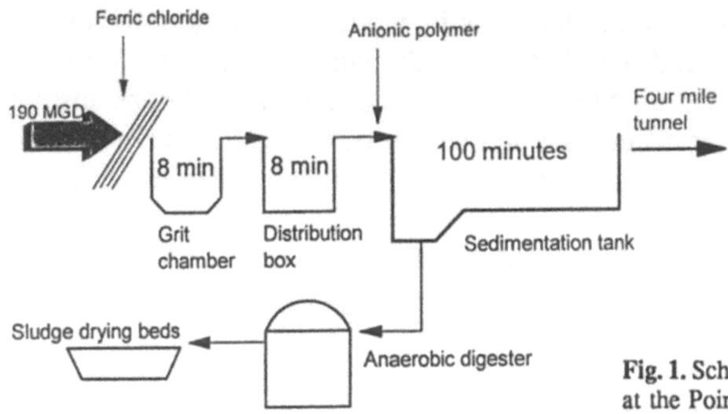

Fig. 1. Schematic layout and retention times at the Point Loma plant

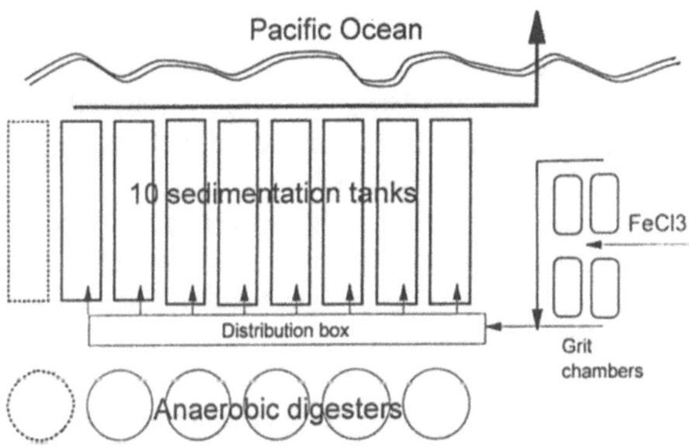

Fig. 2. Present configuration of the Point Loma plant

The present construction activities at the Point Loma plant will involve the addition of two more sedimentation tanks and the connection of an extended outfall tunnel. The addition of two more sedimentation tanks will increase the basins by 25 % and with a new influent distribution system, balance the flow better. Unfortunately, the construction is interfering with the present operation of the plant. Figure 2 is a schematic diagram of the Point Loma plant.

During dry weather periods approximately 15 MGD from Tijuana in Mexico is also treated at the Point Loma plant. The wastewater from Mexico is being considered for treatment in a separate waste treatment plant closer to the border. The process that will probably be used at this treatment plant is advanced primary treatment. This plant will improve the situation at Point Loma.

3. Tunnel Extension

The lack of secondary treatment at the Point Loma plant, and no disinfection of the effluent has caused sanitary problems during certain periods. The old outfall tunnel, 2.5 miles long, ended near the Point Loma kelp beds, a recreational area for divers and fishermen.

Since 1993, the extended outfall tunnel is in service, discharging the effluent four miles from the shore and 320 feet below the surface. Intensive testing by the City of San Diego's Ocean Monitoring Program and The Scripps Institute of Oceanography have selected the site and design carefully. The environmental impact on the marine life in the kelp beds will be negligible when the extension is in use.

The cost of this tunnel extension exceeds $45 million.

The area around the collecting system of San Diego is shown in Figure 3.

4. Clean Water Act Regulation

The Clean Water Act was passed as a U.S. federal law in 1972, and states that all waste treatment plants in the United States must meet an effluent standard of maximum 30 mg/l of total suspended solids and BOD. This rule is referred to as the 30/30 rule. San Diego, and many other US cities with ocean outfalls, applied for a waiver and this was approved by the EPA in 1981.

In 1987 the City of San Diego voted to withdraw the waiver application and to come into compliance with the Clean Water Act. The city also committed itself to an extensive water reclamation and reuse program and the San Diego Clean Water Program was established to carry out these goals.

Following San Diego's withdrawal of its waiver application, the EPA filed a law suit against the city for violation of the Clean Water Act. Between 1987 and 1990, San Diego developed a plan to come into compliance with the law. This program included upgrading the collection system to prevent spills, complying with bacterial standards in the kelp beds, upgrading the Point Loma plant to secondary treatment and constructing six water reclamation plants. In 1990, the cost of this plan was estimated to be approximately $2.4 billion or a rate payment of $360 per household per year.

In January 1990, the EPA and San Diego entered into a consent decree that would put San Diego's plan on a legally enforceable schedule for compliance. At the same time many scientists suggested that upgrading the Point Loma plant to secondary treatment would not benefit the environment. Oceanographers from Scripps Institute at the University of California San Diego testified that the BOD discharge from Point Loma would do absolutely no harm to the ocean environment. The focus should be put on solids removal.

The federal judge whose approval of the consent decree was also required was troubled by the many scientists who indicated that Point Loma's secondary treatment would not improve the environment. The cost for this upgrade was estimated at $1 billion and the favorable impact on the ecology in the kelp beds was questioned.

The judge deferred his decision on the consent decree until mid-1993. In the interim he instructed San Diego to conduct tests at the Point Loma plant to determine if it could improve the solids removal efficiency of the chemically enhanced primary treatment process. This testing program, which did not include

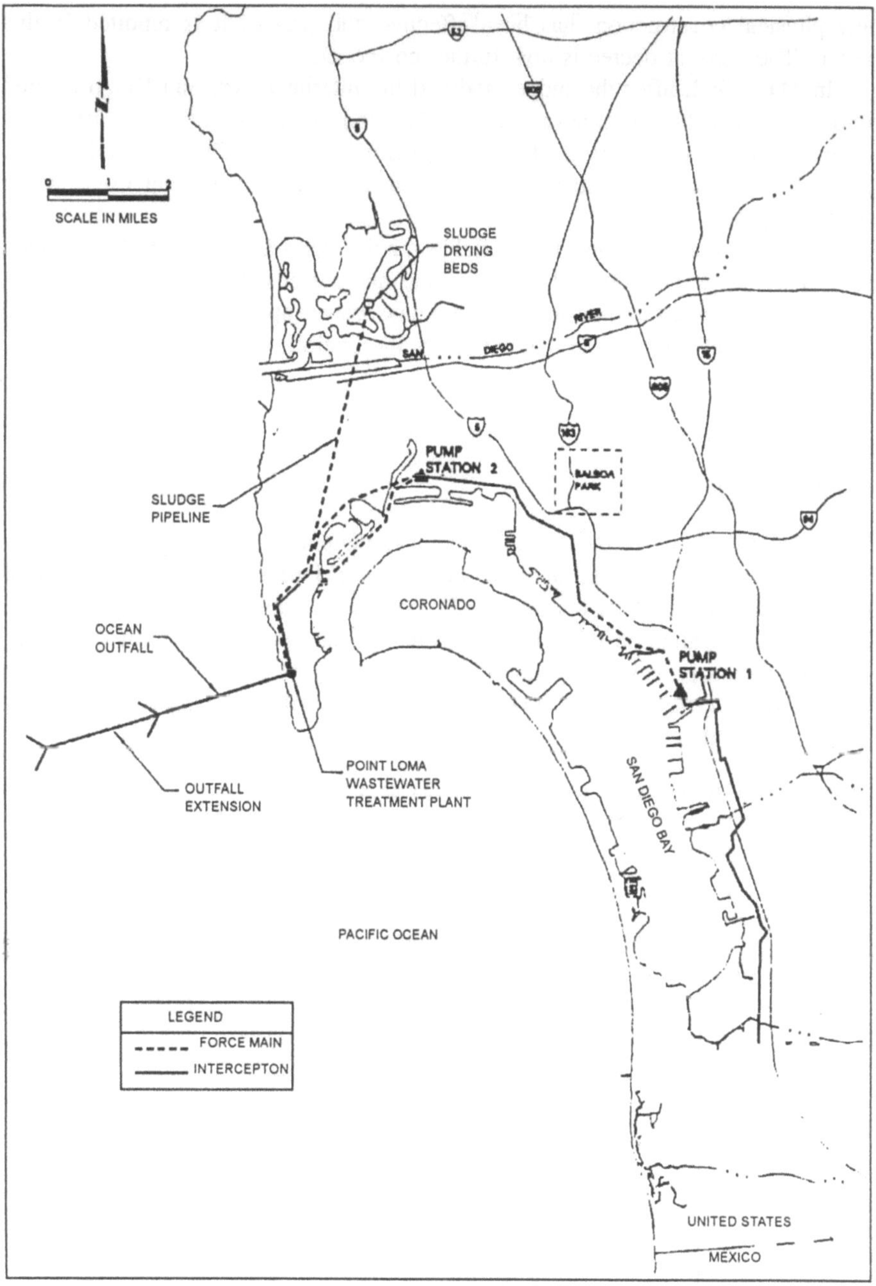

Fig. 3. San Diego's sewerage system and tunnel outfall

any physical construction, has been effective and part of it is reported in this report. The consent decree is now further postponed.

In May 1993, after the judge modified his interim order, San Diego entered into a new scaled down program termed "the consumer alternative". A 19 month extension on all secondary treatment projects was granted. This $1.3 billion program does not include a secondary treatment upgrade of the Point Loma waste treatment plant.

Presently, many options are available for the Point Loma plant. A continued effort may solve the technical problems and make it possible to comply with the Clean Water Act. Other possibilities are a new waiver application or trying to find legislative relief through the upcoming reauthorization of the Clean Water Act.

5. Test Program

5.1 Jar Test Evaluation

The standard chemical program at the Point Loma plant is a dosing of 40 ppm dry base ferric chloride and 0.2 ppm anionic polymer. Different jar test studies were done to evaluate other possible chemical products. The selection of polymer is done by full scale evaluation of different products from vendors. To further improve the coagulation process, poly-aluminium chloride (PAC) was looked upon as an alternative to ferric chloride. Several jar tests were carried out with the following significant results.

Table 1. Results from initial jar tests with ferric chloride and PAC

Product	Dose mole/m^3	Dose ml/m^3	Turbidity NTU	Removed turbidity	Cost $/m^3	Cost/ rem. turb.
Blank	0	0	145	0	0	0
Ferric chloride[1]	0.1	32	62	83	0.0036	8.59
	0.2	64	26	119	0.0059	13.11
	0.3	96	23	122	0.0107	22.46
	0.4	128	17	128	0.0143	28.92
	0.5	160	14	131	0.0178	36.37
PAC[2]	0.1	24	38	107	0.0060	56.07
	0.2	48	9.4	135	0.0120	88.52
	0.3	72	6.4	139	0.0180	129.87
	0.4	96	5.2	140	0.0240	171.67
	0.5	120	4.2	141	0.030	212.92

[1] Ferric chloride with 12 % Fe and density 1.4
[2] PAC, poly-aluminium chloride with 9 % Al and density 1.25

The effect of the added ferric chloride was immediate. Already at a dose of 0.2 mole/m^3 the turbidity was reduced by 60 %. The result was further improved at higher dosages but a very good result could not be obtained. Turbidity values lower than 14 NTU could not be obtained with ferric chloride, even at dosages higher than 0.3 mole/m^3.

The effect of the PAC addition was also immediate and a very clear water could be produced. At higher dosages of PAC, 0.3 mole/m^3 and above, turbidity levels were usually around 5 NTU. The good results from the PAC addition could not be obtained with the ferric chloride application.

PAC was more expensive to use than ferric chloride. The removal of turbidity at the higher PAC additions was significantly more expensive than the removal with ferric chloride. Since ferric chloride was very effective at low dosages but could not reach low turbidity readings, this led to the discussion of a double dosage of ferric chloride and PAC. The purpose was to use ferric chloride to remove larger particles and PAC to remove the colloidal material. The results from the experiments are shown in Figure 4.

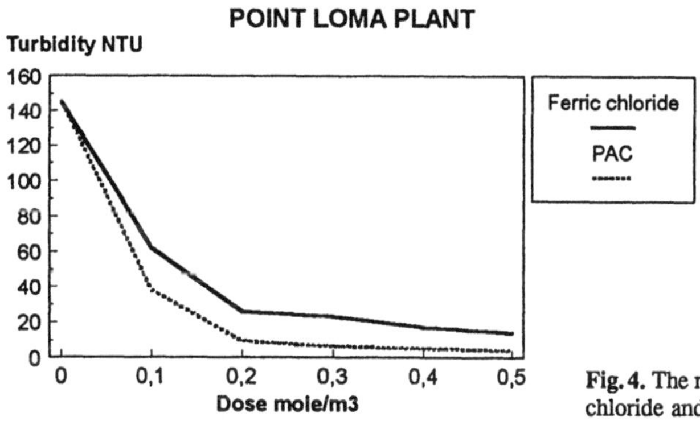

Fig. 4. The results from jar tests with ferric chloride and PAC

Experiments with the double dosage of ferric chloride and PAC had the following results.

Table 2. The results from a combined dosage of ferric chloride and PAC

Products (Tab. 1)	Dose mole/m^3	Turbidity NTU	Turbidity removed
Blank	0	335	0
Ferric chloride	0.1	62	273
Ferric chloride	0.2	26	309
PAC	0.1	38	297
PAC	0.2	9.6	325
Ferric + PAC[1]	0.1	12	323
Ferric + PAC	0.2	4.3	331

[1] 50 % of molar dosage from ferric chloride and 50 % from PAC

It can be seen that the combined dosage was more effective than the single dosage of either product alone. This means that the two products ferric chloride and PAC supported each other and that they could be utilized in a dual system.

This lead to a new application that was both possible to apply in practice and had economic feasibility. It was therefore decided to test this system in full scale operation together with the normal use of 0.2 ppm anionic polymer.

5.2 Full Scale Trial

5.2.1 Effect on Floc Formation and TSS-removal. As an initial approach to the full scale trial, 24 hour testing was done on the influent water. A variation in treatability of the water was noted, namely, that higher dosages were needed during early morning hours and in the late afternoon. The results are shown in Figure 5.

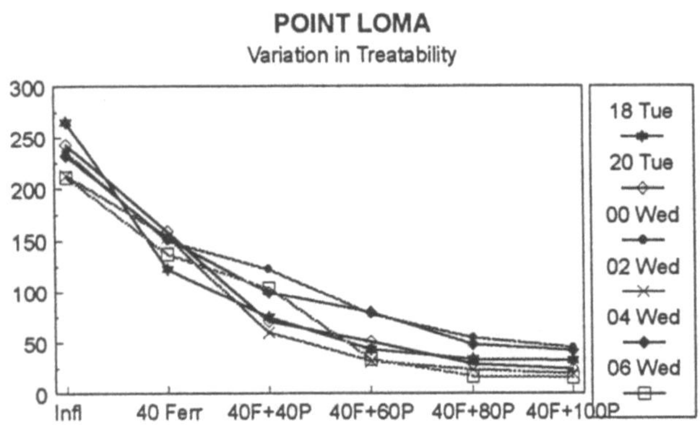

Fig. 5. The variation in treatment results over a period (numbers in the legend indicate hours of the day)

Based on the variations in the course of a day, the following dosage regime was established.

- 35 ppm dry base ferric chloride or 0.22 mole/m^3 at all times
- 60 ppm liquid PAC (9 % Al) or 0.20 mole/m^3 during easy-to-treat periods
- 80 ppm liquid PAC (9 % Al) or 0.26 mole/m^3 during hard-to-treat periods
- 0.2 ppm anionic polymer during easy-to-treat periods
- 0.25 ppm anionic polymer during hard-to-treat periods.

The dosages were adjusted to the incoming flow that is dependent on the number of pumps in operation at the main pumping station prior to the plant. During a two week period only ferric chloride and anionic polymer were dosed on certain days and PAC was applied only on selected mid-week days. The shift was always done at midnight so that the result could be analyzed through the routine sampling that was done at the plant. A flow proportional sampling box takes continuous samples from the final effluent and the results are presented after triple analyses.

The average results from the trial can be seen in Table 3.

The reduction of TSS increased significantly during the trial when the ferric chloride application was combined with the PAC application. This result was also confirmed visually and through turbidity measurements on the effluent.

The results from the jar tests indicated again that PAC is more effective in removing smaller particles than ferric chloride. Ferric chloride and PAC together remove colloidal material down to a smaller floc size than only the ferric chloride addition alone will do.

Table 3. Average results from the trial can be seen in Table 3

Process	Final TSS mg/l	Reduction in TSS %	Total BOD mg/l
Only ferric + polymer	40	84	100
Ferric chloride + PAC + polymer	32	88	80

5.2.2 Results on Floc Removal and Sludge Production. It was obvious that ferric chloride and PAC together formed a better floc and that more impurities in the wastewater were precipitated. This was also shown during the trial since the BOD values decreased from 100 mg/l with only ferric chloride to 80 mg/l when PAC was dosed. This was more significant than the TSS improvement. Some of the BOD reduction was due to dissolved BOD being precipitated with the PAC application.

The floc which is formed must also be removed. Increased sludge blanket levels were observed when PAC was dosed. This later obstructed the operation. Due to the high overflow rate, the Point Loma plant is sensitive to high sludge blanket levels. The reason for the higher sludge blankets may be one or more of the following:

- Increased floc formation produces more sludge.
- The high dosage of chemicals produces a chemical sludge that must be removed.
- The sludge generated from PAC has a lower density than the ferric sludge.

If a consistent good result should be maintained with the combined dosage of ferric chloride and PAC, the sludge removal capacity must be increased at the plant. This cannot be done today since the anaerobic digesters are hydraulically used to full capacity. A better sludge removal system would consistently improve the result.

A better adjustment of the chemical dosage will result in lower chemical sludge production. The adjustment can be made by correlating the dosage to the variation in treatability and to the influent TSS.

5.2.3 Enhanced Treatment Test. Since high sludge blanket levels existed and the possibilities for their removal at the Point Loma plant were limited, a system for the production of smaller sludge volumes was investigated. It was found that if the coagulants ferric chloride and PAC were combined with a cationic polymer, the dosage of the inorganic materials could be decreased.

This meant that the total number of moles of inorganic salts could be less but the same good result was obtained with the combination of a cationic polymer.

The number of iron moles from ferric chloride and aluminium moles from PAC could be reduced by 20 % and 30 %, respectively. This was obtained using 1.5 ppm cationic polymer.

A study was done to investigate the effect on sludge volume. Several jar tests with different types of treatment were correlated in one graph. The treatment consisted of ferric chloride alone, ferric chloride and PAC, and the inorganic salts combined with a cationic polymer. The results are presented in Figure 6.

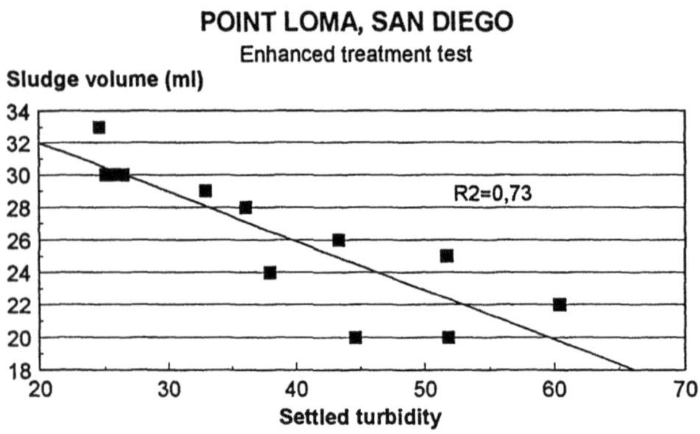

Fig. 6. Correlation between sludge volume and turbidity removal

There was a good correlation between removed turbidity and sludge volume, $r^2 = 0.73$, but there was no single treatment factor that could be shown to be effective in reducing the sludge volume. The different treatments were scattered around the line in the graph and the effect of the cationic polymer could not be seen in the sludge volume readings.

The test was later repeated whereby some jars also included anionic polymer. All jars that received the treatment with anionic polymer were found to have the lowest sludge volume, as seen in Figure 7. It can therefore be assumed that this treatment will lower the sludge volume.

The enhanced treatment tests resulted in a recommendation to use a cationic polymer dosed in conjungtion with the inorganic salts. The dosage levels of the ferric chloride and PAC should be reduced by approximately 25 %. Prior to the sedimentation basins, an anionic polymer should be dosed.

The system with ferric chloride + PAC + cationic polymer and later an anionic polymer will probably reduce the TSS levels in the effluent and produce a smaller sludge volume.

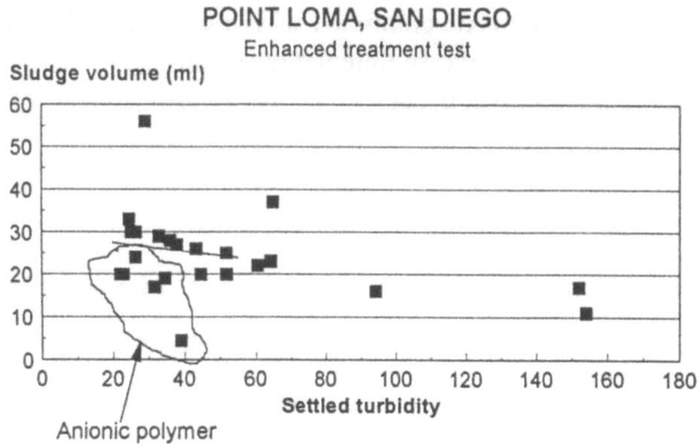

Fig. 7. Results from various tests on sludge volume and removed turbidity

6. Conclusions

The testing program at the Point Loma plant has shown the following:

The ferric chloride dosage at the plant is very effective and small dosages ($\sim$ 0.1 mole/m^3) have a good effect on TSS reduction. Smaller particles and colloidal material cannot be removed with ferric chloride even at very high dosages ($>$ 0.5 mole/m^3).

The combination of both ferric chloride and PAC reduces the colloidal material and results in very good floc formation. Turbidity levels are very low with the combined dosage and the TSS reduction is improved.

Full scale tests have indicated that the application of PAC increases the TSS reduction by 3–5 mg/l. During a testing period, the solids reduction was increased significantly from 84 to 88 % with the use of PAC.

The influent water at the Point Loma plant has a variation in treatability over the 24 hour period. During early mornings and late afternoons, the water is more difficult to treat and higher dosages are necessary.

During periods with PAC dosage and good floc formation, high sludge blanket levels were recorded. The high sludge blankets may be a result of better solids removal, overdose of inorganic chemicals and/or lower sludge density.

To improve the result consistently at the Point Loma plant, it is necessary to install a better sludge removal system.

If the inorganic coagulants ferric chloride and PAC are combined with a cationic polymer, the dosage of the inorganic salts can be reduced by approximately 25 %. This had no effect on sludge volume but should be beneficial since less chemical sludge is produced.

If the application of ferric chloride, PAC and cationic polymer is combined with an anionic polymer, smaller sludge volumes are recorded in lab tests.

In conclusion, therefore, solids removal at the Point Loma plant can be improved with the following means:

1. Combined dosage of ferric chloride and PAC
2. Adjusted dosage over the 24 hour period

3. Reduced levels of ferric chloride and PAC combined with a cationic polymer
4. The use of an anionic polymer prior to the sedimentation tanks
5. The installation of better sludge removal capacity

References

NOAA, Natinal Oceanic and Athmospheric Administration: 50 Years of Population Change Along the Nation's Coast. Rockville, Maryland 1990

EPA, US Environmental Protection Agency: National Water Quality Inventory. Report to Congress, EPA 503/9-92/006, Washington DC 1992

EPA, US Environmental Protection Agency: Response to Information Request for the National Research Council. Water and Technology Board, Office of Water, Washington DC 1992

Handbook on Water Treatment. Kemira Kemwater, Helsingborg, Sweden

Hansson, B., et al.: Full Scale Trial at the Point Loma Treatment Plant. Report to the City of San Diego, Metro Waste Water Division, San Diego

Rosén, B.: Trouble Shooting in Water and Waste Water Treatment. Management of Urban Water Supply and Waste Water Treatment, Gothenburg 1993

Bengt Hansson
Environmental Systems AB
P.O. Box 1294
S-280 60 Broby
Sweden

Alan Langworthy
The City of San Diego
4918 North Harbour Drive
San Diego, California 92106
USA

Industrial Waste Management: The Case of the Tannery Industry

D. Petruzzelli, R. Passino, M. Santori, and G. Tiravanti

Abstract

120 000 t/y of dry sludges are produced as a result of the precipitation techniques commonly applied for metals control in the 40 Mm^3/y of wastewaters discharged by the tannery industry in Italy. Different approaches for tannery waste management, based on "conservative environmental technologies", are proposed in the paper, together with a comparative cost-benefit evaluation. Details on a demonstration campaign run using an ion exchange pilot plant (10 m^3/d) at a real installation are also given in the paper.

Introduction

The tannery industry is blamed for the environmental impact related to the presence of metals (Cr, Al, Fe) in the wastes. 50 % of the feed chemicals to the production processes, essentially Cr(III) salts, are lost and discharged in the effluent waters. 40 $M m^3/y$ ($\approx$ 100 $M m^3/y$ in Europe) of liquid wastes (500 mg/L Cr(III)) are produced annually in Italy, and this leads to the formation of 120 000 t/y dry weight of sludges (300 000 t/y in Europe) (1–5 % Cr content) as a result of the currently applied wastewater technologies [1].

Conventional technologies for metals control in liquid effluents are based on chemical precipitation, usually as insoluble hydroxides, using lime or sodium hydroxide. The operation, however, leads to the formation of noxious sludges that must be disposed of properly in controlled landfills, to avoid the environmental impact associated with the potential leaching of the metals [2].

In some areas (Italy and Germany) tannery sludges were also applied to land as soil amendants, assuming the low solubility of the chromic salts and thus their essential immobility in the water-soil environmental compartments [3]. This latter practice should be strongly discouraged in view of the evidence that chromic species are easily oxidized to chromate in the soil environment [4]. Chromates are more environmentally mobile due to the higher solubility of their salts and, as a consequence, the impact on living organisms is stronger because of the ability of these anions to permeate the cellular membranes [5].

Due to the environmental persistence of the metals, the latest trends in their control are based on "conservative technologies" allowing for the possibility of

R. Klute / H.H. Hahn (Eds.)
Chemical Water and Wastewater Treatment III

removing and recycling metals to the production lines of the industrial activity of origin, or to other activities. Ion exchange as a "conservative" technology allows for removal of metals from the liquid phase at the strictest concentration limits imposed by the enforced legislations, and their recovery from the spent regeneration eluates [6–8].

Three examples of conservative technologies applied to the solution of the enviromental problems related to the liquid and solid effluents of the tannery industry are reported in this paper.

An economic evaluation for each solution is also reported, together with a final comparative cost-benefit evaluation of the proposed solutions, as well as with respect to the currently applied practice based on landfill disposal.

Description of the Processes

Three processes for solving the environmental problems related to the tannery industry are described.

Process 1: Acid Leaching and Reprecipitation of Metal-Containing Tannery Sludges

Table 1 reports the average composition of the mixed sludges, including both the inorganic fraction from the preliminary metal precipitation operation and the organic fraction from the biological wastewater treatment operation. Based on their qualitative characteristics, reference mixed sludges are classified as "hazardous" in the legislation enforced in most of the industrialized countries.

Table 1. Average composition of mixed tannery sludges from 5 different Italian installations [1]

Parameter	Sample 1 %	Sample 2 %	Sample 3 %	Sample 4 %	Sample 5 %
Dry residue (105 °C)	38.1	4.2	35.8	32.1	26.2
Organics	39.5	65.7	56.2	69.1	48.2
Inorganics (550 °C)	60.5	34.3	43.8	30.9	51.8
TOC	22.9	38.2	24.2	38.8	28.0
N_{tot}	0.4	2.8	0.4	0.6	1.8
N_{org}	4.0	6.3	3.3	6.0	4.6
P_{tot}	0.03	0.01	0.01	0.03	0.01
Cr(III)	2.9	1.3	2.7	4.6	2.1
Al(III)	0.7	0.1	0.8	1.2	0.9
Fe(II)	7.2	16.2	0.6	3.3	4.1
Ca	15.0	1.9	12.0	12.2	9.1

The first treatment process is based on an improved management of the precipitation operations, currenty used for metals control, which are based on sulphuric acid extraction of the mixed organic-inorganic sludges, their selective separation and reprecipitation from the acid extract [9]. The inorganic fraction of the composite sludges, containing only the precipitated metal hydroxides, is properly disposed of in controlled landfills, whereas the residual detoxified organic sludge fraction can be safely applied to land as soil amendant.

Figure 1 shows a conceptual scheme of the process for the acid leaching of the mixed sludges.

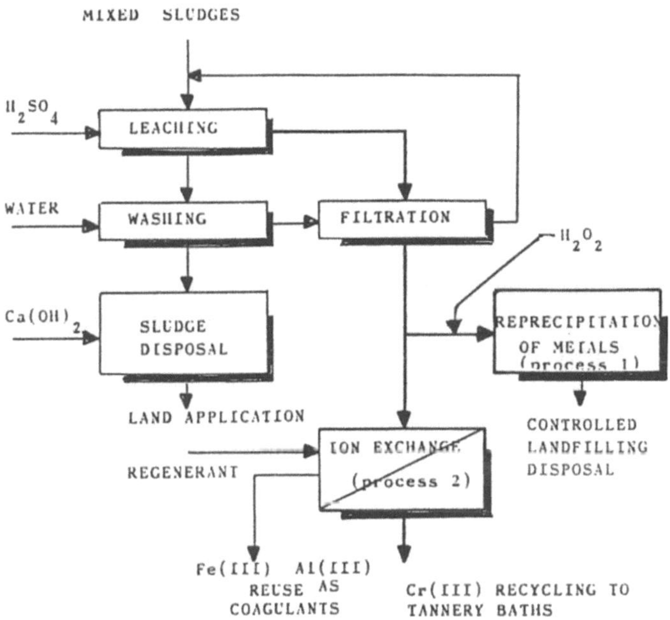

Fig. 1. Conceptual scheme of the acid leaching operations from tannery sludge. Two alternatives for the leachate treatment (reprecipitation of the metals (process 1) and ion exchange (process 2) are also included

As shown in Figure 1, the proposed solution is based on the following steps: leaching of the sludges, washing of the residues, filtration and land application of the organic detoxified residues. The figure includes also the metals reprecipitation step; an ion exchange treatment step, also potentially applied to the resulting sludge, will be discussed later (process 2).

Different acids (H_2SO_4, HNO_3, HCl) at different concentrations were tested as leachant solutions with essentially equivalent behaviour; 1 M sulphuric acid was preferred to minimize the concentration of the calcium ion (one of the major components in the sludge) in the leachate due to the formation of insoluble $CaSO_4$. Excess dosage of sulphuric acid (60 % with respect to the stoichiometric value), pH $\leq$ 1, and a solid to liquid ratio of 1 : 20 was necessary for the metal extraction efficiency to be greater than 85 % from the sludges. Relatively low residual quantities of metals (0.1–0.2 % Cr) were detected in the leached residue, which can be safely disposed of in agriculture as soil amendant.

Table 2 reports the typical composition of the acid leachate obtained from sample 2, reported in Table 1.

Table 2. Typical composition of the sulphuric acid leachate from the tannery sludge*

Parameter	Concentration mg/l	Parameter	Concentration mg/l
Cr(III)	450	$SO_4^{=}$	6000
Al(III)	300	Cl	1000
Fe(II)	400	$PO_4^{=}$	150
Ca^{2+}	500	N_{total}	500
Mg^{2+}	350	TOC	1500
Na^{+}	850	pH	1.0

* Obtained from sample 2 in Table 1

By adding the appropriate stoichiometric quantity of NaOH to this solution, the quantitative reprecipitation of the Cr, Al and Fe hydroxides can occur, which makes it possible to dispose of these metals safely in controlled landfills. The amount of this latter mineral sludge is, on the average, only one hundredth of the original composite organo-mineral sludge, with the related economic revenues associated with the landfill costs, which are usually applied on a gravimetric basis.

Process 2: Acid Leaching and Ion Exchange Separation and Recovery of Metals from the Leachate of Tannery Sludge [9]

The first step of process 2 is the same as the acid leaching procedure previously described in process 1 (see Fig. 1); the acid leachate of the mixed sludge is then sent to an ion exchange process for selective separation of the metals. The oxidation of ferrous to ferric species is obtained by the stoichiometric addition of hydrogen peroxide to the acid leachate before treatment.

The process utilizes a weak base anion resin, selective for ferric species [12], followed by a strong cation sulphonic resin for the separation and recovery of chromium and aluminium. Figure 2 shows the conceptual scheme of the ion exchange section.

The acid leachate (see Table 2) is eluted onto the resin columns where the ferric species are retained by complexation onto the weak base aminophenolic resin [12], whereas the aluminium species are retained by the cation resin, and chromic species break through immediately from the column. The effluent solution from the ion exchange sections, containing essentially chromic sulfate (about 75 % of the chromium initially present in the solid waste), is ready for recycling to the production lines as chrome tanning bath, after increase of the metals concentration.

The resins are regenerated by using 5 BV (Bed Volumes) of 1 M H_2SO_4 and 10 BV of a 50 % mixture of 1 M H_2SO_4 and 2 M Na_2SO_4, respectively. The spent

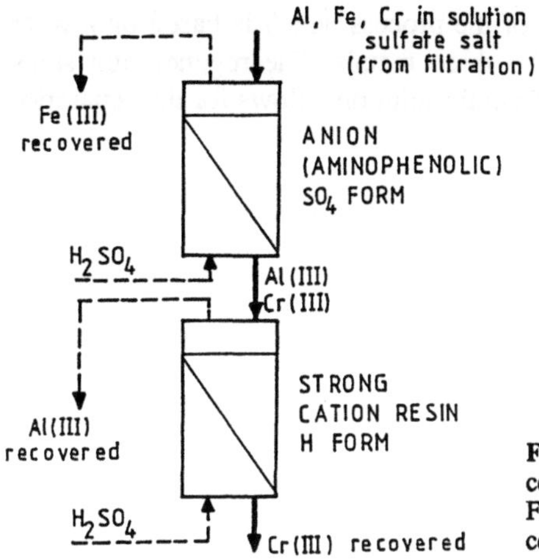

Fig. 2. Conceptual scheme of the ion exchange process for separation and recovery of Cr(III), Al(III), Fe(III) from acid leachate of tannery sludge (process 2)

resin regeneration eluates, containing alum and ferric sulphate, can either be recycled directly as flocculating agents in the general wastewater treatment operations or, after precipitation of the metals by lime or NaOH addition, the resulting limited quantities of sludge can safely be disposed-of in controlled landfills.

Process 3: Ion Exchange Separation and Recovery of Metals from "Segregated" Wastewaters from Tannery Operations. The IERECHROM Process [10]

The third process, namely IERECHROM (Ion Exchange REcovery of CHROMium), is applied to that "segregated" fraction (10 % of the total volume) of the tannery effluents (spent chrome tanning baths plus leather washing waters) containing the massive amount of the metals lost from the tanning operation. Table 3 reports the average composition of the reference wastewaters.

Table 3. Average composition of the "segregated" wastewaters from the tannery operations

Parameter	Concentration mg/l	Parameter	Concentration mg/l
Cr(III)	180	Cl^-	850
Al(III)	0.5	$SO_4^=$	450
Fe(III)	0.2	TOC	120
Ca^{2+}	300	pH	3.5
Mg^{2+}	120	Cond.	5.7 mS/cm

Figure 3 shows a conceptual scheme of the process which is based on a weak electrolyte carboxyl resin selective to the three metals. The regeneration steps, carried out sequentially with oxidants and acidic solution, allows for the separation and recovery of the metals.

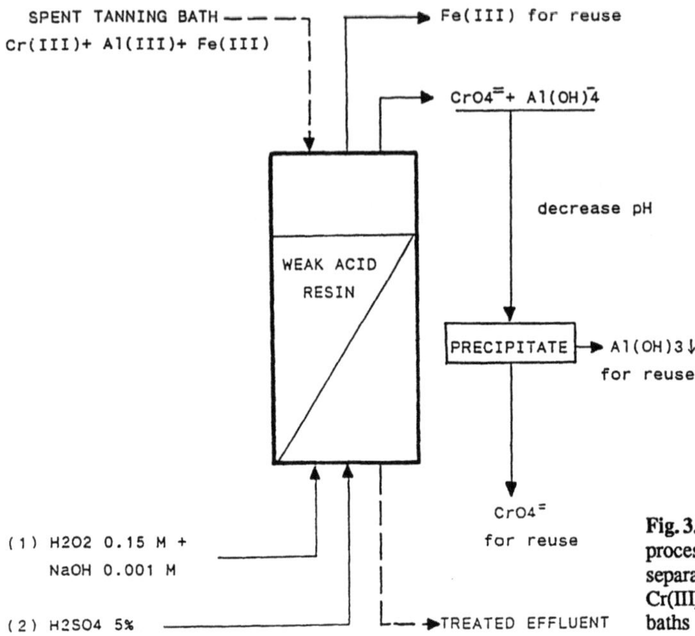

Fig. 3. IERECHROM process for selective separation and recovery of Cr(III) from spent tanning baths

In ion exchange phenomena, the advantages gained during the operation in terms of selective retention of the ionic species, are paid for during the regeneration step when a strong chemical potential is necessary to overcome the high free energy of the interaction of the ionic species with the charged functional groups. In the case of the reference wastewaters, regeneration of the resin by using conventional acidic or alkaline solutions failed. Efficient regeneration levels were obtained only by using unconventional chemicals such as hydrogen peroxide in alkaline media. By using 10 BV of the mixture 0.15 M H_2O_2, 0.5 M NaOH, 1 M NaCl, trivalent chromium and aluminium species were internally (in the resin phase) oxidized to chromate and hydrolyzed to aluminate, respectively. These latter anionic species were immediately rejected by the carboxylate resin groups by the Donnan effect [11]. Aluminate and chromate species from the spent alkaline regeneration eluate were then separated by pH neutralization of the solution, which induced the precipitation of $Al(OH)_3$. A final polishing step of the resin, with 5 BV of 1 M H_2SO_4, was necessary for the elution of the ferric species.

Figure 4 shows a picture of the 10 m^3/d pilot plant assembled for the demonstration campaign run at the EUROPEL tannery installation in Solofra, Naples, S.W. Italy [10]. The demonstration campaign was concluded satisfactorily, confirming the technical feasibility of the process. Figure 5 shows the typical perfor-

Fig. 4. Picture of the IERECHROM pilot plant treating 10 m^3/d of segregated tannery wastewaters

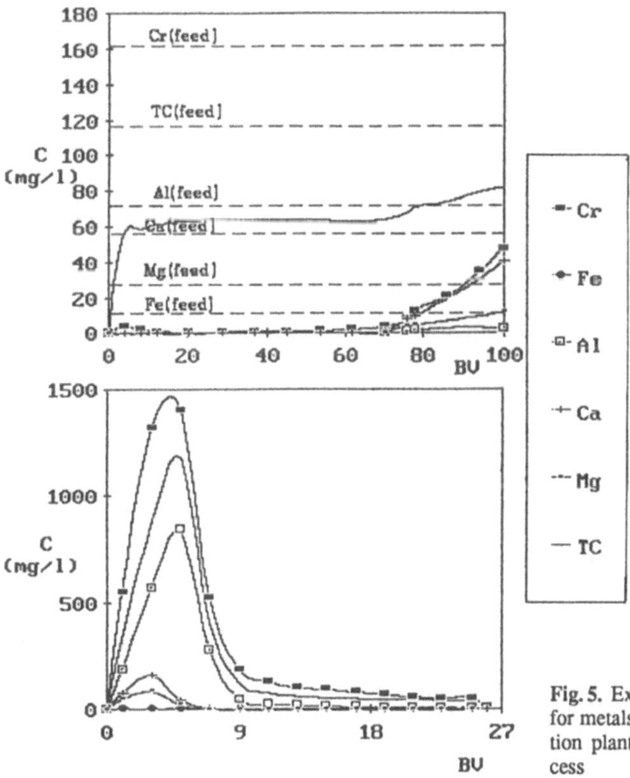

Fig. 5. Exhaustion-regneration curves for metals in the 10 m^3/d demonstration plant of the IERECHROM process

mances of the carboxyl resin; about 100 bed volumes (ave. Cr-leakage < 1 mg/l) of metal free waters were produced during the exhaustion step. The column was then regenerated with ≈ 10 BV of alkaline hydrogen peroxide solution, thus recovering, for the tannery production lines, about 90 % of the chromium initially present in the segregated wastewaters. Satisfactory results were also obtained by the use of chromium sulfate, recycled from the above process, for the preparation of fresh chrome tanning baths.

Technical-Economic Comparison of the Proposed Solutions

Generally speaking, it is difficult to make a detailed cost benefit evaluation of environmental processes without including the economic revenue associated with the minimized impact due to the avoided discharge of the waste into the environmental compartments. However, this parameter is difficult to evaluate and, for this reason, it was not included in this study. In the following a comparative economic evaluation of the proposed solutions is reported.

Table 4 reports the summary of the cost items for different processes and, for comparison, the current Italian costs for the practice of tannery sludge landfill disposal (conventional treatment).

Table 4. Comparative economic evaluation for the three proposed solutions (December 1993)

Item	Process 1 *	Process 3 **	Process 3 ***
Investment costs (pumps, valves, reactors, reservoirs, filters, electric boards)			
a) Acid leaching US$	235 000	235 000	–
b) Leachate treatment US$	65 000	176 000	–
Total investment costs US$	300 000	411 000	60 000
Amortization (10 % in 10 years) US$	30 000	41 000	6 000
Unitary investment costs US$/t	7.5	10.3	3.2 US$/m^3
Operating and maintenance (O & M) costs (chemicals, electric power, labour)			
a) Acid leaching US$/t	39	39	–
b) Leachate treatment US$/t	11	126	–
Total O & M costs US$/t	50	165	6.0 US$/m^3
TOTAL (Investment + O & M) US$/t	57.5	175.3	9.2 US$/m^3
GRAND TOTAL US$/t	90	210	9.2 US$/m^3

Conventional treatment costs: 180 US$/t****

* Installation treating 4000 t/y wet (80 %) sludge; including metal reprecipitation from leachates, land application of detoxified residue, landfilling of metal hydroxides.

** Installation treating 16 000 m^3/y of leachate; the grand total includes the leaching operation, the ion exchange and land application operations.

*** Installation treating 1875 m^3/y of "segregated" wastewaters; operating, total and grand total costs are reported on a volumetric basis (US$/m^3).

**** Metal precipitation of the overall tannery wastewaters, followed by controlled landfilling of the resulting mixed sludges.

With reference to process 1, the unitary investment costs for an installation treating 20 t/d of wet (80%) mixed sludges, (compositions reported in Table 1) are around 7.5 US$/t (amortization 10% in 10 years), whereas the operating and maintainance (O & M) costs are around 50 US$/t, for an overall (investment plus O & M) cost amounting to 57.5 US$/t. The grand total by including the leaching operation, the reprecipitation of the metals, their controlled landfilling, and land application of the detoxified sludges amounts to 90 US$/t (Dec. 1993). The low technological content of the process means that most of the overall costs are for labour and O & M, whereas the investment costs are associated mainly with the filtration system.

With reference to process 2, for a full scale installation having the potential of 80 m^3/d (corresponding to about 20 t/d of leached sludges), the unitary investment costs amount to about 10.3 US$/t (same amortization as before), whereas the O & M costs are about 165 US$/t. The overall unitary cost of process 2, by including the acid leaching of the sludges, the ion exchange operation, and the disposal expenses for the land application of the detoxified residues in agriculture, would amount to 210 US$/t.

For process 3, the costs are expressed on a volumetric basis because the process refers specifically to the "segregated" liquid effluents from the tannery operations. For the 10 m^3/d demonstration plant actually designed and realized, the investment costs were around 3.2 US$/$m^3$, whereas the operating costs were around 6 US$/$m^3$; the overall costs amounted to about 9.2 US$/$m^3$ of treated segregated wastewater.

A comparison of the data for the three proposed processes with the conventional treatment costs (metal precipitation of the overall tannery wastewaters, followed by landfilling of the resulting mixed sludges) (180 US$/t) indicates that process 1 costs are about one half lower (90 US$/t) and process 2 costs on the same order of magnitude (210 US$/t) as conventional treatment costs. It is interesting to note that the costs for process 3 are about 2.5 times lower than conventional treatment (see Table 5). As already mentioned, these comparisons do not include

Table 5. Comparison of the costs between conventional technology (precipitation) and process 3 on a volumetric basis

Wastewater	Technology		
	Conventional (precipitation of metals and landfilling)	Advanced, Process 3 (ion exchange recovery of metals)	
	mixed	segregated	residual
Volume [m^3]	1	0.1	0.9
Resulting sludges [kg/m^3]	20	–	18.5
Sludge disposal costs [US$/$m^3$]	3.5*	–	0.6**
Treatment costs [US$/$m^3$]	–	0.92	
Total costs [US$/$m^3$]	3.5	1.5	
Cost Ratio	2.5	1	

* Assuming controlled landfill disposal costs of 180 US$/t wet (80%) weight.

** Assuming agriculture disposal costs of 30 US$/t wet (80%) weight.

economic revenues from the recycling operation of the chromic sulfate recovered to the tannery operations and from the minimized environmental impact related to avoiding the introduction of metal sludges into the environment.

Conclusions

The process based on metals leaching of tannery sludges is attractive, as a "conservative" environmental technology, because it allows detoxification of the sludges (they can be applied safely in agriculture, at lower costs).

After the leaching operation two possible alternatives are available for the treatment of the metal-containing extracts (i.e., reprecipitation of the metals, according to process 1, or ion exchange separation and recovery of metals, according to process 2).

As illustrated, the metals reprecipitation process leads to the formation of minimized amounts of inorganic sludges to be disposed-of properly in controlled landfills. The overall costs, including the leaching operation, are about 90 US$/t. The process competes favourably with the common practice of landfill disposal of the whole sludges (180 US$/t).

The ion exchange process 2 allows for the recovery of about 75% of the overall chromium currently lost in tannery sludges; it can be recycled to the tanning operations. The cost, including the leaching operations, of process 2 is around in 210 US$/t, still competitive with the landfilling practice.

The costs resulting from the treatment of the segregated wastewaters from the tannery operations (the IERECHROM process) could be at least 2.5 times lower than the currently used landfill disposal. It is likely that, with the progressive saturation of landfills, on the one hand, and the rising costs of metals, on the other, the IERECHROM process will be even more attractive in the future.

Acknowledgements. The assistance of Mrs. G. Bagnuolo, N. Limoni, C. Longobardi and M. Stramaglia is gratefully acknowledged in running part of the experimental work. The authors wish to thank Mr. G. Laera and Miss N. Rapanà, working under contract for the Ministry of the Environment, for their untiring cooporation during the demonstration campaign at the tannery installation. Financial assistance from the UNIC (Unione Nazionale Industria Conciaria) is also acknowledged.

References

[1] Simoncini, A.: Fanghi di Conceria: Quali Alternative di Utilizzazione? AES *8* (1987) 36–43
[2] Patterson, J.W.: Metals Separation and Recovery. In: Metal Speciation Separation and Recovery, J.W. Patterson and R. Passino (Eds.). Lewis, Chelsea, MI, USA, 1987
[3] Adriano, D.C.: Chromium. In: Heavy Metals in Natural Waters, R.S. Di Santo (Ed.). Springer, Heidelberg 1984
[4] Nakayama, E., Kuwamoto, T., Tsurubo, G., Fujinaga, T.: Chemical Speciation of Chromium in Seawater. Effects of MnO_2 and Reducible Organic Materials on the Redox Processes of Chromium. Anal. Chim Acta *130* (1981) 401–404
[5] Heary, L.E., Ray, D.: Kinetics of Cr(III) Oxidation to Cr(VI) by Reaction with MnO_2. Environ. Sci. Technol. *21* (1987) 1197–1193

[6] Petruzzelli, D., Tiravanti, G, Liberti, L., Passino, R.: Specific Resins for Metal Ion Separation. The Cr(III)/Al(III)/Fe(III) System. In: Recent Developments in Ion Exchange, P. Williams and M.J. Hudson (Eds.). Elsevier Sci. Pub., London 1990
[7] Petruzzelli, D., Passino, R., Santori, M., Tiravanti, G.: Selective Ion Exchange for Chromium Recovery from Tannery Wastes. In: Separation Chemistry Techniques for Radioactive Wastes and Other Specific Applications, L. Cecille, M. Casarci, and L. Pietrelli (Eds.). Elsevier Appl. Sci., London 1991
[8] Petruzzelli, D., Alberga, L., Passino, R., Santori, M., Tiravanti, G.: Exhaustion-Regeneration Behaviour of Carboxylic Resins. The Cr(III)/Al(III)/Fe(III) System. React. Polym. *18* (1992) 95–105
[9] Italian Ministry of the Environment: Recupero di Cromo da Rifiuti Solidi (Fanghi) e Liquidi (Bagni Esausti) di Concerie. Final Report, Bari, Sept. 1993
[10] Petruzzelli, D., Tiravanti, G., Santori, M., Passino, R.: Chromium Removal and Recovery from Tannery Wastes. Laboratory Investigation and Field Experience on a 10 m^3/d Demonstration Plant. XVII Biennial Conference IAWQ, Budapest, Hungary, July 24–30, 1994
[11] Helfferich, G.; Ion Exchange. McGraw-Hill, New York 1962, p. 134
[12 Hodgkin, J.H., Eibl, R.: Ferrin Ion Chelation by Aminophenol Resins. React. Polym. *4* (1986) 285

D. Petruzzelli and G. Tiravanti
Istituto di Ricerca sulle Acque, CNR
Via De Blasio 5
I-70123 Bari
Italy

R. Passino and M. Santori
Istituto di Ricerca sulle Acque, CNR
Via Reno 1
I-00185 Rome
Italy

Stormwater Treatment

Characterization and Advanced Treatment of Urban Stormwater from a Separate System

B. Heinzmann

Abstract

New investigations have shown that stormwater from a separate sewer system is an important pollution factor. The concentration of several parameters is higher than in the treated wastewater and partially exceeds the minimum requirements concerning wastewater discharge into the receiving waters. The load of several parameters in stormwater is equal to wastewater both in the investigated drainage area as well as in the western part of Berlin. With the upgrading of the wastewater treatment plants in Berlin, the significance of stormwater as a factor for the load of receiving waters increases. Therefore it is important to ensure a good treatment of the stormwater. The preliminary tests have shown that for coagulation and flocculation of the very weakly buffered and very soft stormwater, polyaluminium chloride (with a share of cationic flocculant aid), which reacts less acidic, is especially suitable. The optimum dose is 0.06 mmol/l Al^{3+}, whereby the pH value has never fallen below 6. In the pilot plant investigations it was possible to work with this constant dose of polyaluminium chloride because the pH value and the acid capacity during the stormwater flow remained relatively unchanged. The best advanced stormwater treatment process is coagulation and flocculation in a pipe designed for both microfloc and macrofloc formation and for separation by filtration. In this case, in particular phosphate, organic compounds, filterable solids and anthropogenic substances, e.g. lead (a high portion of which comes into receiving waters in Berlin by stormwater) and copper could be removed. These results indicate that, in general, this process functions more efficiently than a stormwater tank where particles can only settle.

1. Introduction

Although Berlin has an abundance of water bodies, the flow rate and the velocity of discharge are very low. These receiving waters are loaded by treated sewage. In addition, they are polluted by the overflow from the combined sewer system and are recipients of stormwater from the stormwater drain in separate system. The quality of the receiving waters has been progressively deteriorating, especially after heavy rainfalls. This special water situation, which is comparable to

R. Klute / H.H. Hahn (Eds.)
Chemical Water and Wastewater Treatment III

the river and lake system in the states Brandenburg and Schleswig-Holstein, requires not only attention to particulate but also to dissolved compounds such as phosphate. Meanwhile many national and international investigations were carried out and results on the quality and the characteristics of urban stormwater from a separate system were obtained and the sources of pollution were determined. But the transfer and the use of these data is difficult because the drainage areas are so different. Consequently, exact data and knowledge about the characteristics and quality of stormwater from the separate system in Berlin are not available. Until now it has not been possible to quantify the load of the receiving waters due to the stormwater from the storm sewer. In order to determine the rate of flow and the quality, and to characterize the urban stormwater, flow measurements were carried out and volume-proportional samples were taken in the storm sewer of a 32.39 ha large drainage area in Berlin in the years 1989 and 1990. This was the basis for determining annual average concentrations and loads in the investigated drainage area and estimating the annual loads into the receiving waters in both the investigated drainage area as well as the western part of Berlin. Hence, a comparison with treated wastewater was made and a scientific evaluation of stormwater from the separate system was possible.

The quality of stormwater from a storm sewer could be improved through advanced treatment. The removal of the phosphorus load in stormwater is also very important as far as running projects for the restoration of lakes and taking measures against eutrophication in Berlin (for example Hässelbarth [1]) are concerned. Hence, experiments should be carried out for the advanced treatment of stormwater from a separate system.

2. Methods

Firstly, a suitable drainage area for the investigations had to be located. Criteria for the choice were:

- Both the drainage area as well as the storm drainage must be large enough, so that samples from the stormwater near the outflow of the storm drainage system can be taken over a long period of time.
- The drainage area must be representative of the area of 160 km^2 in the western part of Berlin which is drained in the separate system.
- An investigation period of two years was chosen to take into account meteorological variations in different seasons.
- In order to measure the height of precipitation and the duration of the precipitation period, a Hellmann recording pluviometer was installed at the Tegel water works only 500 m from the investigated drainage area.
- Furthermore it was necessary to determine the size of the built-up area of the investigated drainage area.

It was necessary to determine if coagulation and flocculation of very soft stormwater with a very low acid capacity is possible without decreasing the pH below 6, or whether the addition of caustic soda for neutralisation would be necessary.

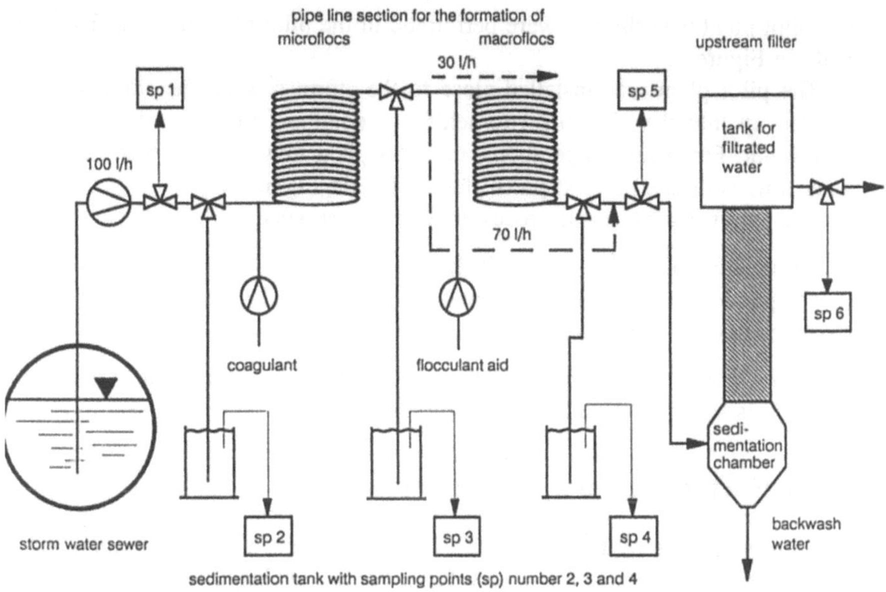

Fig. 1. Schematic diagram of the pilot plant

Furthermore the concentration and the removal efficiencies of different parameters using several treatment processes should be determined and a comparison should be made with the efficiency of stormwater treatment in a stormwater tank.

For the preliminary tests, a standardized jar test procedure was used and the optimum dosage of different coagulants and flocculant aids was found. The experiments were performed with nine different coagulants and six different flocculant aids. Criteria for the choice were:

- improvement of stormwater quality (high removal efficiencies and low concentrations, if possible) after the treatment,
- low decrease of acid capacity and pH value,
- low coagulant dose,
- low residual concentration of iron and aluminium.

Because of the logistics of taking samples by hand in the preliminary tests, stormwater after the first flush was used frequently. These samples are more often less polluted than the first flush.

After these experiments the most suitable coagulants with the optimum dose were used for advanced treatment by coagulation and flocculation in a pipe, followed by separation of the complete stormwater flow in pilot plant experiments. The pH value should not be less than 6. The efficiency of the sedimentation and filtration separation technique was also investigated. Furthermore the optimum length of the pipe line section for the formation of macroflocs was determined.

The pilot plant experiments were performed in the pilot plant illustrated schematically in Figure 1.

The pilot plant was installed close to the storm sewer. The stormwater was pumped up into the pilot plant to achieve a continual inflow of 100 l per hour stormwater. For parallel operations, the pilot plant may be used such that stormwater flow is divided in two parts (dashed line in Figure 1).

The following treatment processes were investigated:

- sedimentation (1 hour) = SW-Sed,
- coagulation-flocculation-sedimentation (1 hour) = C-F-Sed and
- coagulation-flocculation-filtration = C-F-Fil,

where SW is stormwater inflow to the pilot plant, C is coagulation, F is flocculation, Sed is sedimentation and Fil is filtration.

3. Characteristics of Urban Stormwater

New investigations by Heinzmann [2] have shown that the concentration and the load of several parameters are very high. For an evaluation of stormwater it is important to compare these parameters with the minimum requirements concerning wastewater discharge into the receiving waters and with treated wastewater of a sewage plant. Table 1 shows the minimum requirements [3 (*) and 4 (**)] and the frequency of excess for several parameters. The concentrations of lead and zinc in the urban stormwater are about 1.3 to 3.5 times higher than the concentrations of the sewage inflow of the sewage treatment plant Ruhleben. The stormwater has concentrations of filterable solids about 4 to 5 times higher and organic compounds and total phosphorus up to 2.3 times higher than the highly efficient sewage treatment plant Ruhleben under dry weather conditions. Furthermore the load of several parameters in the stormwater is equal to wastewater both in the investigated drainage area and in the western part of Berlin. Table 2 shows a comparison of

Table 1. Minimum requirements [3 (*) and 4 (**)] and frequency of excess

Parameter mg/l	Minimum requirements mg/l	Frequency of excess in % 1989	1990
Filterable solids	20 (*)	77	70
COD	75 (**)	62	36
NH_4-N	10 (**)	6	0
NO_3-N	5 (**)	0	4
P_{total}	1 (*) (**)	32	16
Aluminium	3 (**)	15	13
Copper	0.5 (*)	0	0
Iron	3 (**)	24	15
Lead	0.5 (*) (**)	10	0
Zinc	2 (*) (**)	42	24

loads from stormwater from the separate system and of treated wastewater released into the receiving waters in the western part of Berlin. A large portion of filterable solids, organic compounds, phosphorus, copper, iron, aluminium and particularly lead and zinc enter the receiving waters via stormwater from the separate system, but otherwise total nitrogen is not a pollution factor. In comparison with former investigations, the concentration of lead decreased, since the consumption of leaded gasoline has decreased. One significant source of the high zinc concentration in stormwater is the eave troughs in the investigated drainage area. This means that the real load of zinc in the western part of Berlin is lower than the estimation in Table 2.

Table 2. Comparison of loads of stormwater (SW) from a separate system and of treated wastewater (WW) into the receiving waters in the western part of Berlin in 1990 in t/a

Parameter	Annual load 1990 in t/a		
	Stormwater load	Wastewater load	Coefficient SW : WW
Filterable solids	2016	2574	0.78
COD	3254	11812	0.27
DOC	472	2983	0.16
Total phosphorus	16	81	0.2
Total nitrogen	170	5013	0.03
Aluminium	38.6	55	0.7
Copper	1.1	1.8	0.61
Iron	57.6	68.4	0.84
Lead	2.3	< 0.76	–
Zinc	42.9	9.4	4.5
Adsorbable organic halogens	1.2	14.8	0.08

Consequently, zinc, lead, copper, aluminium, iron, organic compounds, phosphorus, filterable solids and particularly suspended substances with adsorbed anthropogenic pollutants found by Hahn [5], i.e. heavy metals and organic micropollutants (e.g. PAC and PCB), in the storm drainage are an important pollution factor. The significance of the stormwater for the quality of the receiving waters increased with the upgrading of the sewage treatment plants. Hence, it is necessary to ensure a good treatment of the stormwater in the storm sewer before it will be discharged into the lakes and rivers.

Table 3 shows characteristics of urban stormwater flow important for coagulation from the storm sewer in the separate system in Berlin. In general, the stormwater is very soft with a very low acid capacity. As a rule, the colloids and the suspended particles have a negative charge. The concentration of salt is mostly very low. The concentration of solids and the turbidity fluctuates. Depending on the season, the stormwater may contain high concentrations of particulate and dissolved organic substances.

Table 3. Anual average, minimum and maximum concentrations of several parameters in stormwater in Berlin, 1990

Parameter	Units	Annual average	Minimum	Maximum
			of the average concentration of stormwater flow	
pH			6.5	7.5
Acid capacity	mmol/l	0.360	0.22	0.62
Conductivity	μS/cm	105	52	260
SO_4^{2-}	mg/l	29	12	75
Calcium	mg/l	13	5.6	48.7
Magnesium	mg/l	1.24	0.47	4.59
UV-Absorbance at 254 nm	1/m	19	7.50	258
BOD	mg/l	28	5	184
COD	mg/l	113	16	662
Turbidity (1 h)	FTU	53	6	205
Filterable solids	mg/l	70	7	446

4. Cosiderations on the Coagulation of Stormwater

Since stormwater is very soft, it is possible to transfer the experience of colloid stability, particle destabilization and coagulation gained from very soft waters over to stormwater. In soft water the colloids are very stable, because the operational range of the electrostatic repulsion is higher than in hard water. Calcium and magnesium have a destabilizing effect on the colloids before the coagulants are added. Ødegaard et al. [6] have shown that at equal colloid and particle concentrations, soft water contains more colloids and particles with a diameter of < 1 μm and less with a diameter of > 1 μm than hard water. A large portion of the fine colloids has already been coagulated in hard water. Therefore, as a rule, the coagulant dose for destabilization is lower than in soft water. In soft water and in cases in which the pH value is less than 7, the working pH range for a good coagulation is very small. Furthermore, extra care should be taken with coagulation and flocculation experiments in the winter time due to the low temperature of the stormwater (0–10 °C).

Because of this special stormwater characteristic it seemed important to investigate, above all, the effect of prepolymerized metal salts, especially polyaluminium chloride, in advanced stormwater treatment. Advantages of polyaluminium chloride use are lower alkalinity consumption, high removal of particles, wider working pH range for an optimum coagulation and flocculation, less sensitivity to low temperatures and lower residual aluminium concentration. These advantages, particularly the low consumption of alkalinity and the decrease of the pH value can be intensified by using a mixture of polyaluminium chloride and cationic polymer.

5. Results of the Preliminary Tests

The coagulation experiments with the soft and weakly buffered stormwater flow were carried out in a jar test apparatus. The best results were achieved with a pre-polymerized aluminium chloride which has a share of flocculant aid with a cationic charge (low molecular weight acrylamide copolymer with quaternary aminocarboxylic ester). This coagulant Sachtofloc 46.6 (SF) causes the system to become less acidic than traditional coagulants. The optimum coagulation dose was 0.06 mmol/l Al^{3+}. This corresponds to a dose of 0.18 mg/l cationic flocculant aid. The pH value never fell below 6 using this coagulant and dose, hence, good results were obtained. Large amounts of phosphorus, organic compounds and filterable solids were removed and the concentration of anthropogenic pollutants, e.g. lead and copper, were also decreased to a high degree. Zinc elimination was very low. Heavy metal removal was only slightly better using a higher coagulant dose of 0.1 mmol/l Al^{3+}. The residual aluminium concentration in the jar test procedure after the sedimentation step was higher than 0.5 mg/l. The reason for this may be that the kinetics of aggregation are too slow for the destabilization mechanism of charge neutralization by adsorption [7]. Thus flocs which have a poor sedimentation property are partly formed.

A prerequisite for good coagulation is the destabilization of the colloids in the stormwater. It is effected by adsorption of the soluble-hydrolysis species on the colloids, i.e. classic charge neutralisation and the mechanism of adsorption coagulation. Sometimes the mechanism of sweep coagulation came into the process. A proportional dose of coagulant is required concerning particle concentration by the mechanism of adsorption coagulation. There is a proportional relationship between the concentration of particles with a negative charge and coagulant demand. If the dose is too low the colloids cannot be completely destabilized. The zeta potential, as a criterion for the colloid stability and destabilization, remains negative. In several cases, a destabilization successfully took place when the zeta potential was approximately zero and good treatment results could be achieved. The concentrations of filterable solids, total phosphorus and turbidity decreased considerably. This is shown in Figure 2 and 3 in the test with the coagulant SF on October 30, 1989. A coagulant overdosage may cause a restabilization of the colloids and particles. The zeta potential is positive, therefore giving poor results. Furthermore it can happen that the range for the precipitation will be reached due to a high coagulant dose, but the formation of flocs is not possible because the pH is going down. This is above all the case with traditional coagulants (for example ferro chloride sulphate), which react much more acidically and so the pH value of the stormwater decreases considerably. Therefore the residual concentration of iron and aluminium increases. This requires a neutralization with a caustic soda solution. The pH decrease is less when using polyaluminium chloride because of the high degree of hydrolysis and the low dose. Furthermore the working pH range for a good coagulation is much higher. The probability to reach the range of adsorption coagulation and sweep coagulation is much higher in cases using polyaluminium chloride than in cases using traditional coagulants.

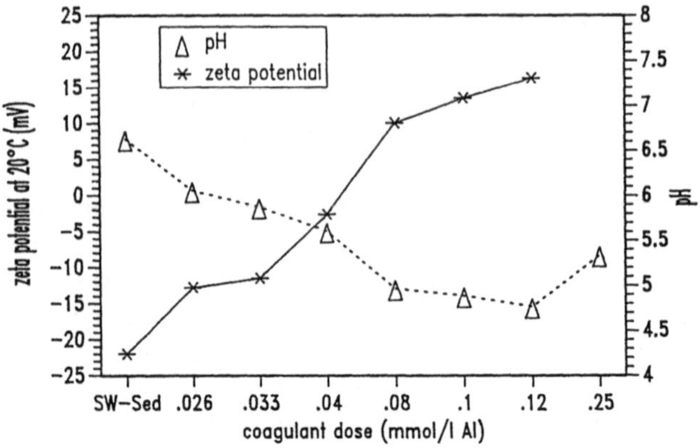

Fig. 2. Zeta potential and pH value depending on the dose of polyaluminium chloride on October 30, 1989

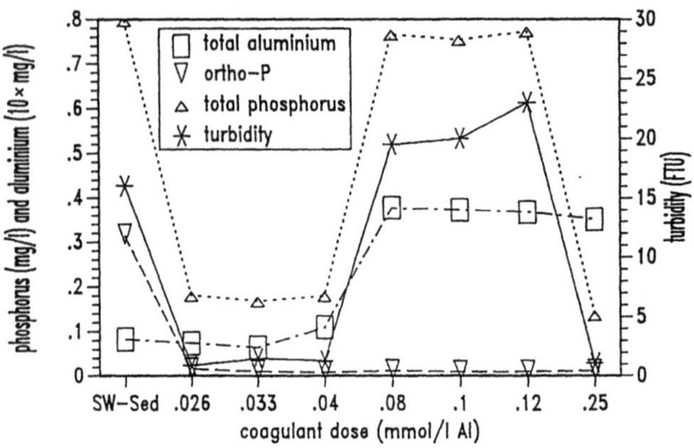

Fig. 3. Concentrations of total aluminium, phosphate (ortho-P), total phosphorus and turbidity depending on the dose of polyaluminium chloride on October 30, 1989

6. Practical Experience at a Pilot Plant

During the stormwater flow in the storm sewer, the level of pH and acid capacity remained relatively unchanged. Larger variations could not be observed. Hence, it was possible to work with a constant coagulant (polyaluminium chloride) dose of 0.06 mmol/l Al^{3+} in the pilot plant investigations and operation control was not required. This is shown in Figure 4.

In Figure 5, the results of the various treatment processes are compared with those of the untreated stormwater flow (SW) for May 6, 1991. An advanced stormwater treatment via pipe flocculation with the coagulant SF was successfully implemented. pH values of less than 6 were never reached, hence, the addition of caustic soda was not necessary. The acid capacity was more than 0.1 mmol/l during this treatment test. The concentration of the filterable solids was reduced to below 10 mg/l and, by the stormwater treatment through coagulation-flocculation-sedimentation (white triangles in Fig. 5), the turbidity was always lower than 10 FTU. The concentration of the filterable solids was lower than 2 mg/l and the turbidity was always below 5 FTU after stormwater treatment by coagulation-

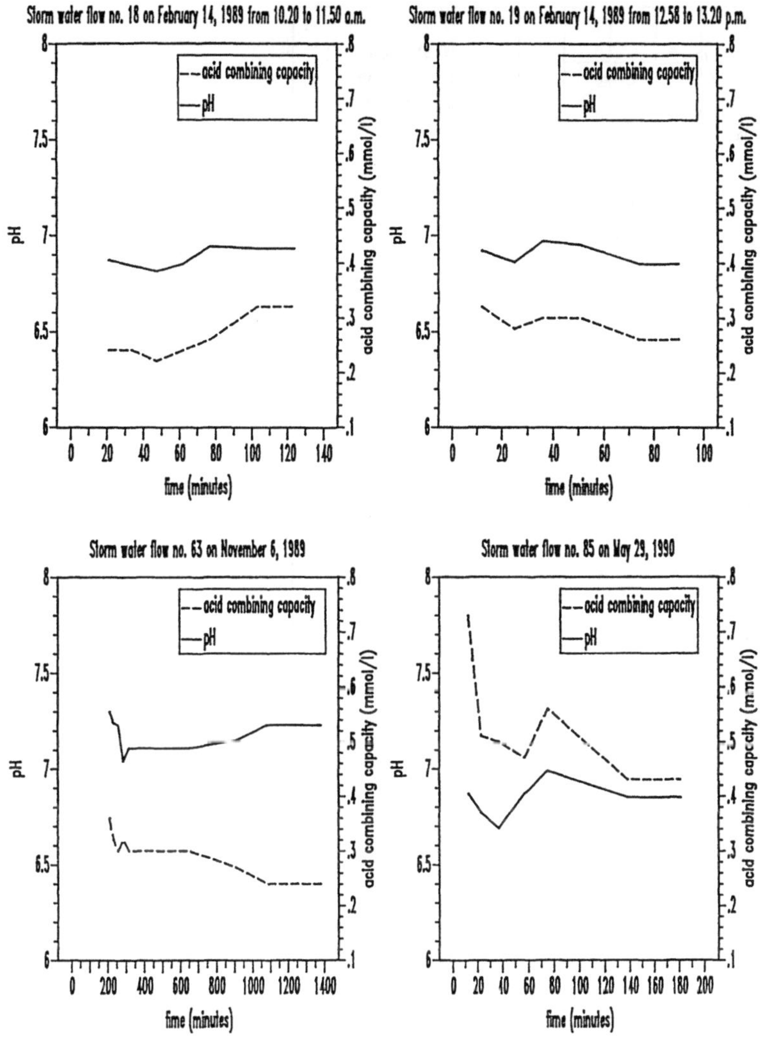

Fig. 4. Time course of acid capacity and pH value

flocculation-filtration (black points in Fig. 5). This represents a considerable enhancement. Both advanced stormwater treatment processes are definite improvements over the 1 hour sedimentation in a sedimentation tank (black squares). These results were verified in other tests.

Figure 6 shows the results of the tests on the different treatment processes with regard to filterable solids and turbidity depending on the inflow concentrations. The removal efficiency can be determined from the lines in the graph. In the pilot plant investigations, the best results were achieved using the advanced stormwater treatment process coagulation-flocculation-filtration (black points). The process coagulation-flocculation-sedimentation (triangles) did not run as well. Both

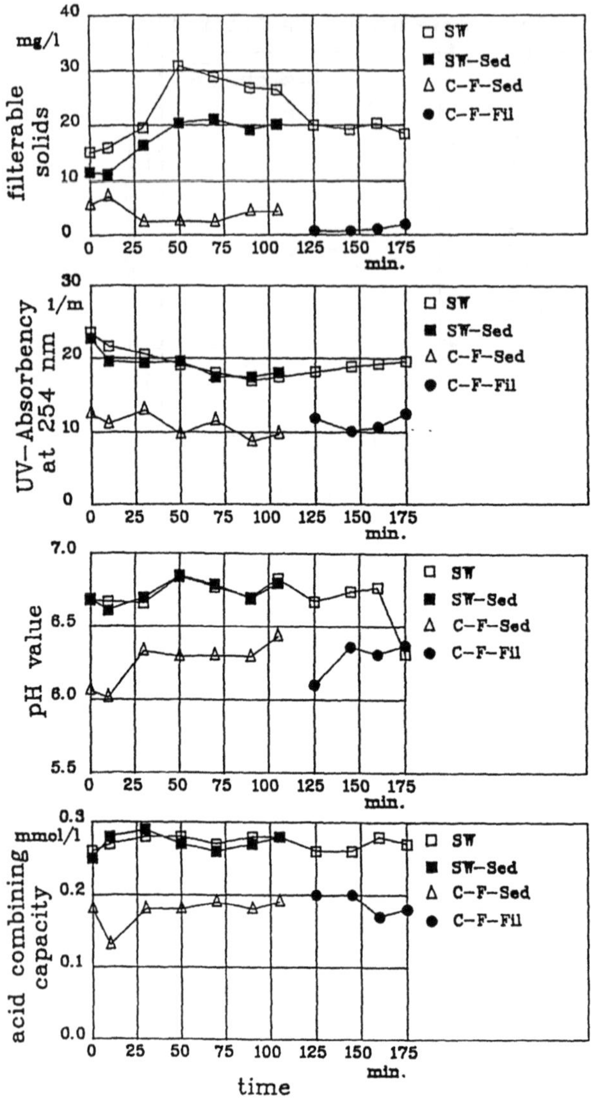

Fig. 5. Stormwater treatment results in terms of filterable solids, UV-absorbance at 254 nm, pH and acid capacity on May 6, 1991

processes, however, achieved better results than treatment of the stormwater by one hour sedimentation (black squares). The poor results of the treatment process C-F(8m)-Sed (black triangle) indicate that a pipe line section of 8 m, i.e. a flocculation time (flow time) of 1 minute is to short for the formation of macroflocs. The results of the process C-F(50m)-Sed show that a flocculation time of approximately 6 minutes is sufficient. However, additional stormwater experiments with a high content of organic compounds have shown that the flocculation time should really be longer than 10 minutes.

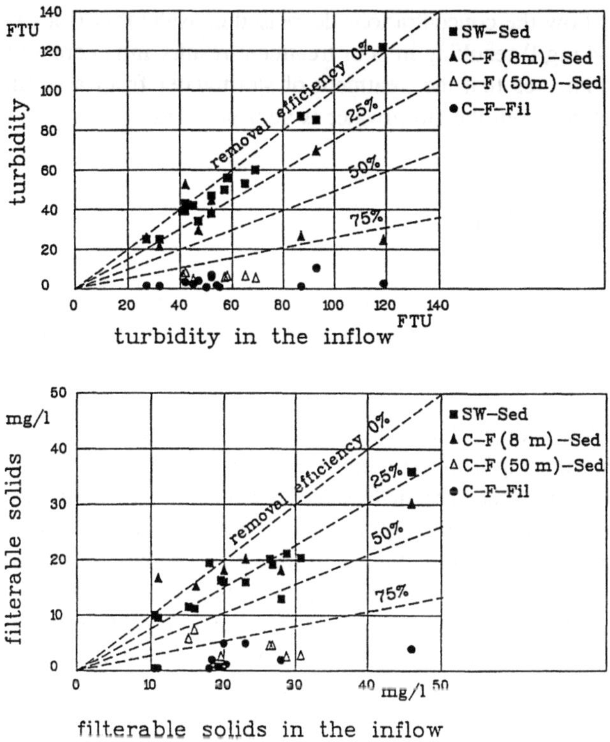

Fig. 6. Removal efficiencies and concentrations in terms of turbidity and filterable solids in the treatment processes depending on inflow concentrations

The dry period before the rain on March 19, 1991 was very long. Therefore the first flush of the stormwater was very polluted, resulting in high concentrations of all parameters which are given in Table 4. Through the one hour sedimentation, a considerable decrease in the concentrations of the parameters filterable solids, COD and lead was achieved. Coagulation with the coagulant PAX (polyaluminium chloride), however, resulted in an even greater improvement (Table 4). The phos-

Table 4. Treatment results (concentration (con.) and removal efficiency (rem.)) at the onset of stormwater flow on March, 19, 1991

	Filterable solids		COD		Pb		Zn	
	con. mg/l	rem. %	con. mg/l	rem. %	con. mg/l	rem. %	con. mg/l	rem. %
SW	197	–	256	–	0.17	–	5.44	–
SW-Sed	85	57	36	86	0.09	47	4.92	10
C-F-Sed	45	77	18	93	0.04	76	4.08	25

phate concentration in the stormwater (0.01–0.45 mg/l P) was also reduced by the pipe flocculation to below 0.007 mg/l P during the complete run-off. On the other hand, the results for zinc were poor and its removal efficiency was very low. At the end of this stormwater flow the concentrations decreased considerably (for example filterable solids 26–4 mg/l) resulting in poor treatment results and decreased removal efficiency. This indicates that the treatment of stormwater from a storm sewer is particularly effective when the stormwater is very polluted.

7. Discussion and Conclusions

These experiments prove that advanced stormwater treatment functions far better than the one hour sedimentation and treatment in a stormwater tank where particles can only settle. Higher removal efficiencies and lower concentrations were achieved for several parameters. For example, in the stormwater flow the filterable solids were reduced to a level lower than 5 mg/l and the concentration of total phosphorus was lower than 0.2 mg/l. Investigations made by Terzioglu et al. [8] at the stormwater tank near the Dianasee in Berlin show that the concentration of filterable solids was approximately 53 mg/l and the concentration of the total phosphorus was approximately 0.55 mg/l. Figure 7 shows a comparison of the average removal efficiencies in terms of filterable solids, turbidity, total phosphorus and COD as a result of the treatment processes with values obtained from the literature (Terzioglu et al. [8] and Weibel et al. [9]). In the pilot plant experiments the one hour sedimentation had a low removal efficiency in comparison to the data from the literature, because the investigated stormwater had poor sedimentation characteristics. Furthermore the reason for the good results with the stormwater tank at Dianasee is also the very low surface loading of 0.0094 m^3 per m^2 and hour during the investigation. The time span includes the rainy season and the dry periods between rains. Since the last stormwater flow was in the stormwater tank, the sedimentation process could continue to work.

The phosphate concentration was reduced in all these tests to a level lower than 0.025 mg/l phosphorus. Anthropogenic pollutants, such as the heavy metals lead (a large part reaches the water bodies through stormwater) and copper, were always removed to a high degree. In contrast, the removal of the heavy metal zinc was very low, due to its solubility in this pH range and to the poor adsorption on the new flocs. Under these circumstances, only measures which prevent the release of zinc into the environment can help, for example, the replacement of zinc in eave troughs with plastic material.

The best coagulant was found to be polyaluminium chloride which has a share of cationic flocculant aid. Therefore a combined effect of coagulant and flocculant aid was achieved. A further advantage of using this coagulant is that just one feeding device for the chemical dosage is required. In all these tests with this coagulant, the pH value and the acid capacity decreased slightly. A pH value lower than 6 was not observed. Therefore, the residual concentration of dissolved aluminium was always lower than 0.15 mg/l after the treatment. Furthermore, the concentration of total aluminium could only be decreased through coagulation-flocculation-filtration using the coagulant SF.

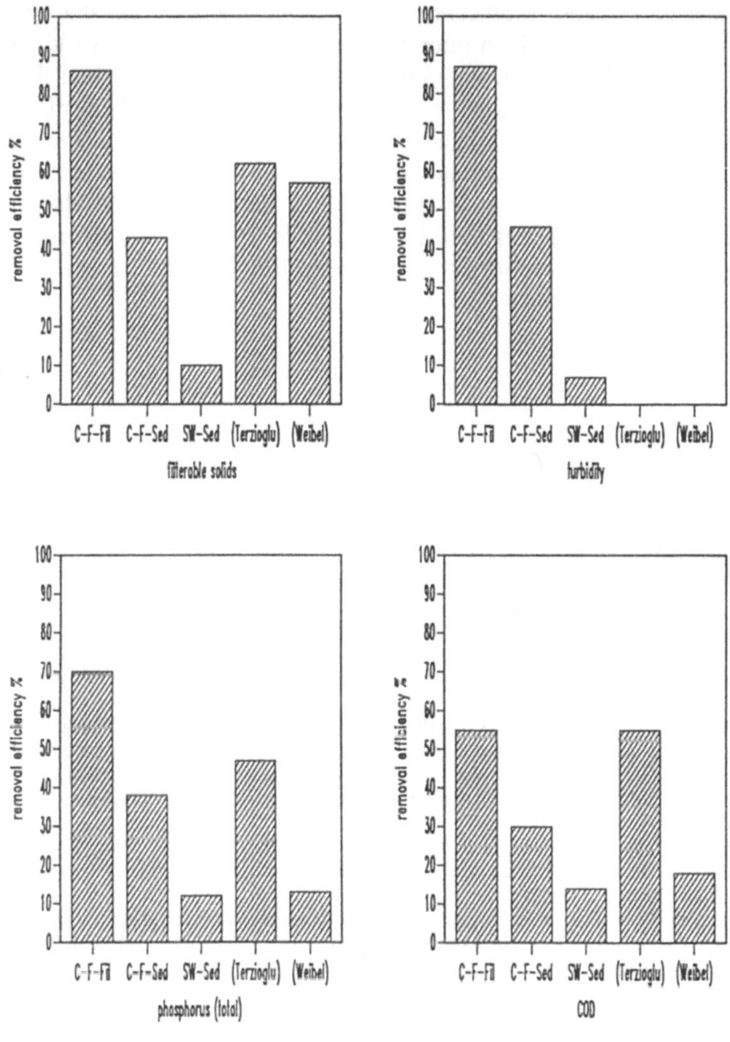

Fig. 7. Comparison of the average removal efficiencies in terms of filterable solids, turbidity, total phosphorus and COD as a result of the treatment processes with literature values (Terzioglu et al. [8] and Weibel et al. [9])

For a good formation of macroflocs, a flocculation time of at least 6 minutes (pipe line section 50 m) and a dosage of a flocculant aid (anionic charge) is required when working with the coagulant PAX. An improvement of the sedimentation through the coagulant's 5 % SiO_2 component could not be observed. In comparison to the coagulant SF, the pH value and the acid capacity were more reduced. Finally, it can be pointed out that the flocculation time for both coagulants should be longer than 10 minutes. An important prerequisite for the good formation of the macroflocs is sufficient time, especially for separation through sedimentation.

The best advanced treatment process works with the coagulant SF and flocculation in a pipe designed for both microfloc and macrofloc formation and for separation by filtration. In general, this process is definitely more efficient than a stormwater tank. Thus, an advanced stormwater treatment process is available for the improvement of stormwater quality. This technology should be utilized in cases where the receiving waters are particularly loaded by the stormwater from the separate system, for example at Müggel Lake in Berlin. It should also be utilized where urban stormwater is used for seepage for groundwater replenishment and for the treatment of storm run-off from intensively used traffic areas (for example intersections and motor-ways).

References

[1] Hässelbarth, U.: A Concept for the Conservation and Restoration of Lakes: The Example of Lake Tegel in Berlin. In: Chemical Water and Wastewater Treatment, A. Grohmann, H.H. Hahn and R. Klute (Eds.). Fischer, Stuttgart New York 1985, pp. 243–255

[2] Heinzmann, B.: Beschaffenheit und weitergehende Aufbereitung von städtischen Regenabflüssen. In: Fortschritt-Berichte VDI. VDI-Verlag, Reihe 15, Nr. 113 (1993)

[3] Eidgenössische Verordnung über Abwassereinleitungen vom 8. Dezember 1975. Der Schweizerische Bundesrat (Stand am 1. Juli 1990)

[4] Der Bundesminister für Umwelt, Naturschutz und Reaktorsicherheit: Allgemeine Verwaltungsvorschrift v. 08.09.1989 über Mindestanforderungen an das Einleiten von Abwasser in Gewässer – Rahmen-AbwasserVwV – Anhang 1 und 40. Herausgegeben vom Bundesminister des Inneren 40. Jahrgang (1989)

[5] Hahn, H.H.: Niederschlagsbedingte Schmutzstoffbelastung der Gewässer – Beitrag der verschiedenen Belastungspunkte und der möglichen Schadstoffgruppen erneut unter die Lupe genommen. Präsentation eines BMFT-Verbundprojektes: Schadstoffe im Regenabfluß aus städtischen Gebieten. Schriftenreihe des Institutes für Siedlungswasserwirtschaft der Universität Karlsruhe, Bd. 58 (1990), pp. 63–83

[6] Ødegaard, H., Fettig, J., Ratnaweera, H.C.: Coagulation with Prepolymerized Metal Salts. In: Chemical Water and Wastewater Treatment, H.H. Hahn, and R. Klute (Eds.). Springer, Berlin Heidelberg New York 1990, pp. 189–220

[7] Jekel, M.: Huminstoffe im Flockungsprozeß der Wasseraufbereitung. Veröffentlichungen des Bereichs und des Lehrstuhls für Wasserchemie und der DVGW-Forschungsstelle am Engler-Bunte-Institut der Universität Karlsruhe, H. 26 (1985)

[8] Terzioglu, K., Peter, A., Sarfert, F.: Regenklärbecken am Dianasee in Berlin. Ein Pilotprojekt im Zusammenhang mit dem Sonderprogramm zur Verbesserung der Wassergüte der Berliner Seen. Korrespondenz Abwasser *34* (12) (1987) 1296–1302

[9] Weibel, S.R., Weidner, R.B., Christianson, A.G., Anderson, R.J.: Characterization, Treatment, and Disposal of Urban Stormwater. Proceedings 3rd IAWPR Conference Munich (1966), pp. 329–343

Bernd Heinzmann
Berliner Wasser Betriebe
Buddestr. 33
D-13507 Berlin
Germany

Dissolved Air Flotation: Efficient Removal of Micropollutants from Stormwater Runoff

R. Pfeifer and H. H. Hahn

Abstract

Stormwater runoff washes off a considerable amount of pollutants from the surfaces of urban catchment areas. Most of the pollutants are transported particle-bound and it is well known that solids of a small grain size (suspended solids) serve as their host carrier. In the traditional design, the stormwater sewer of a separate system discharges into an intermediate sedimentation tank at the outfall of the receiving water to reduce the pollutants load. Due to the extended sedimentation time of small particle, this type of stormwater treatment facility is incapable of removing finer, polluted solids fractions. A demonstration project has shown that chemical coagulation and liquid/solid separation by dissolved air flotation results in an efficient removal of pollutants and allows recovery of the natural waters.

Introduction

In separate systems, surface water is collected in its own sewer system (isolated from municipal sewerage) and is transported from the urban catchment area towards the receiving water. At the outlet of the catchment areas, where stormwater treatment plants are located, strong variations in the amount of water and the transported pollutants load were observed during runoff events. Intensity and type of the pollutants load depend on the prevailing utilisation of the catchment area. A large share of traffic surfaces in the catchment area increases the amount of inorganic and organic micropollutants (like heavy metals and substances known asmineral-oil-type hydrocarbons (MOTH)), which are predominantly transported adsorbed to solids. It is above all the fine-particulate solids fraction which is highly polluted. Research in the transport of mineral-oil-type hydrocarbons in stormwater runoff has shown that the solids fraction of a particle size of $d < 63$ μm can carry 32 % of the total load. As to heavy metals, it has been shown that the same particle fraction carried 53 % of the total lead load [1, 2]. Micropollutants typically accumulate in aqueous systems, build deposits, hardly degenerate biologically and can harm the ecological system. To reduce the stormwater-related introduction of fine-particulate solids into the receiving water, the use of suitable liquid/solid separation methods must be considered ([3]). Subsequently, the results of a continuously operating flotation plant (pilot plant) shall be discussed. Then,

R. Klute / H.H. Hahn (Eds.)
Chemical Water and Wastewater Treatment III

these results will be compared with the sampling results observed in an existing (full-scale) stormwater sedimentation tank, located in Karlsruhe/Grünwinkel.

Description of the Plant

Research work was performed at the outlet of the Karlsruhe/Grünwinkelseparate system. Before reaching the receiving water, the stormwater runoff is lead through a subterranean stormwater sedimentation tank which is permanently filled. A container, which is located above the tank, accommodates the treatment plant.

The plant will automatically start to operate during a runoff event. Measuring the water level in the sewer allows to determine the beginning of stormwater runoff while turbidity measurements make it possible to determine the concentration of pollutants. The plant will be loaded as soon as turbidity exceeds a predefined limit (10 FTU). Loading will be stopped as soon as turbidity sinks below the above limit. Thus, the plant will only be fed with the heavily solids/pollutants-loaded runoffvolume. The plant was designed for a load of $Q = 1\ m^3/h$.

Figure 1 presents a flowchart of the pilot plant. During a stormwater event a submerged pump in the sewer will feed the plant with a split flow from the sewer runoff. First, the split flow passes a bar rack and a grit chamber to remove coarse solids and sand. Then, the coagulation and the flotation reactors will be fed. Part of the pretreated stormwater runoff can optionally be stored and processed later.

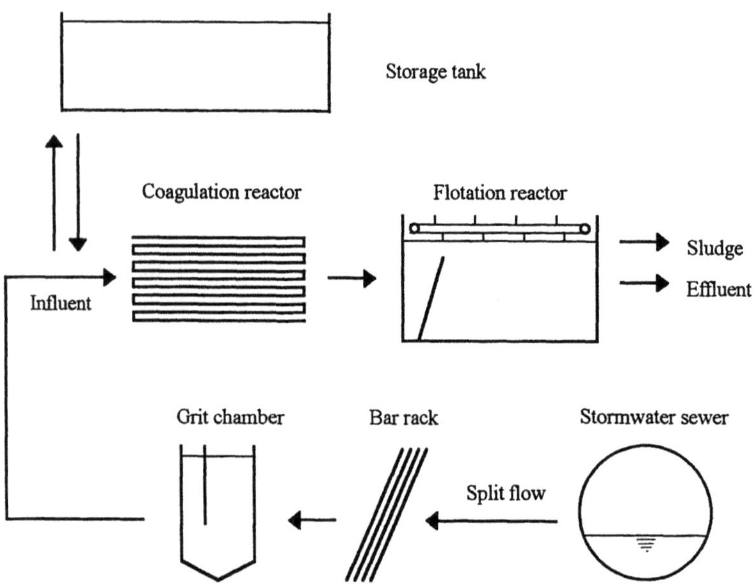

Fig. 1. Flowchart of the pilot plant

A pipe reactor was built for coagulation; it consists of several parallel straight pipe stretches (each of a length of l = 1.75 m) and elbows (of 180 °C each) in slopes. This reactor has a total length of 40 m and a steady diameter of d = 40 mm. Each elbow radius is r = 80 mm. The reactor's dimensioning and construction was checked and confirmed by tests.

The flotation reactor's effective surface is $A = 0.30\ m^2$. It works according to the air dissolved flotation method. A split flow of the plant's effluent is introduced into a saturator where air is dissolved under a pressure of 3 bars. The reflux is injected into the reactor and is released via a pin valve in close vicinity to the plant's influent. The released water is mixed with the stormwater runoff before entering the reactor.

Coagulation of Stormwater Runoff

In the design of the pilot plant the main objective was to achieve a liquid/solid separation of the particulate impurities by flotation with an as small as possible dosage of chemicals. One was striving for a simple solution refraining from the use of polymer coagulants. For this purpose, bench-scale research in coagulation and flotation. Apart from the impact of the dosage of the coagulant, the impact of the size of the mean velocity gradient and the reaction time on the coagulation results was focused on. Mean velocity gradient G [s^{-1}] and reaction time t [s] are used to describe turbulence relationships during coagulation and to transfer bench-scale results to reactors of a larger scale (scale-up). The product of G and t is referred to as the Camp number (Ca [-]).

Batch tests were carried out in a flotation testing device. The same reactor was used as a flotation reactor. This avoids any modification of the solids aggregates formed during the coagulation phase by transferring the sample later into the flotation cylinder. Ferric(III) chloride was used ascoagulant. The process conditions were maintained steady for the flotation (saturated water ratio R = 30 %, flotation time t_{flot} = 5 minutes). The results were judged with regard to an optimum removal of solids by flotation.

Coagulation Process Conditions. Bench-scale tests have shown that relatively small flocs (mean particle diameter d = 300 μm) of a compact shape and an increased shear stability are particularly suitable for flotation. The desired configuration of solids aggregates was obtained by an appropriate setting of the mean velocity gradient and reaction time without coagulants. The reduction of the G value to $G = 60\ s^{-1}$, which was steadily maintained for 120 s, following a rapid mixing phase with an increased G value ($G = 1100\ s^{-1}$, t = 15 s), produced optimum process conditions. Upon completion of the flotation phase, a stable flotation layer had formed. The scum did not release any particles. An extension of the reaction time at a G value of $G = 60\ s^{-1}$ did not improve the removal of solids. Quite to the contrary, it produced instability in the flotation layer.

Based on the bench-scale results, a pipe reactor was built to allow the formation of flocs of a compact shape and an increased shear stability. The dimensioning of the pipe reactor was also based oninformation published in the literature [4, 5]. The above described reactor type was built (cf. Fig. 1). The mean velocity gradient was calculated on the basis of the observed head loss at a loading of $Q = 1\ m^3/h$ in the straight pipe stretches and in the elbows. The mean velocity gradient is $G = 70\ s^{-1}$ in the straight pipe stretches rising to $G = 500\ s^{-1}$ in the elbows. At a total reaction time of $t = 185$ s, a Camp number of Ca = 32 300 will result. This value lies in the range of values typical for pipe coagulation [6].

Investigations at the pilot plant showed favourable results for the removal of solids. The flotation layer formed a stable scum allowing easy removal. One must assume that this modification of turbulences during coagulation results in the formation of compact solids aggregates of an increased shear stability in the selected pipe reactor. Due to their configuration, these solids aggregates facilitate adhesion of air bubbles.

Dosage of Chemicals. The impact of the dosage of chemicals on the removal of solids is depicted in Figure 2. The turbidity values of the original sample ("Influent") and the values observed after coagulation/flotation ("Effluent") are depicted in correlation to the concentration of chemicals for the batch tests. Process conditions for coagulation and flotation were maintained steady while the dosage of ferric(III) chloride was varied. All tests used samples of stormwater runoff collected from various events at the beginning of runoff. A fast flotation velocity was observed in numerous tests. In these tests the optimum removal of solids was achieved after a flotation time of 2 to 3 minutes. In Figure 2 turbidity values observed after coagulation/flotation ("Effluent") are depicted, which were measured after a flotation time of 5 minutes. It can be noticed that already a small dosage of 4 mg Fe^{3+}/l resulted in a turbidity reduction of 84 %. A dosage of 10 mg Fe^{3+}/l reduced turbidity by 94 %.

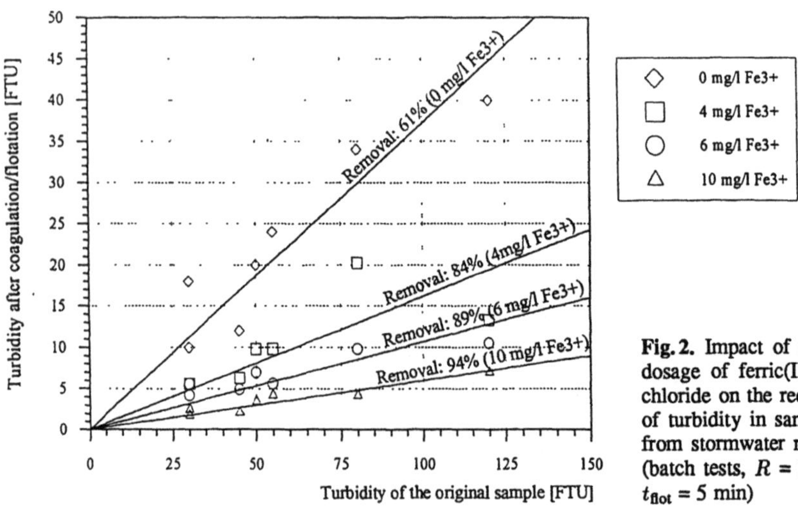

Fig. 2. Impact of the dosage of ferric(III) chloride on the reduction of turbidity in samples from stormwater runoff (batch tests, R = 30 %, t_{flot} = 5 min)

The formation of flocs could be clearly observed during the tests at any time. The pH value did not sink below pH 6 after chemicals were added. On the basis of the tests conducted with the testing device, the dosage was set to 5–10 mg Fe^{3+}/l in the continuous operation of the pilot plant.

The general results of the tests proved a favourable flotability of solids in stormwater runoff. An appropriate setting of the energy input facilitates an optimum removal of solids and a stable flotation layer. Polymers are not required. Without prior coagulation, up to 55 to 65 % of the suspended solids and COD could be removed by flotation. This favourable flotability is due to an increased amount of surfactants (e.g. mineral-oil-type hydrocarbons and tensides) and the natural particle distribution characterised by a large share of the 100 to 200 μm particle fraction.

Removal of Pollutants by Coagulation/Flotation

During a stormwater event the pilot plant will automatically start into operation. Figure 3 shows time-related sampling results of a selected event for influent and effluent, respectively, of the flotation plant. Start of the runoff event was registered at 12:25 p.m. as a result of the observed water level. Sewer runoff reached the 6 cm mark at 12:30 p.m. At this minimum water level the submerged pump transports the split flow from the sewer to the pilot plant. Sampling of the influent and effluent was done at intervals of 10 minutes.

The turbidity and COD values depicted in Figure 3 show a course typical for stormwater runoff. The start of the runoff event is characterised by an increased load of solids and pollutants, resulting in increased turbidity and COD values. As the event continues, these values will decrease. After a process period of 70 minutes loading is automatically stopped (at 13:40 p.m.), as the turbidity value in the plant's influent sinks below the limit of 10 FTU. The plant's effluent values are extremely low from the very start on. The turbidity mean value is 4 FTU with only very small variations. The COD mean value is 10 mg/l. The effluent values are also largely stable. Mass balancing makes it possible to determine a turbidity removal of 82 % and a COD removal of 78 %.

The particle size distribution in the plant's effluent allows to determine the removal of particulate bound micropollutants. Volume distributions in the influent and effluent are analysed. Figure 3 shows the median values (d_{50}). The analysis includes particle sizes from 5 μm to 300 μm. The plant's influent is characterised by a broad spectrum of particles at the start of loading, formed by the particle fraction that was easily washed off at the beginning of a stormwater event. As the event continues, the particle sizes decrease and the distribution across the spectrum narrows. The effluent shows small particle sizes from the beginning on. Figure 4 depicts the entire particle spectrum observed in the effluent. The cumulative frequency was entered according to particle diameter. It becomes obvious that the distributions are limited to a narrow size range. The diameter d_{50} concurs with the mean value of 16 μm for all effluent samples.

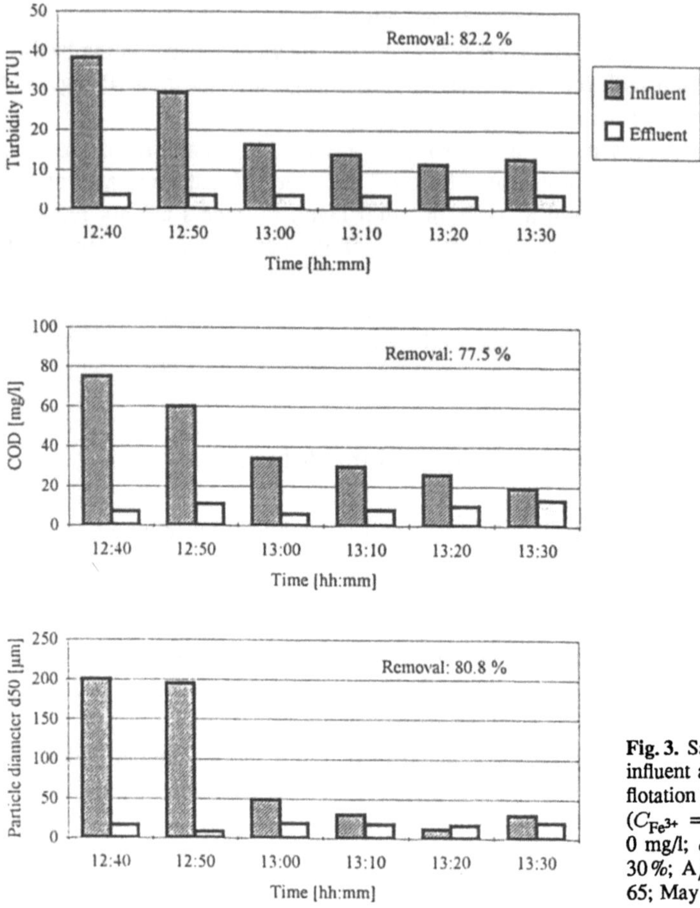

Fig. 3. Sampling results of influent and effluent of the flotation plant ($C_{Fe^{3+}}$ = 5 mg/l; C_{Poly} = 0 mg/l; q = 5 m/h; R = 30 %; A/S = 0.07; Event 65; May 11, 1993)

Certain parameters from the influent and effluent sampling in the flotation plant are listed in Table 1. These concentration parameters represent load-weighted mean values as a result of six individual observed values. It can be noticed that on average 97 % of the suspended solids could be removed. The effluent's low particle concentration is reflected in the low remaining turbidity value. This makes it possible to remove the predominantly solid-bound pollutants efficiently. Lead, for instance, was reduced by 93 %, copper by 83 % and the sum parameter mineral-oil-type hydrocarbons (MOTH) was reduced by 75 %.

Sampling Results of the Stormwater Sedimentation Tank

Event-related samples were taken from the Karlsruhe/Grünwinkel sedimentation tank in order to document its performance. The samples confirm that stormwater runoff can vary drastically in quantity and quality, i.e. the amount and the pollution load. Transport and removal of pollutants shall be explained in the

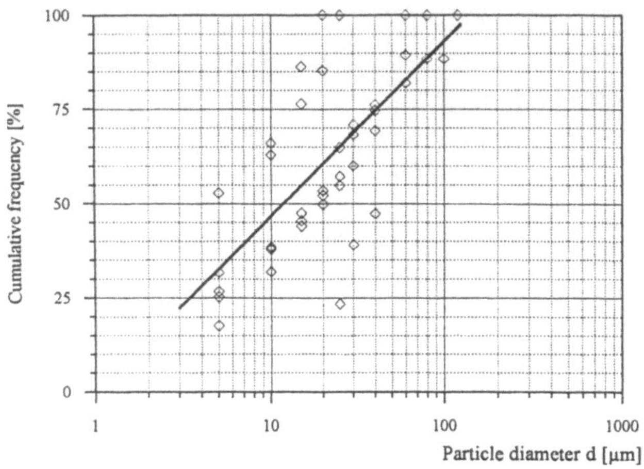

Fig. 4. Distribution of particles in the plant's effluent (based on six individual measurements)

Table 1. Removal of pollutants from strom water runoff after coagulation/flotation

	Influent[1]	Effluent[1]	Removal [%]
Turbidity [FTU]	22.3	4.3	81
SS [mg/l]	57.9	1.6	97
COD [mg/l]	40.6	9.7	76
Lead [μg/l]	42.9	3.0	93
Cadmium [μg/l]	1.0	0.2	80
Zinc [μg/l]	520.0	240.0	54
Copper [μg/l]	76.2	13.0	83
MOTH [μg/l]	11.0	2.8	75

[1] Load-weighted mean

following on the basis of a typical runoff event. Figure 5 includes the sampling results for the suspended solids and observed water level parameters for influent and effluent of the sedimentation tank, respectively. Computer-controlled sampling was done at 5 minute intervals. The effluent was sampled in parallel to the influent.

The influent shows a high solids load at the start of the event, which reaches its peak of 150 mg/l after a short while (15 minutes after the start) and decreases rapidly as the event continues. The peak concentration of solids precedes the runoff peak. Sampling results of the effluent show a lower concentration of suspended solids (maximum: 82 mg/l) than in the influent. The values found in the effluent do not vary so drastically. Their variation spectrum is, however, smaller than that of the influent. Solids and pollutants are removed, but amount and distribution of the solids and pollutants load in the influent do affect the effluent as can be seen in the concentration curve. In this stormwater event suspended solids were reduced by 34.5 %.

Table 2 lists the removal rates for certain parameters in the stormwater sedimentation tank, which were determined according to load balances. For comparison, Table 2 also includes the rates obtained by coagulation/flotation (cf. Table 1).

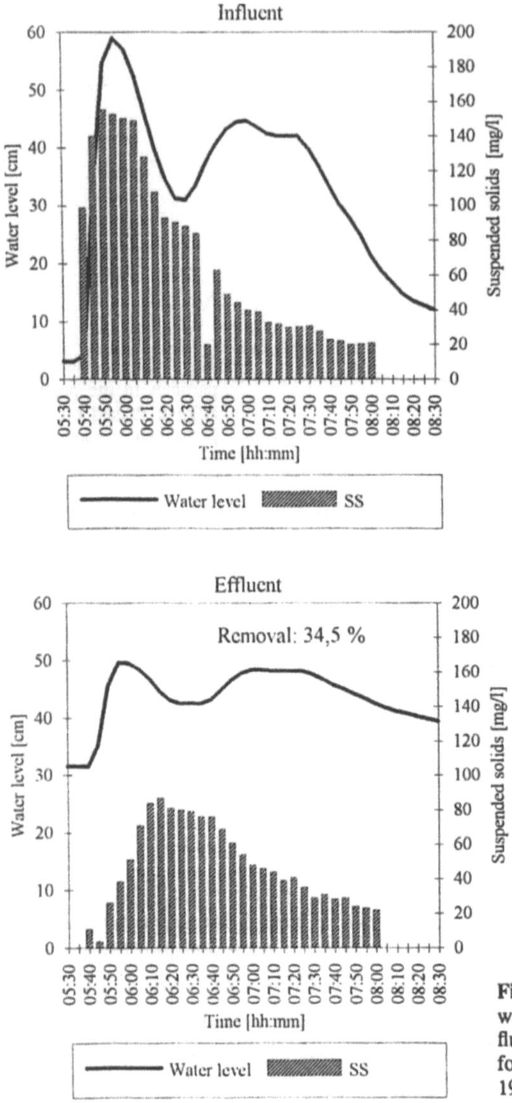

Fig. 5. Simultaneous sampling of the stormwater sedimentation tank's influent and effluent; the graphics show the results obtained for suspended solids (Event 68; May 14, 1993)

The sampling results confirm that an average of 45–50 % of the solids can be removed in the sedimentation tank. These findings are reflected by the reduction rates of turbidity and suspended solids. At the same time, solid-bound pollutants are also removed. Measured as COD, the reduction rate for these pollutants is 41 %. One must assume that the removed solids are above all rapidly sedimenting large solids fractions. More specific investigations on the distribution of particle sizes confirm that the effluent can be highly polluted by fine-particulate solids fractions ($d < 300\ \mu m$). This explains the quite insufficient reduction of heavy metals and mineral-oil-type hydrocarbons (MOTH). A comparison of the reduction rates observed in the stormwater sedimentation tank with the values observed after treatment by coagulation/flotation proves that particularly highly polluted solids fractions, which are characterised by an unfavourable sedimentation behaviour, can be efficiently removed by liquid/solid separationviaflotation. This solids fraction cannot be removed in the sedimentation tank.

Table 2. Removal of pollutants in the stormwater sedimentation tank (SST) and after coagulation/flotation

	SST (full-scale plant) %	Flotation (pilot-scale plant) %
Turbidity	49	81
SS	45	97
COD	41	76
Lead	36	93
Cadmium	39	80
Zinc	13	54
Copper	27	83
MOTH	27	75

Conclusions

The treatment of stormwater runoff by coagulation/flotation produces a stable effluent quality despite varying solids loads. The removal of the finer grain fractions in the effluent is of particular importance. This facilitates a sharp increase in the reduction ofmicropollutants. Stormwater sedimentation tanks experience drastic variations in load due to stormwater runoff, which in turn varies in quantity and quality. Thus, flow conditions favourable for sedimentation (plug flow) can only be established in a very limited way. This results in a relatively low removal of pollutants. Stormwater sedimentation tanks are no suitable means to remove finc-particulate solids. We recommend the use of flotation plants for the treatment of stormwater runoff, if the introduction of micropollutants is detrimental to aqueous systems and impairs the reclamation of drinking water. This can, in particular, apply to impounded rivers or lakes.

References

[1] Chebbo, G., Musquere, P., Bachoc, A.: Solids Transferred Into Sewers. Proceedings 5th Int. Conf. of Urban Storm Drainage, Osaka 1990, pp. 885–890

[2] Hahn, H.H., Pfeifer, R.: The Contribution of Parked Vehicle Emissions to the Pollution of Urban Runoff. Proceedings 4th Int. Conf. Highway Pollution, 1992, 217–226

[3] Pfeifer, R., Hahn, H.H.: The Need for Advanced Treatment of Stormwater Runoff from Separate Sewer Systems. In: Innovations in Flotation Technology, P. Mavros and K.A. Matis (Eds.). NATO ASI Series, 208, Kluwer Academic Publishers, Dordrecht 1992, pp. 521–525

[4] Rodman, D.J.: Investigation Into Hydraulic Flocculation with Special Emphasis on the Removal of Algae. Ph.D.-Thesis, Water Research Center, Medmenham 1982
[5] Grohmann, A.: Flocculation in Pipes: Design and Operation. In: Chemical Water and Wastewater Treatment, H.H. Hahn and R. Klute (Eds.). ISWW Series *40*, University of Karlsruhe, Karlsruhe 1985, pp. 113–131
[6] Zabel, Th.F.: Flotation in Water Treatment. In: Innovations in Flotation Technology, P. Mavros and K.A. Matis (Eds.). NATOASI Series *208*, Kluwer Academic Publishers, Dordrecht 1992, pp. 431–454

Rüdiger Pfeifer and Hermann H. Hahn
University of Karlsruhe
D-76128 Karlsruhe
Germany

High Speed Settling of Stormwater with Microsand

E. Guibelin, F. Delsalle, and P. Binot

Abstract

Biology does not afford treatment of CSO's (Combined Sewer Overflow) waters under fair technical conditions. Due to limited upflow velocities during the settling phase, the usual physico-chemical processes do not afford treatment of such high flow rates under fair economic conditions.

The ACTIFLO process, using recycled microsand as a ballast to increase the settling speed of flocculated particles, overcomes this economic problem through lowering by a factor of 10 (approx) the lamellar area required to treat a given water flow.

Removal of 80 % of TSS, 60 % of BOD, 15 % of NTK and 80 % of P is achieved with a consumption of about 60 g $FeCl_3$, 0.8 g polyelectrolyte and 3 g of microsand per m^3 of CSO's water, within a total area of only 0.03 m^2/m^3 influent per hour.

Plant start-up is achieved within about 15 minutes.

Microsand weighted flocculation appears to be a good way for disposing of the COS's water from large and medium urban catchments economically.

Microsand and Lamellae: Why?

Physico-Chemical Treatment of Stormwater: An Efficient Complement to Biological Treatment

Medium and large rain storm events generate in combined sewer networks very large water and pollutant flow.

Depending upon the peak flow duration and the kind of biological process used (high, medium or low load activated sludge processes), biological treatments could handle stormwater flows ranging between two and four times the average dry weather flow.

Above those flow levels, some more polluted waters may be stored in stormwater buffer tanks and redistributed to the biological treatment plants.

The size of stormwater tanks is, however, limited by the high cost of land in urban areas as well as by the requirement to treat stored waters within one to two days to avoid fermentations and odours.

R. Klute / H.H. Hahn (Eds.)
Chemical Water and Wastewater Treatment III

The main alternative, to avoid or limit large and polluting combined sewer overflow discharges, is to settle polluting charges, before discharge to receiving waters.

Although settling characteristics of CSO's waters are proving fairly good, for organic pollutants and heavy metals as well, (Chebbo, 1992), the cost of the required settling area is prohibitive using classical static settling.

Addition of a flocculation step results in an improvement of the treated water quality through coagulation of the colloidal portion of the pollution, whilst addition of a lamellar settling device allows for a reduction, by a factor of 5 to 10, of the horizontal settling area.

Optimization of the Flocculation and Settling Step

In such a flocculation and lamellar settling process, water quality is closely related to the quality of the chemical treatment, to be optimized for each specific case through careful testing and selection of the best suited chemicals, chemicals dosing and flocculation conditions.

Investment costs are closely related to the size of the horizontal settling area, i.e. to the settling velocity of the flocculated particles.

Thanks to M. Stokes, we have known for some time that settling velocity of a spherical particle is strongly dependent upon its size and volume weight (the larger it is, the quicker it is).

The first research on the classical flocculation process involved a quick flash-mixing followed by a long (say 20 minutes) flocculation step in order to end up with large flocculated particles. Such processes allowed for Hazen velocities of about 1.2 m/h in lamellar settling-tanks.

Further developments favoured smaller and more compact flocculated particles, through longer retention time in several flash mixers, followed by a much shorter (say 6 minutes) residence time in a flocculation tank. Such processes resulted in better economics (flocculation residence time shortened by a factor of two, Hazen velocity increased from 1.2 to say 1.8 m/h).

A radical change in the settling step economics requires a correspondingly radical change in the allowed Hazen velocity, however. 1.2 m/h to 1.8 m/h is good. 10 m/h minimum are required to make the financing of CSO's physico-chemical treatment acceptable for large or medium urban catchments (i.e. for flow rates above 5 or 10 m^3/s).

A good way to increase the settling velocity of organic flocculated particles is to stuff them with weighty (d = approx. 2.65 T/m^3) microsand particles. This process has already been used by OTV for potable water treatment for years, allowing for upflowvelocities up to 50 m/h.

Microsand Weighted Settling in Potable Water Treatment

Former Microsand Weighted Processes

Microsand weighted settling was first used by OTV for physico-chemical treatment of surface water before the filtration step. A mixture of raw water, microsand, and reagents is settled in the Cyclofloc with a upflow velocity up to about 8 m/h.

Addition of lamellae and a fluidized bed area resulted into a upflow velocity increases to about 15 m/h in the Fluorapid process (Fig. 1).

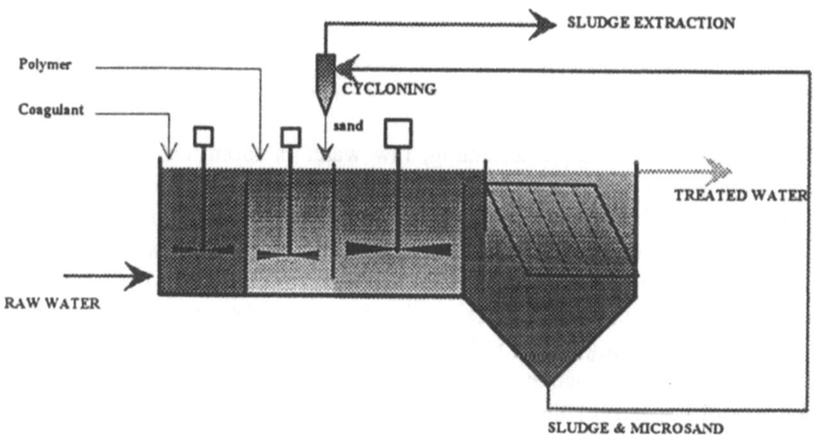

Fig. 1. Actiflo process

Actiflo Process for Surface Water Treatment

In the Actiflo process, the coagulation and flocculation steps are carried out together by means of mixing with sand particles (dia. = 60 micrometers) in two specific tanks, ahead of the settling step.

Raw water destabilized with a metallic salt such as $FeCl_3$ (concentration say 20 g/m^3) enters the injection tank (2 min. retention time) where it is mixed with microsand (about 2 kg/m^3 of water) together with a polyelectrolyte (about 0.1 g/m^3).

Then, the forming flocs are introduced into the maturation tank (6 min. retention time) where the flocs are slowly grown to the optimum size. The flow of water and grown flocs is then directed to the lamellar settling area, with an upflow velocity of approx 40 to 50 m/h. Settled waters are extracted from the top of the settling area through perforated weirs.

The mixture of sand and settled sludges is circulated from the bottom of the settler to a set of hydrocyclones, where sand is separated from sludge and recycled to the injection tank, while extracted sludge is sent to sludge treatment.

Almost all the sand is recycled and the actual microsand consumption is quite marginal (about 2 g per m^3 of raw water).

This process produces a good and stable quality of settled water, with TSS lower than 5 mg/l and turbidity lower than 2 NTU including flooding period increases of TSS from 50 mg/l to 500 mg/l.

After numerous applications of the Cyclofloc process, 10 Actiflo plants are presently in operation, with good results.

Actiflo Process for Stormwater Treatments

Since stormwater pollution appears to have good settling characteristics, the above result of using Actiflo for drinking water strongly suggested that it also be tested on combined sewers effluent. A 250 m^3/h prototype unit was built and installed in the SIAAP effluent treatment plant in Colombes, in the western suburb of Paris. The prototype was tested from the end of 1992, with the following configuration and the following results.

After screening and rough desanding, raw water is submitted to a first step of coagulation where colloids are destabilized by ferric chloride. The coagulated water enters a second tank of flash mixing, where flocculation is initiated through the addition of an anionic polyelectrolyte. In this step, microsand, which acts like a ballast, is also added. Coagulated suspended solids are bound to the microsand grains through polyelectrolyte bridges. In the third tank, the so-called maturation tank, conditions of mixing allow optimal growth of the floc. For example, a high velocity gradient would provoke a breakdown of the floc. On the contrary, a slight agitation would not permit the pelletization of the flocs. In all tanks, sizes and shapes of mixers, homothetic ratios, rotating and crosswise velocities as well as residence times have been optimized with the aim of obtaining a large number of flocs of high density tanks to the microsand. In accordance with Stokes' law, these particles settled very quickly on the large surface developed by the lamellae in the settling tank, where a laminar counterflow settling takes place.

The settled water is recovered by means of weirs before being released into the river. As for the settled flocs, they are collected by means of a scraper or pyramidal pits before being injected into hydrocyclones. In these devices, sludge is separated from microsand. The sludge comes out from the overflow and is afterwards treated by a thickener. Clean microsand is extracted from the hydrocyclone before being recycled ahead in the flash mixing tank. Thus the loss of sand is very low.

Prototype Design

Hydraulic flow entering the prototype unit can be adusted up to 250 m^3/h. Total volume of the coaxially stirred coagulation tank, injection tank and maturation tank is about 30 m^3 (i.e. about 7.5 min. retention time under maximum flow conditions). The settling area is made of PVC lamellae with a 55 ° slope. The horizontal surface of the settling area was first set at 6 m^2, but was soon reduced to 1.9 m^2 to allow high upflow velocities while coping with the 250 m^3/h hydraulic limitation.

The sand and sludge mixed flocs are collected in the pyramidal pit shaped bottom of the settler and extracted discontinuously through electrovalves. Sludge and sand mixture collected in a buffer tank is directed through a pump to a hydrocyclone which separates the sand (recycled to the injector tank) from the sludge.

An effective size of 135 μm was chosen for the microsand in order to limit sand losses at high speeds.

The following parameters were modified during the trials:

1. Residence time in the flocculation tanks.
2. Upflow velocity on the lamella surface.

These two parameters could be tested independently through variations of input flow and lamella surface.

3. Dosage of polyelectrolyte and ferric chloride.
4. Microsand rate, calculated as the ratio of weight of microsand recycled in the injection tank to the raw water volume entering the same tank during a given time.
5. Recycling flow rate.

The main parameter considered in this paper was the upflow velocity expressed as ratio of input flow by the horizontal settling area starting from 6 and going down to 1.9 m^2.

The main dependent variables recorded were the following:

- TSS removal
- COD removal

  and if required:
- BOD removal, total phosphorus and nitrogen as TKN
- Concentration of sludge evacuated from the hydrocyclones and ability of this sludge to settle.

Results and Discussion

TSS Removal

Figure 2 describes total suspended solids removed as a function of applied load of total suspended solids, for several upflow velocities.

Up to the 130 m/h maximum upflow velocity tested with the minimum 1.9 m^2 lamellar device, it is clear that the good TSS removal efficiency (about 80 % on the average), is almost independent of the upflow speed.

Settling problems appear to be solved up to high speeds with the microsand weighted flocculation technology, as long as the flocs are stuffed with enough micro-sand.

Optimization of depollution efficiency is linked to optimization of the flocculation conditions (chemical and hydraulic conditions, which largely depend upon the quality of influent waters).

For a specific water sample, jar tests have demonstrated that optimal chemical conditions are similar to those in classical lamellar settling.

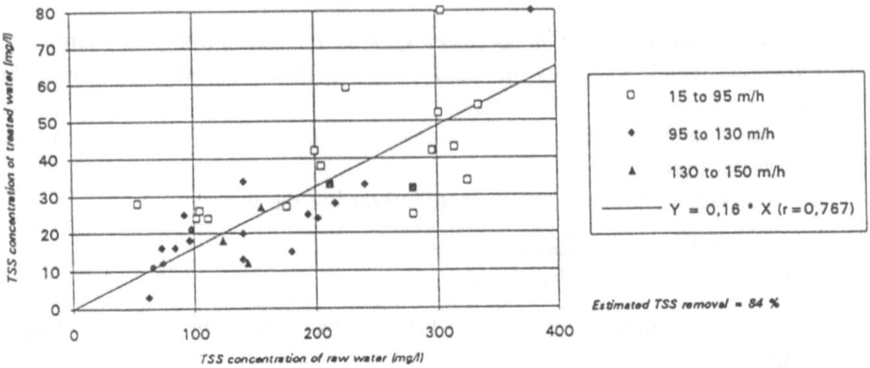

Fig. 2. TSS removal with Actiflo process as a function of upflow velocity (37 daily samples in Colombes pilot)

COD and BOD Removal

Figures 3 and 4 describe COD and BOD loads removed as a function of applied load for several upflow velocities.

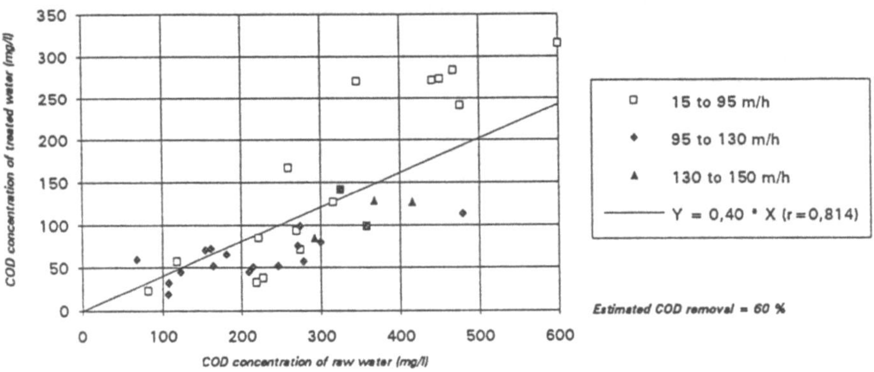

Fig. 3. COD removal with Actiflo process as a function of upflow velocity (37 daily samples in Colombes pilot)

Here again pollution efficiency proves to be almost independent of upflow velocities (within the tested range of velocities).

With the Colombes waters, BOD removal averaged about 70 % of applied BOD load, while COD removal was limited to a 55 % average.

This could be linked to a total average of 77 % insoluble BOD in Colombes raw water, against 63 % insoluble COD.

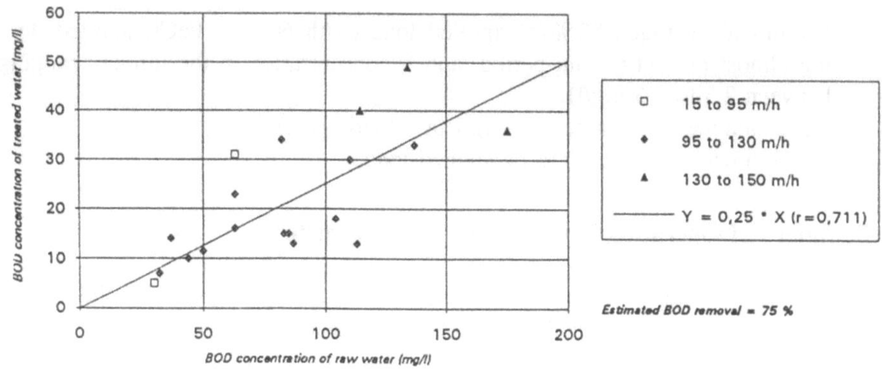

Fig. 4. BOD removal with Actiflo process as a function of upflow velocity (20 daily samples in Colombes pilot)

Total Phosphorus and TKN

- Figures 5 and 6 describe P and KN loads removed as a function of applied loads.

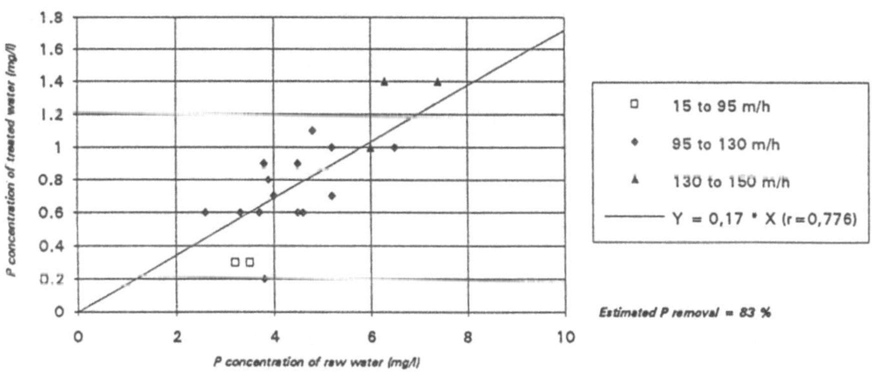

Fig. 5. Total phosphorus removal with Actiflo process as a function of upflow velocity (20 daily samples in Colombes pilot)

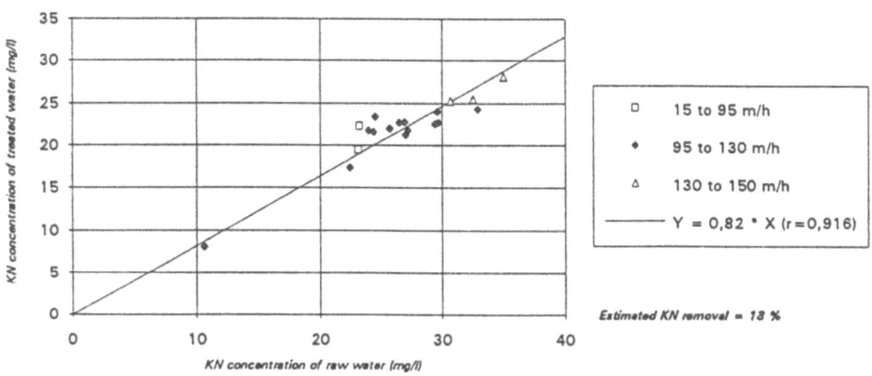

Fig. 6. Kjeldahl nitrogen removal with Actiflo process as a function of velocity (20 daily samples in Colombes pilot)

- P removal averaged 82 % of applied load (with 60 mg/l $FeCl_3$ applied during almost the entire trial period, and P concentration in the influent ranging between 2.5 to 7.5 mg/l).
- All P samples were under 1.5 mg/l in settled waters.
- KN removal averaged 17 % of applied load.

Summary of Operating Performance and Parameters

At the present stage of the experiments, and based upon the Colombes water quality, the following removal efficiencies have been secured (Figs. 7, 8):

Parameter	Average removal %
TSS	80
COD	55
BOD	69
Total P	82
Total KN	17

with operating conditions set as follows:

Parameter	Rate
Upflow velocity	Up to 135 m/h
Total residence time in coagulation and flocculation tanks	Less than 10 min
Ferric chloride	50 to 100 g/m^3 (usullay 60 g/m^3)
Polyelectrolyte	0.5 to 1 g/m^3 (usullay 0.8 g/m^3)
Recycled microsand rate	2 to 8 kg/m^3
Power consumption	Less than 30 Wh/m^3

Sludge was extracted from the hydrocycloning stage with a concentration in the range of 2–8 g/l.

Extracted sludge demonstrated a good thickening ability. Through in-line addition of polymer to sludge and a simple thickening on the lamellae, a concentration of 25 to 30 g/l was easily obtained.

Prototype Start Up and Shut Down

Since microsand is a chemically inert material, the shut down procedure requires only the operation of the sludge and sand pump for some time, in order to clean the sand from the sludge and to store it in the injection tank. If the shut down

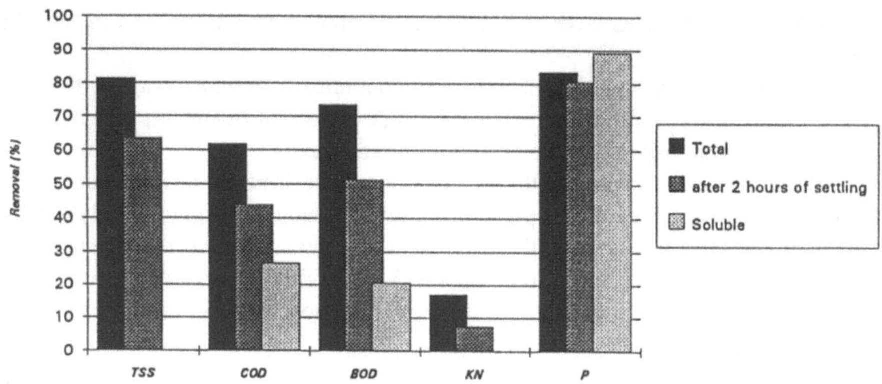

Fig. 7. Average removal of the main pollution parameters (20 to 40 daily samples in Colombes pilot)

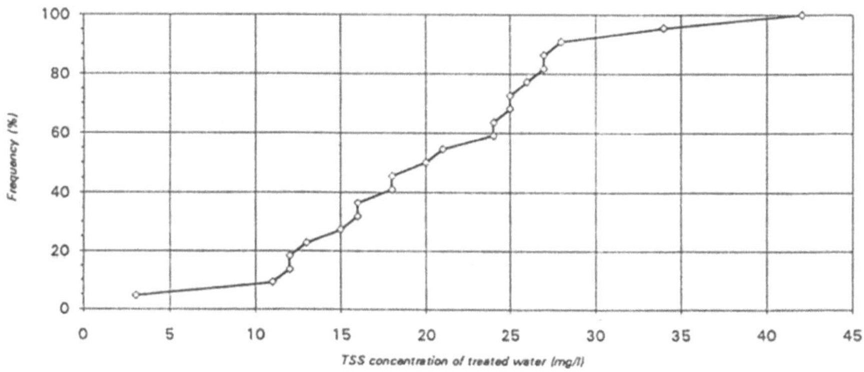

Fig. 8. Frequency of TSS concentration outflow with Actiflo process (22 daily samples in Colombes pilot – $TSS_{influent}$ = 60–200 mg/l)

period is expected to last some time (say, more than a few days), chlorination of water avoids fermentation problems.

Start up of the prototype is carried out at full load within about one retention time (i.e. TSS removal 80 % within about 20 min).

Conclusion

The microsand weighted settling technology has been used for years for potable water treatment.

Applied to combined sewers stormwaters, this compact technology allows TSS, COD, BOD and P removal close to those observed on classical lamellar settling physico-chemical treatments, with the settling area decreased by approx a factor of 5 (or by a factor of 50 compared to classical static settling physico-chemical treatment).

That means flows as large as 5 m^3/s can be applied to a 14 m × 14 m settling area.

Such technology could constitute an economically feasible way of treating the large stormwater flows untreatable through biological treatment, especially in urban areas with large water flows and high land costs.

The microsand-weighted settling devices could be installed downstream of the CSO's protecting the treatment plant or the sewage network, being run only during medium and large rain events (therefore with low overall chemical consumption).

The same devices could also be operated in dry time, downstream of the final clarifiers, acting as final polishers and for post-dephosphatation. Preliminary tests on this configuration are promising.

Bibliography

Argaman, Y.A. Kaufman, W.J.: Turbulence and Flocculation. J. Div. Sanit. Eng., Pro. Am. Soc. Civil Eng. *96* (SA2) (1979) 223–241

Argaman, Y.A.: Pilot Plant Studies of Flocculation. JAWWA *63* (12) (1971) 775–777

Bourdon, F., Jarosz, J., Ottavioli, J., Mallen, A.: Optimizing Operating and Control of Physico-Chemical Lamella Settling. IAWPRC – IWSA Specialized Conference – Coagulation, Flocculation, Filtration, Sedimentation and Flotation in Water and Wastewater Treatment, Jönköping, Sweden, April 24–26, 1990

Chebbo, G., Saget, A., Bachoc, A.: Elements pour le dimensionnement des ouvrages de dépollution des rejets urbains par temps de pluie. AFITE la pollution par les eaux pluviales Qualité-Collecte-Solution d'avenir, Lyon, 4–5 nov. 1992, pp. 17–26

De Dianous, F., Pujol, E., Druoton, J.C.: Industrial Application of Weighted Flocculation Development of the Actiflo® Clarification. In: Chemical Water and Wastewater Treatment, H.H. Hahn and R. Klute (Eds.). Springer, Berlin Heidelberg New York 1990, pp.

Guibelin, E.: Le procédé Actiflo – Décantation lamellaire par flocs lestés des eaux de surface. Tribune de l'eau, Nov./Déc. 1990 (special Pollutec) pp. 3–6

Janssens, J.G., Buekens, A.: Theoretical Analysis and Practical Application of the Kinetic Model for Flocculation in the Interpretation of Jar Tests. Aqua *2* (1987) 91–97

Morissey, S.P., Harlemans, D.R.F.: Retrofitting Conventional Primary Treatment Plants for Chemically Enhanced Primary Treatment in the USA. In: Chemical Water and Wastewater Treatment II, R. Klute and H.H. Hahn (Eds.). Springer, Berlin Heildelberg New York 1992, pp. 401–416

Ostrowski, M.: Storm Water Treatment in the Federal Republic of Germany. AFITE la Pollution par les eaux pluviales Qualité - Collecte - Solutions d'Avenir, Lyon, 4–5 Nov. 1992, pp. 141–149

Sibony, J.: Clarification with Microsand Seeding, a State of the Art. Water Research *15* (1992) pp. 1281–1290, 1981

Tabuchi, J.P.: Les impacts des rejets urbains de temps de pluie. Agence de l'Eau Seine Normandie, 24 Juin 1991

E. Guibelin, F. Delsalle, and P. Binot
O.T.V. (Omnium de Traitement et de Valorisation)
11, Avenue Dubonnet
F-92407 Courbevoie
France

Sludge Handling

Chemical Treatment – Consequences for Sludge Biosolids Handling

P. Balmér

Abstract

The factors affecting sludge production rates at wastewater treatment plants are discussed. Methods are presented for the estimation of additional sludge production when chemical precipitation is employed. Sludge production rates at full scale plants employing chemical precipitation are given. When using iron or aluminium salts sludge production increases by 22–35 %. At plants with lime precipitation the sludge production can increase by 100 % or more. Experiences with thickening, stabilization and dewatering of sludges from chemical precipitation plants are discussed. With chemical precipitation the release of phosphorus to the water phase during anaerobic digestion is small, otherwise chemical precipitation has no essential impact on sludge treatment. It is concluded that an increase in sludge production is the only negative consequence of chemical precipitation on sludge treatment. Recovery of chemicals may change this in the future.

Sludge – Management's Main Headache

In most countries the treatment and disposal of wastewater is well regulated. When a consent to dispose of wastewater is granted and the effluent is in agreement with the standards, the operation of a treatment works is, despite all the problems normally encountered, a relatively straightforward task.

Sludge treatment operations and especially sludge disposal are often much more troublesome. The reasons for this are that sludge is more difficult to treat than wastewater and many more parties are involved in sludge disposal. In addition to environmental authorities, local public health authorities, the local solid waste disposal organisations and farmers and their organisations are often involved. The public often express strong opinions about sludge disposal. Many wastewater treatment plant (WWTP) managers have experienced that sludge disposal methods that seemed to be reliable, canbe closed almost overnight because of political decisions, public opinion pressure or new regulations.

Another important aspect of sludge treatment and disposal is cost. The operating cost for sludge treatment and disposal is often a major part of the total

R. Klute / H.H. Hahn (Eds.)
Chemical Water and Wastewater Treatment III

operating budget. The trend is also that costs connected to disposal are rapidly increasing.

When chemical treatment of wastewater is being considered, it is therefore of importance not only to evaluate the treatment efficiency, cost of chemicals and the necessary treatment plant volumes, but also the effect on sludge production rates, sludge treatment, and sludge disposal.

Sludge Production Rates

Calculation of Sludge Production

The sludge production at a treatment plant depends on:

- the degree of treatment
- the degree of degradation of organics
- the addition of chemicals to the treatment process

The amount of dissolved organics and, especially, the amount of suspended solids in the effluent from the treatment plant have a strong effect on the net sludge production. Substantial reductions in net sludge production can easily be "achieved" if the per capita effluent flow or the effluent suspended solids concentrations are high (Fig. 1).

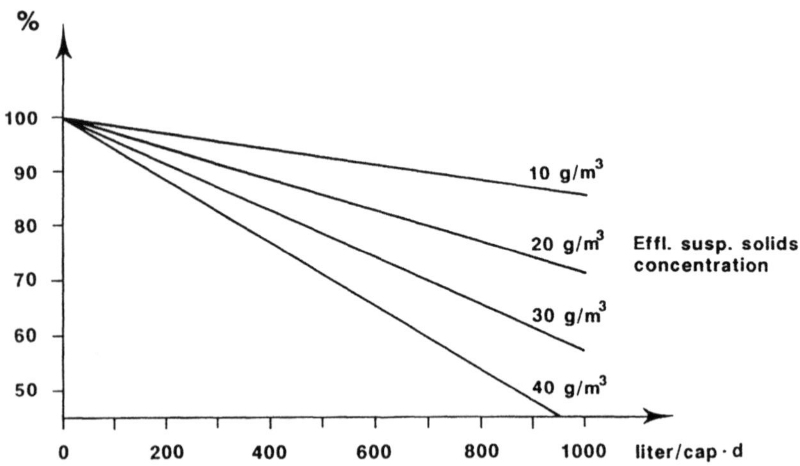

Fig. 1. Net sludge production as a percentage of gross sludge production as a function of per capita wastewater flow and final effluent suspended solids concentration. (Gross sludge production of 70 g DS per capita and day is assumed)

During biological treatment, organics are removed to some degree. The degradation in biological wastewater treatment is, however, quite small in relation to the total amount of organics removed, as illustrated in Figure 2. The main part of the organics degradation takes place in the digestion.

When precipitation chemicals are added to wastewater treatment processes, sludge production increases as the chemicals form phosphates and other precipitates.

Balmér and Fredriksen (1975) collected extensive data on sludge production on a pilot scale. With this data, Balmér (1974) showed that the increase in sludge production can be calculated stoichiometrically by assuming that the phosphates in the wastewater are removed as $AlPO_4$ and $FePO_4$, respectively, and that any surplus precipitant is removed as $Al(OH)_3$ or $Fe(OH)_3$. Any improvement (deterioration) in suspended solids removed, due to the chemical precipitation, has to be added to (deducted from) the stoichiometrically calculated amounts.

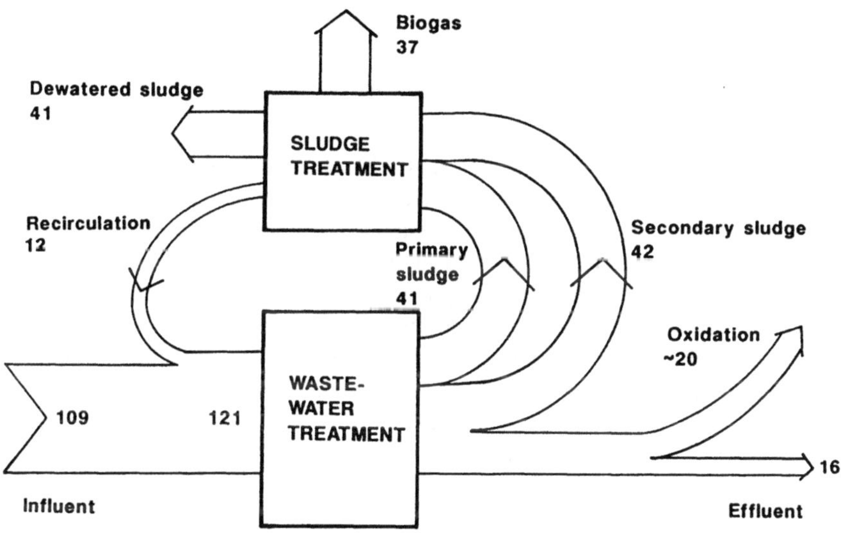

Fig. 2. COD balance for the Rya WWTP in Göteborg, Sweden. All data given as tons COD per day

When lime is used as a precipitant, sludge production estimation is more complicated. The phosphate precipitate (calcium hydroxylapatite) is small (less than 20 % in most cases) in relation to the total precipitate. The main components of the precipitate are calcium carbonate and unutilized lime. Balmér (1984) showed that the requirement of alkali for the necessary pH rise in lime precipitation was closely related to the molar sum of the ammonia and bicarbonate content of the wastewater (Fig. 3). The lime requirement was about 50 % higher than the alkali requirement. This indicates that not all lime is dissolved.

The precipitation of calcium carbonate can be estimated from the initial hydrogen carbonate concentration, the pH and the calcium carbonate solubility product. It must, however, be noted that the solubility products in standard handbooks are too low.

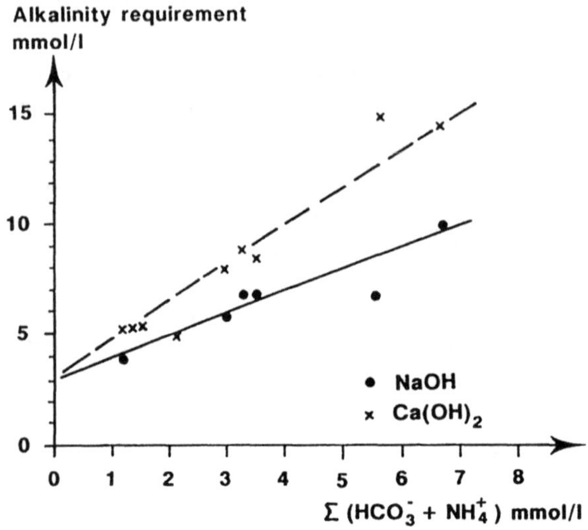

Fig. 3. Sodium hydroxide and lime requirements to reach pH 11.4 in wastewaters are the molar sum of bicarbonate and ammonia ions

A value of 0.2×10^6 (Wiechers et al. 1980) can be recommended for practical applications. In addition to the calculated amounts of calcium carbonate the undissolved lime in the sludge has to be estimated. This can be done by assuming that the amount of lime used in excess of the alkali requirements can be found as undissolved lime in sludge. In lime precipitation magnesium hydroxide is also formed. The amount is, however, negligible in relation to the total amount of sludge formed. Balmér (1984) made a large number of lime precipitation experiments and found that the amount of sludge calculated by the procedure described above agreed with the actual amount formed within ±20% for 80% of the data.

Full Scale Experiences

In a recent survey Balmér and Mattsson (1993) collected information on sludge production and the consumption of chemicals from wastewater treatment plants with full biological treatment, chemical precipitation and sludge digestion. If the amount of sludge formed from the addition of chemicals is calculated, as discussed above, the fraction of the total sludge production due to the chemicals can be calculated. The results of these calculations are shown in Figure 4. With normal amounts of precipitation chemicals, 30–40 moles per capita and year, the fraction of chemical sludge as dry solids amounts to 18–26% when iron is used as a precipitant. This means that chemical precipitation will increase the amount of sludge at a treatment plant by 22–35%. The median sludge production at wastewater treatment plants with full biological treatment and chemical precipitation with iron salts was about 22 kg DS/cap/year after digestion. Only 20% of the sludge production data deviated more than 20% from this figure. In all calculations it has been assumed that one population equivalent (pe)equals 70 g BOD7 per day, i.e. the approximately 60 g BOD5 per day.

With alum as a precipitant, the increase in sludge production should be slightly lower as indicated in Figure 4. However, alum is used mostly in post-precipitation. In post-precipitation the alum doses are often quite high since a considerable amount of the alum is an excess dose in order to reduce the pH to achieve good precipitation and floc formation. In the survey of full scale plants, sludge production of plants employing alum precipitation was in fact higher than at plants using ferrous or ferric iron precipitation. In an unpublished study from 1980 the author collected sludge production data from plants using alum in post-precipitation mode. The average sludge production rate at these plants was 30 kg DS/cap/year. When alum or prepolyimerized aluminium salts are used in pre-precipitation, full scale experience shows that the doses can be kept low and thus considerably lower sludge production rates are to be expected.

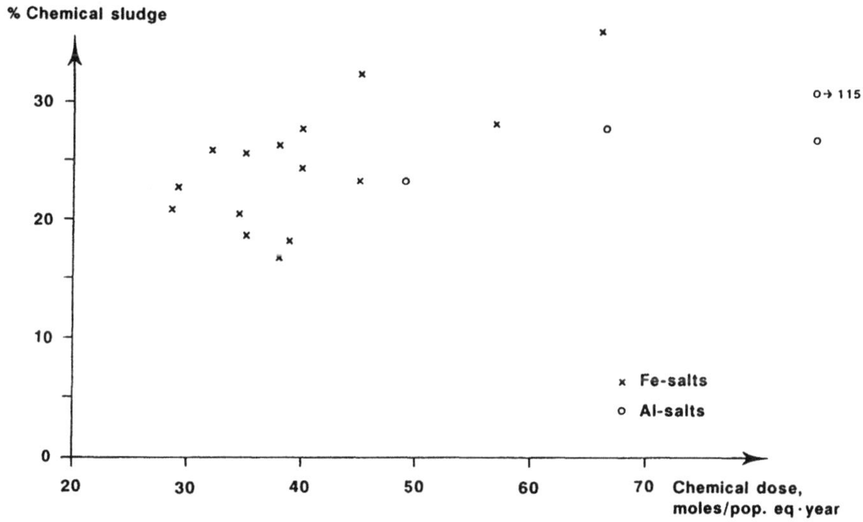

Fig. 4. Chemical sludge production as a percentage of total sludge production at wastewater treatment plants with primary and secondary treatment and sludge digestion verses dosage

Considerably higher sludge production rates can be expected with lime precipitation. Balmér (1984) collected full scale data from plants employing post precipitation with lime. At plants with low alkalinity in the secondary effluent caused by nitrification, the lime dose was 200 g/m^3 or lower while at plants without nitrification the lime dose was 250–500 g/m^3. At the low lime dose plants the sludge production was 30–40 kg DS/pe/year and 50–100 kg DS/pe/year at the high lime dose plants.

Ødegaard (1992) reported full scale experiences from over 50 Norwegian plants using chemical precipitation of raw and presettled wastewater. With alum the sludge production was 35.7 ± 14.9 kg DS/(capita · year) and with ferric iron 21.3 ± 13.5 kg DS/(capita · year). Most of these plants did not stabilize the sludge.

Sludge Treatment

To Mix or Not to Mix

When chemicals are added prior to primary treatment or prior to secondary treatment, the chemical precipitate is thoroughly mixed with the primary and waste sludge, respectively.

When chemicals are used in a separate tertiary step (post-precipitation) it is possible to treat the chemical sludge separately from the primary and secondary sludges. Several post-precipitation plants in Sweden were initially designed for separate handling of the chemical sludge. In most cases this meant that the chemical sludge was thickened separately and mixed with the digested primary and secondary sludges before dewatering.

Most of these plants have abandoned this practice for two main reasons. Firstly, a considerable amount of the "chemical" sludge consists of precipitated suspended solids that have escaped from the secondary settlers. The chemical sludge is therefore often not inert and may need digestion or some other type of stabilization. Secondly, it is difficult to achieve a homogeneous mixture when primary and secondary sludges are mixed with chemical sludges just prior to dewatering. This makes it difficult to achieve stable operation of the dewatering equipment. If the chemical sludge is mixed with the primary/secondary sludges in the digester, inhomogeneity problems are solved.

Thickening

Chemical sludges from post-precipitation with alum or ferric iron are in most cases difficult to thicken. Their thickening properties are often as bad or even worse than waste activated sludge (Fredriksen 1974). Chemical sludges from the precipitation of raw wastewater are considerably easier to thicken.

Practical experiences in Norway and Sweden indicate that concentrations of 3–5 % are generally achieved.

If lime is used as a precipitant the thickening properties are excellent. In some cases the opposite problem has arisen: the lime sludge thickens so well that it is difficult to pump it out from the settler sludge pits and from the thickeners.

Anaerobic Digestion

The general experience is that digestion of mixed primary and secondary chemical sludges from ferrous, ferric and alum precipitation works well. The increased mass of sludge may affect the retention time in the digester. Gas production appears unaffected. On the other hand, there are a number of plants reporting increased gas production. In most cases these plants employ pre-precipitation. Pre*precipitation removes more organic material in the primary settlers and this sludge is degraded anaerobically instead of aerobically, the net result being an increase in gas production.

With ferrous and ferric iron as a precipitant there has been concern that phosphorus could be released under the reduced conditions that prevail in an anaerobic digester. Practical experience has shown however that the dissolution of phosphorus during anaerobic digestion is negligible. An advantage with sludge from ferrous and ferric ion precipitation in the digester is that hydrogen sulphide is effectively bound by the iron.

There is limited information on the anaerobic digestion of sludges from the precipitation of raw or presettled wastewaters. Recently an anaerobic digestion plant for sludges from ferric iron precipitation of raw wastewater was put into operation at a large wastewater treatment plant outside Oslo. Sagberg (1994) reports good operational results.

Aerobic Digestion

At small treatment plants aerobic digestion is often preferred to anaerobic. Aerobic digestion also works well with sludges from chemical precipitation. Eikum et al. (1974) studied aerobic digestion of mixed primary chemical sludges from precipitation with alum. The degradation was similar to what could be expected from aerobic stabilization of primary sludges. Recently Paulsrud and Nedland (1994) reported on thermophilic aerobic digestion of mixed primary chemical sludges from chemical precipitation with ferric iron. Also in this case the results were in agreement with what has been achieved with primary/secondary sludges.

Dewatering

In dewatering, the dry solids in the sludge cake, the solids recovery and the consumption of conditioning chemicals are of concern.

Solids recovery well above 90 % is generally achieved with proper conditioning when mixed primary and secondary chemical or mixed primary chemical sludges are dewatered.

Information on sludge cake dry solids was collected by Balmér and Mattsson (1993). The average cake dry solids was 24 %. The sludges were digested, mixed primary/secondary chemical sludges. The average polymer consumption at these plants was 2.8 kg/ton DS with values ranging from 1.4 kg to 5.2 kg/ton DS. The dewatering equipment used was decanter centrifuges and continuous band filters. The data base was too small to draw any conclusions on the relative performance of centrifuges and band filters. The data on polymer consumption and sludge cake dry solids are better than in the 1980 survey by the author.

Ødegaard (1992) reported sludge cake concentrations of 21.1 ± 2.7 % DS at plants with chemical precipitation of raw or presettled wastewater using alum as a precipitant and 26.6 ± 6.4 % DS when ferric iron was used. The polymer consumption was 2.86 ± 0.97 kg/ton DS.

There exist no good data where the sludge from the same plant with and without chemical precipitation can be compared. The data that exist are from different time periods and, as the performance of dewatering equipment and conditioning chemicals have improved considerably during the last 20 years, such comparisons are not relevant.

Disposal

Sludges from chemical treatment plants are valuable in agriculture since the phosphorus concentration is high, 2.5 % P as an average in Sweden (Sjöqvist and Wikander-Johansson, 1985). To what extent this phosphorus is available to plants is still under debate.

A factor of increasing importance for sludge disposal is the concentration of heavy metals in the sludge. Most important is, of course, to ensure that there are no heavy metal containing industrial discharges to the sewerage system, but the heavy metal content in the precipitation chemicals is also important. There are several examples of severe heavy metal contamination of sludge from low quality chemicals. With the high quality chemicals now on the market, the heavy metal contribution from the chemicals is normally negligible.

The heavy metal of greatest concern when sludge is disposed of to agriculture is cadmium. The quality of a phosphorus fertilizer can be expressed as the Cd/P-ratio. The successful control of industrial discharges, the high quality of precipitants and the high concentration of phosphorus in the sludge have, at most Swedish treatment plants, resulted in a sludge with a Cd/P-ratio that is as good as or even better than commercial fertilizers.

The only drawback of chemical treatment from a disposal point of view is the increased volume to be disposed. If the emerging technology of recovery of chemicals (Göransson and Karlsson 1994) is successful this drawback can be eliminated.

References

Balmér, P.: Slamproduksjon ved kjemisk felling. (Sludge Production from Chemical Precipitation.) Vann *9* (1974) 33–38 (in Norwegian)

Balmér, P., Fredriksen, O.F.: A Pilot-plant Scale Evaluation of Potential Precipitants in the Secondary Precipitation Process. Wat. Res. *9* (1975) 721–727

Balmér, P.: Erfarenheter med fällning av kommunalt avloppsvatten med kalk. (Experience from Precipitation of Municipal Wastewater Using Lime.) Chalmers Univ. of Technol., Dep. of San. Eng., Publ. 3:84 (1984) (in Swedish)

Balmér, P., Mattson, B.: Kostnader för drift av avloppsreningsverk. (Wastewater Treatment Plant Operation Costs.) VA-Forsk Rapport 1993-15. ISBN 91-88392-66-X

Eikum, A.S., Carlson, D.A., Paulsrud, B.: Aerobic Stabilization of Primary and Mixed Primary-Chemical (Alum) Sludge. Wat. Res. *8* (1974) 927–935

Fredriksen, O.F.: Fortykking av kjemisk slam. (Thickening of Chemical Sludges). Vann *9* (1974) 39–42

Göransson, J., Karlsson, I.: (1994). Beneficial Use of Sludge from Sewage Plants and Water Works. In: Chemical Water and Wastewater Treatment III, R. Klute and H.H. Hahn (Eds.). Springer, Berlin Heidelberg New York 1994, pp. 341–352

Ødegaard, H.: Norwegian Experiences with Chemical Treatment of Raw Wastewater. Wat. Sci. Tech. *25* (1992) 255–264

Paulsrud, B., Nedland, K.T.: (1994). Full Scale Experiences with Thermophilic Aerobic Digestion of Primary-Chemical Sewage Sludge. In: Chemical Water and Wastewater Treatment III, R. Klute and H.H. Hahn (Eds.). Springer, Berlin Heidelberg New York 1994, pp. 329–339
Sagberg, P.: VEAS, PO Box 245, N-3740 Slemmestad. Personal Communication (1994)
Sjöqvist, T., Wikander-Johansson, E.: Vad innehåller slammet. (Sludge Composition). National Swedish Laboratory for Agricultural Chemistry, Publ. No 52, Uppsala 1985 (in Swedish)
Wiechers, N.H.S., Lowenthal, R.E., Marais, G.V.R.: Lime Treatment of Wastewater: Development and Application of a Simple Graphical Technique for Predicting the Composition of Lime Treated Secondary Effluent. Prog. Wat. Tech. *12* (1980) 347–358

Peter Balmér
Göteborg Regional Sewage Works
Karl IX: s väg
S-418 34 Göteborg
Sweden

Full Scale Experiences with Thermophilic Aerobic Digestion of Primary-Chemical Sewage Sludge

B. Paulsrud and K.T. Nedland

Abstract

A full scale demonstration project with thermophilic aerobic digestion (TAD) of primary-chemical sludge was started in May 1993 at Vårnes wastewater treatment plant (7300 p.e.). The project is partly funded by the Norwegian State Pollution Control Authority. The main objective is to document plant performance and characteristics of treated sludge when handling primary-chemical sludge.

The demonstration plant is a standard two-stage facility from the German manufacturer Fuchs Gas- und Wassertechnik operating with a batch (once a day) feeding of sludge. A comprehensive monitoring programme has been established and this paper presents data on volatile solids reduction, pathogen reduction, sludge dewaterability and sludge nutrient content.

A general conclusion so far is that the TAD process investigated here seems to be appropriate for stabilisation and disinfection of primary-chemical sludge, and that plant performance corresponds well with the long-term experiences from German plants treating biological or primary + biological sludges.

Introduction

In 1991, the Norwegian health and environmental authorities prepared a proposal for new regulations and guidelines for the treatment and disposal of sewage sludge [1]. This proposal passed public hearings in 1991/92, but has not yet been implemented (Jan. 1994) due to ongoing legal discussions. However, the technical requirements have been set and most municipalities are planning to meet standards in a revised version of the first proposal.

The main objective of the new rules is to increase the reuse of sewage sludge and the goal is to recycle 75 % of the total sewage sludge production for beneficial use on land areas within the year 2000 (the figure for 1993 was about 50 % reuse) [2]. In order to achieve this, high quality sludge should be offered to agriculture, green areas and forestry. This means that the sludge should contain only small amounts of heavy metals, and odour nuisance and health risks should be controlled by proper stabilisation and disinfection methods.

R. Klute / H.H. Hahn (Eds.)
Chemical Water and Wastewater Treatment III

Since sludge stability cannot be measured by a universally accepted standard test and monitored with respect to official standards, the Norwegian authorities have decided to approve certain processes or process combinations for stabilisation and disinfection of sewage sludge. This initial approval is based upon the German "ATV-Regelwerk" [3–5], which includes guidelines on design and operation of disinfection processes which have been employed for several years at full scale treatment plants in Central Europe. The processes approved so far are:

- Composting (windrows, aerated static pile, reactor)
- Thermophilic aerobic digestion
- Thermophilic aerobic pretreatment + anaerobic digestion
- Pre-pasteurisation + anaerobic digestion
- Heat drying
- Lime treatment (addition of quick lime to dewatered sludge)
- Long-term (3–4 years) storage of dewatered sludge.

Thermophilic Aerobic Digestion

Thermophilic aerobic digestion (TAD) utilises the metabolic heat produced during the biological oxidation of sludge to raise the temperature of the sludge to above ambient temperature. Depending on the conditions, temperatures in the range of 40 to 70 °C can be achieved. It is therefore essentially a "wet composting" process and, as a result of the high operating temperature, the rate of stabilisation is rapid. Provided oxygen is not limiting, retention periods of only five to ten days may be sufficient for production of stable and virtually odourless sludge. The digester volume requirements are therefore much smaller than those for cold aerobic digestion, or for conventional anaerobic digestion. Operated in the high temperature regime (> 55 °C), the process also satisfies normal disinfection criteria.

The TAD process for sewage sludge treatment was developed almost simultaneously in Germany [6, 7], the U.S.A. [8, 9] and in the UK [10, 11] during the 1970s and both pure oxygen and air aeration systems were studied. Since then, several TAD plants have been implemented in Europe, and particularly in Germany, where about 35 full scale facilities were in operation in 1989 [12]. The German system manufactured by Fuchs Gas- und Wassertechnik is by far the most widely used system, and therefore it is the system for which the greatest amount of full scale operating experience exists [12]. These plants consist of two reactors in series (two-stage system) utilizing aspirator aerators and mechanical foam cutters.

In Norway, both pure oxygen and air aeration TAD processes were tested on a pilot scale during the mid-1980s [13–15], but no full scale plants were built, mainly due to non-existing requirements for stabilisation and disinfection of sewage sludge. This situation changed with the new sludge regulations and the authorities now encourage the "sludge industry" in Norway to come up with demonstration projects. They were particularly interested in those processes already approved for stabilisation and disinfection and with no full scale experience in treating non-biological sludges. Thermophilic aerobic digestion was one of these, and a full scale demonstration plant was established, partly funded by the Norwegian State Pollution Control Authority.

Description of Demonstration Plant

Vårnes wastewater treatment plant serves the centre of the municipality of Stokke in the southern part of Norway. This old primary treatment plant was converted to a modern chemical precipitation plant in 1992 and at the same time a new sludge handling system was incorporated. Prepolymerized aluminium chloride (PAX 21) is used as coagulant. Figure 1 presents the process flow diagram of the plant, including the TAD demonstration facility. Extensive buffering capacity is included before and after the TAD reactors to allow high flexibility in the operation as a demonstration plant. The main purpose of the heat exchanger is to reduce the sludge temperature prior to dewatering, since dewatering of hot, aerobically digested sludge was expected to be a problem.

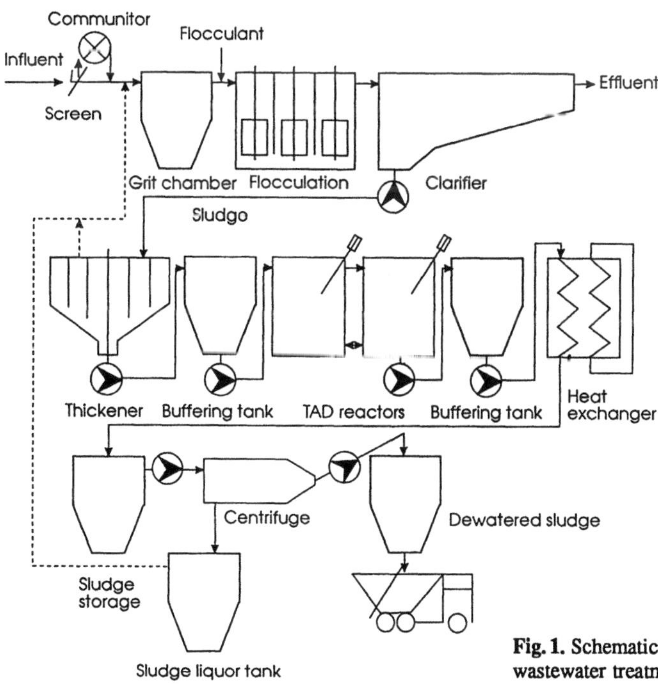

Fig. 1. Schematic flow diagram of Vårnes wastewater treatment plant

The TAD demonstration facility is a standard two-stage Fuchs system, operating with batch feeding (once a day) of sludge. Design values for the wastewater treatment plant and the TAD plant are given in Table 1.

Table 1. Design values for Vårnes wastewater treatment plant

Wastewater treatment plant	Capacity:	7300 p.e.
	Flow:	134 m^3/h
TAD plant	Capacity:	10000 p.e.
	Total solids loading:	1200 kg TS/d
	Volatile solids loading:	720 kg VS/d
	TS concentration in feed sludge:	5% TS
	Volumetric loading:	24 m^3/d
	Retention time:	7 days
	Temperature in reactor 2:	50–65 °C

Sampling, Monitoring and Analysis

A comprehensive sampling and monitoring programme is established to collect performance data, such as volatile solids and pathogen reduction across the TAD system and sludge dewaterability, as well as quality of sludge liquors from full scale dewatering of digested sludge and nutrient and heavy metals content of digested sludge. All sampling points and parameters are summarized in Figure 2 (see dotted boxes and lines).

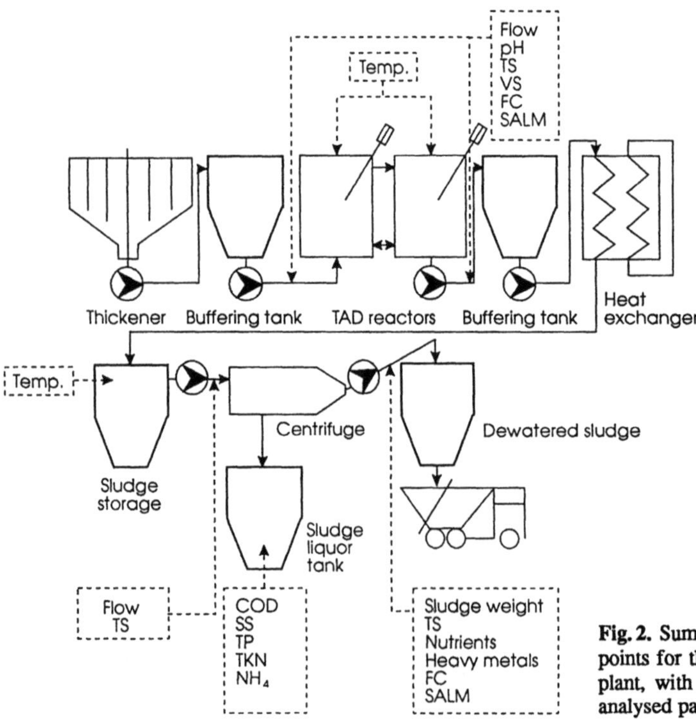

Fig. 2. Summary of sampling points for the sludge treatment plant, with monitored and analysed parameters

Composite samples are taken from each batch of sludge leaving reactor 2 (digested) and entering reactor 1 (raw). This is done by means of automatic solenoid valves controlled by the sludge pumps and a timer. These daily samples are analysed for pH, total solids (TS) and volatile solids (VS). Every week, one set of daily samples is analysed for faecal coliforms (FC) and *Salmonella* (SALM). Temperature is continuously monitored at two levels in both reactors and manual temperature measurements are performed in the storage tank for digested sludge during dewatering periods.

Samples from the sludge dewatering process (feed sludge, sludge cake and sludge liquor) are taken manually during the centrifuge operation which normally lasts for about 6 hours a day, 2 days a week. Once a week, a grab sample of sludge cake is analysed for bacteria. Weekly composite samples are prepared and analysed for total solids (feed sludge and sludge cake) and suspended solids (sludge liquor). Composite samples of the sludge cake are made for every two weeks and analysed for heavy metals and nutrients. One daily sample of the sludge liquor is analysed every two months for COD, total phosphorus (TP), total Kjeldahl nitrogen (TKN) and ammonia.

Results and Discussion

General Operational Experience. The TAD plant was put into regular operation during April 1993 after a long period of problems with the central control system of the treatment plant and with the sludge level sensor in the buffering tank following TAD reactor 2. Normally, a programmable logic controller (PLC) controls the sludge feed and discharge procedures of the TAD plant, but also during the monitoring period reported here (May–November 1993), manual operation was necessary from time to time. There were two major reasons for this:

- Due to some misunderstanding, the PLC was initially programmed to stop raw sludge feeding the next day whenever sludge discharge and feed procedures lasted for more than one hour.
- The sludge level sensor in the buffer tank occasionally produced a false, high level signal to the PLC, which automatically stopped the discharge of sludge from the TAD reactor 2.

From June 28 to August 15, the monitoring programme was stopped since one of the sludge pipes between TAD reactors 1 and 2 was blocked and the reactor had to be emptied. The clogging was probably caused by the special arrangement at this plant: macerating the wastewater screenings and returning them to the plant influent, resulting in a high content of gross particles in the sludge.

Actual Load of the TAD Facility. During the monitoring period, the volumetric loading of the TAD reactors varied between 0 and 25 m^3/d, with an average of 8 m^3/d. This low average load compared to design load (24 m^3/d), is mainly due to the fact that the plant serves a population of only 6000, even though it is

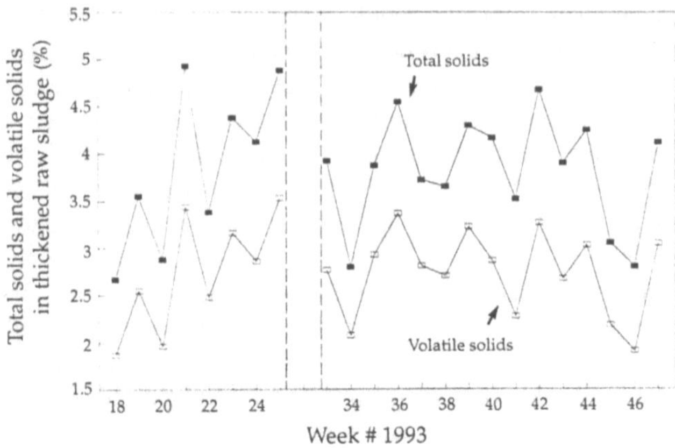

Fig. 3. Variations in feed sludge total solids and volatile solids concentration (weekly averages)

designed for 10000, and that the per capita sludge production is overestimated (120 g TS/(p.e. · d)). The great variations in daily volumetric loading are caused by the aforementioned problems with the automatic operation of the TAD facility.

Organic loading varies even more than the volumetric loading, due to the great variations in total solids and volatile solids content of the thickened raw sludge (see Fig. 3). During some weeks, the average concentration of volatile solids was below 25 kg/m^3 (2.5 %), which is often claimed to be the lower limit for autothermal aerobic digestion [16, 17]. The temperature profile in TAD reactor 2 (see Fig. 4) supports this experience, with falling temperature following the weeks with low volatile solids concentrations in the feed sludge.

Volatile Solids Reduction. There are several methods for calculating VS reduction across aerobic or anaerobic digestion plants [18, 19]. In this case, we have used the "approximate mass balance" method, which defines the fraction of volatile solids reduced (FVSR) as:

$$FVSR = 1 - \frac{V_d}{V_r}$$

V_r Arithmetic average VS content of *raw* sludge added during the period considered
V_d Arithmetic average VS content of *digested* sludge removed during the period considered

The FVSR is presented in Figure 4 as the percentage VS reduction during the TAD process. All values are calculated from 28 days running arithmetic averages of VS content in raw and digested sludge, respectively.

The volatile solids reduction varied between 27 and 48 percent, with an average of 38 percent during the monitoring period. The periods with low VS reduction coincide with low VS concentration of the raw sludge (see Fig. 3) and a corresponding low temperature in the TAD reactors. The solids retention time, however, was fairly constant (20–27 days) throughout the period, with an average of 23 days. This clearly demonstrates the low-loaded mode of operation, and a

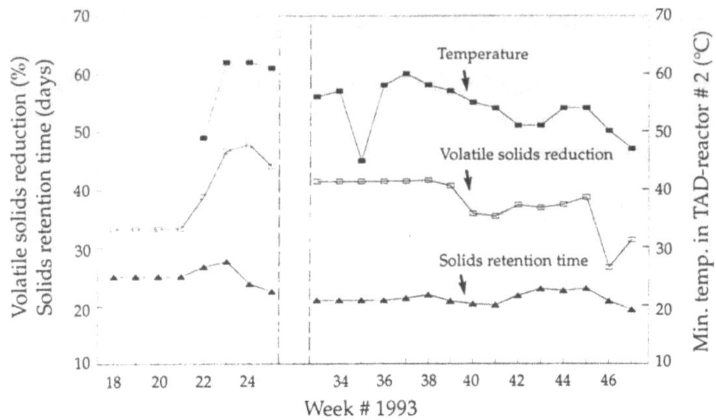

Fig. 4. Volatile solids reduction, solids retention time in the TAD process, and minimum temperature in TAD reactor 2

higher VS reduction would be expected assuming an increased VS content of the thickened raw sludge.

Pathogen Reduction. The content of faecal coliforms (FC) in raw sludge, digested sludge and digested + dewatered sludge is presented in Figure 5. With the exception of two weeks in August with aprobability for cross-contamination of raw and digested sludge during re-start of the process, all samples of liquid digested sludge showed FC values below the new Norwegian standards for sludge disinfection plants (< 2500 faecal coliforms per gram total solids). The high temperature level in TAD reactor 2 (see Fig. 4) and a corresponding high pH-value (average pH = 8.5) explain these results [17].

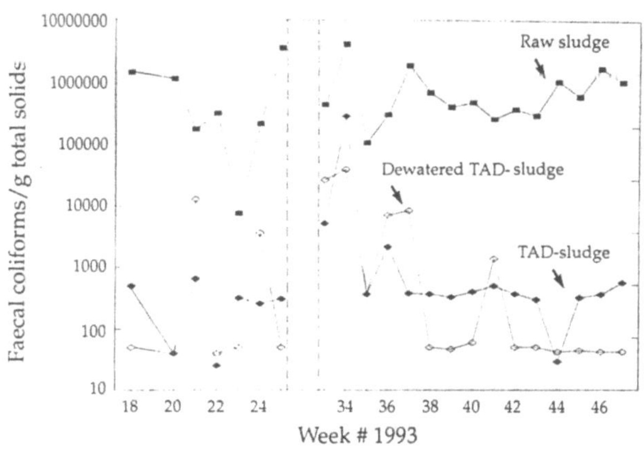

Fig. 5. Variations in faecal coliforms content in raw sludge, digested sludge and dewatered digested sludge

The samples of dewatered sludge, however, show somewhat higher FC values for certain weeks. Possible explanations for this may be recontamination during liquid sludge storage or during the dewatering process, or simply sampling and analytical errors.

All samples were also analysed for *Salmonella* bacteria, but so far only two samples of raw sludge have contained *Salmonella*. *Salmonella* has not been detected in digested sludge samples.

Dewaterability of TAD Sludge. The data recorded from the operation of the decanter centrifuge are presented in Figures 6 and 7.

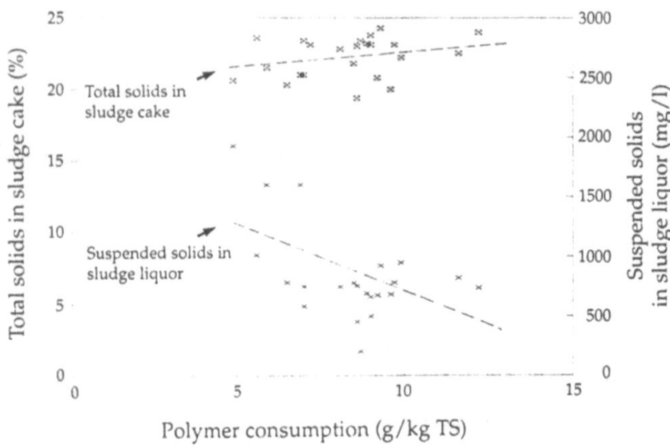

Fig. 6. Cake solids concentration and sludge liquor suspended solids concentration versus polymer consumption

Several combinations of polymer type, polymer dosage and sludge feed rate were investigated during the monitoring period, but so far no "optimum" combination has been found. In Figure 6, all data for polymer consumption (mostly the cationic polymer Zetag 34) are plotted against the corresponding values of sludge cake total solids and suspended solids in sludge liquor (centrate). As expected, the cake solids concentration is almost independent of the polymer dosage, whereas the centrate suspended solids decreases with increasing amounts of polymer up to a level of 8–9 g/kg TS. The average polymer consumption was 8.3 g/kg TS, which is 2–4 times the normal dosage for raw and aerobically digested primary-chemical sludges and somewhat higher than for aerobically (cold) digested primary + waste activated sludge [18].

During commissioning of the centrifuge, cake solids concentrations of 30–35 percent were obtained, but this high solids content caused severe problems with the cake discharge and transportation system. Accordingly, the machine variables were set to produce a more liquid sludge cake and the cake solids content has varied between 19 and 24 percent, with an overall average of 22 percent.

In the design stage, dewatering of high temperature TAD sludge was expected to cause problems and a heat exchanger was incorporated in the process scheme (see Fig. 1). During the different trials with varying polymer dosages, feed sludge rate, etc., feed sludge temperatures also varied due to varying operating conditions of the heat exchanger and the variable retention time in the sludge storage tank.

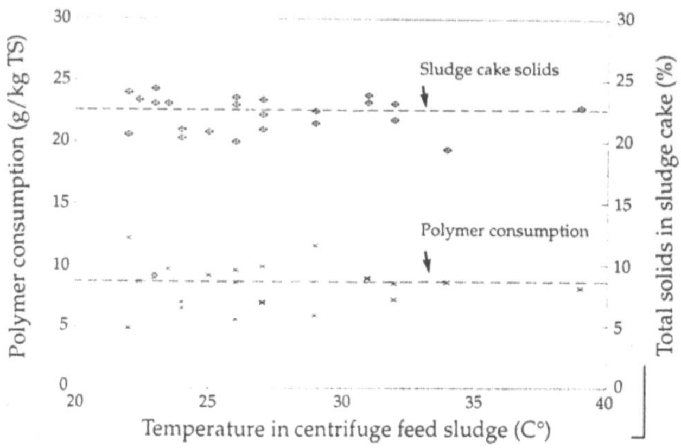

Fig. 7. Polymer consumption and sludge cake solids versus temperature in centrifuge feed sludge

Figure 7 shows that feed sludge temperatures within the range of 20–40 °C have no significant effect on the polymer consumption or the sludge cake solids.

Nutrient Content of Dewatered TAD Sludge. For a beneficial reuse of digested sludge on agricultural land, the nutrient content should be as high as possible. However, Norwegian sludge regulations require dewatering of all sludge prior to disposal to protect the water courses from surface run-off. This means that all digestion processes that solubilise nutrients in the sludge will reduce the nutrient content by dewatering and recycling of the sludge liquor to the treatment plant influent. In the TAD process, some organic nitrogen will be converted to ammonia nitrogen and recycled in the sludge liquor, but nitrification does not occur due to the high process temperature [9].

In Figure 8, the fertiliser value of TAD sludge is compared to those obtained in some Norwegian surveys of raw sludge and anaerobically digested sludge. All values are based upon dewatered sludge and presented as a percentage of total solids in the sludge. The most comprehensive survey from the County of Oppland represents to a large extent plants which produce biological-chemical sludge, whereas the County of Buskerud survey represents mostly primary-chemical treatment plants. The three different plants with anaerobic digestion have either a pre-pasteurisation process (Fredrikstad and Sandefjord) or a thermophilic aerobic pretreatment (Sarpsborg).

Figure 8 shows that the TAD sludge has a total nitrogen and ammonia content equal to or higher than anaerobically digested sludges. The total phosphorus content is even higher than for the raw sludges surveyed (no solubilisation of chemically precipitated phosphorus during digestion), but the potassium content is low.

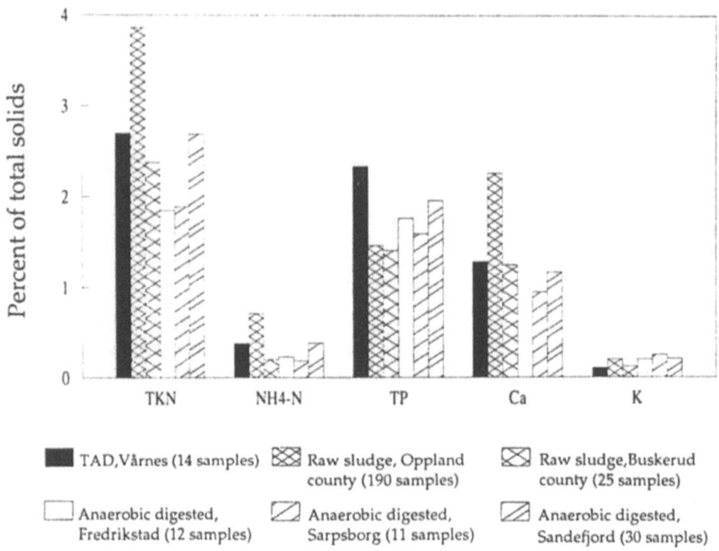

Fig. 8. Comparison of nutrient content in TAD sludge, raw sludge and anaerobically digested sludge

Conclusions

1. Full scale investigations have proven that thermophilic aerobic digestion (TAD) is an appropriate method for stabilisation and disinfection of primary-chemical sewage sludges.
2. In order to attain and sustain temperatures above 50 °C (and preferably above 55 °C) in an autothermal process, the volatile solids content in feed sludge should be at least 25 kg/m^3 (2.5 %). For primary-chemical sludge this means a total solids concentration of at least 3.5–4 % in feed sludge, which can be achieved by a total solids meter controlling the pumping of sludge from the thickener to the TAD facility.
3. A volatile solids reduction of 38 percent (average value) was achieved during the monitoring period (May–November 1993).
4. Faecal coliforms were normally reduced to levels below the Norwegian standard for sludge disinfection processes (< 2500 faecal coliforms per gram total solids). No *Salmonella* was detected in the digested sludge.
5. TAD sludge with temperatures of 20–40 °C can easily be dewatered in a decanter centrifuge, but the polymer consumption (8.3 g/kg TS in average) is higher than for most other sewage sludges dewatered in Norway.
6. The nutrient content in the TAD sludge was equal to or higher than corresponding values for anaerobically digested primary-chemical sludges.

References

[1] Norwegian State Pollution Control Authority and Norwegian Public Health Authority: Regulations and Guidelines for Handling and Disposal of Sewage Sludge, 1991 (in Norwegian)
[2] Norwegian State Pollution Control Authority: Strategy for Disposal of Sewage Sludge, 1992 (in Norwegian)
[3] ATV/VKS-Arbeitsgruppe 3.2.2: Entseuchung von Klärschlamm. Erster Arbeitsbericht, Korrespondenz Abwasser *11* (1986)
[4] ATV/VKS-Arbeitsgruppe 3.2.2: Entseuchung von Klärschlamm. Zweiter Arbeitsbericht, Korrespondenz Abwasser *1* (1988)
[5] ATV/VKS-Arbeitsgruppe 3.2.2: Entseuchung von Klärschlamm. Dritter Arbeitsbericht, Korrespondenz Abwasser *12* (1988)
[6] Loll, U.: Stabilisierung hochkonzentrierter organischer Abwässer und der Abwasserschlämme durch aerob-thermophile Abbauprozesse. Dissertation im Fachbereich Wasser und Verkehr, D 17, TU Darmstadt 1974
[7] Fuchs, H., Fuchs, L., Fuchs, W.: The Exothermic Aerobic Thermophilic Sludge Stabilisation Process. Umwelt *3* (1980)
[8] Matsch, L.C., Drnevich, R.F.: Autothermal Aerobic Digestion. J. WPCF *49* (2) (1977)
[9] Jewell, W.J., Kabrick, R.M.: Autoheated Aerobic Thermophilic Digestion with Aeration. J. WPCF 2 (2) (1980)
[10] Morgan, S.F., Gunson, H.G., Littlewood, M.H., Winstanley, R.: Aerobic Thermophilic Digestion of Sludge Using Air. In: Sewage Sludge Stabilisation and Disinfection, A.M. Bruce (Ed.). Horwood, Chichester, England 1984
[11] Booth, M.G., Tramontini, E.: Thermophilic Sludge Digestion Using Oxygen and Air. In: Sewage Sludge Stabilisation and Disinfection, A.M. Bruce (Ed.). Horwood, Chichester, England 1984
[12] Leonhard, D., Hahn, H.H.: Aerob-Thermophile Schlammstabilisierung in Deutschland. Bericht über eine Studie für den amerikanischen Bereich. Korrespondenz Abwasser *5* (1990)
[13] Paulsrud, B., Haugan, B.-E., Langeland, G.: Aerobic Thermophilic Digestion of Sewage Sludge Using Pure Oxygen. In: Processing and Use of Sewage Sludge, P. L'Hermite and H. Ott (Eds.). Reidel, 1984
[14] Langeland, G., Paulsrud, B., Haugan, B.-E.: Aerobic Thermophilic Stabilisation. In: Inactivation of Microorganisms in Sewage Sludge by Stabilization Processes, D. Strauch, A.H. Havelaar and P. L'Hermite (Eds.). Elsevier, London 1984
[15] Paulsrud, B., Langeland, G.: Aerobic Thermophilic Digestion of Pre-thickened Sludge Using Air. In: New Developments in Processing of Sludges and Slurries, A.M. Bruce, P. L'Hermite and P.J. Newman (Eds.). Elsevier, London 1986
[16] Murray, K.D., Tong, A., Bruce, A.M.: Thermophilic Aerobic Digestion – A Reliable and Effective Process for Sludge Treatment at Small Works. Wat. Sci. Tech. 2 (3/4) (1990)
[17] U.S. EPA, Office of Research and Development: Autothermal Thermophilic Aerobic Digestion of Municipal Wastewater Sludge. Washington DC 20460, EPA/625/10-90/007 (1990)
[18] Fischer, W.J.: Calculation of Volatile Solids Destruction During Sludge Digestion. In: Sewage Sludge Stabilisation and Disinfection, A.M. Bruce (Ed.). Horwood, Chichester, England 1984
[19] U.S. EPA, Office of Water: Technical Support Document for Part 503 Pathogen and Vector Attraction Reduction Requirements in Sewage Sludge. Washington DC, NTIS No. PB93-11069, Natl. Technical Information Service, Springfield, VA 1992
[20] Water Pollution Control Federation: Operation of Municipal Wastewater Treatment Plants, Manual of Practice No. 11 – Vol. III, Second Edition, U.S.A. 1990

B. Paulsrud and K.T. Nedland
Aquateam – Norwegian Water Technology Centre A/S
P.O. Box 6326 Etterstad
N-0604 Oslo
Norway

Beneficial Use of Sludge from Sewage Plants and Water Works

J. Göransson and I. Karlsson

Introduction

Pure water is one of our most important assets today. Therefore there are increasing demands on water treatment. It is our responsibility to return the water to nature in the same condition as we borrow it.

The purification of surface water for drinking water has been done for many years using a metal salt. The metal salt efficiently removes particles and other impurities. The more impurities the surface water contains, the more impurities have to be removed, resulting in more sludge production. The sludge then contains the main metal salt, organic matter and water. The sludge is in some places considered to be a problem but if some of its components can be recovered, it might be an asset.

After we use water, it is polluted with organic matter, phosphorus, nitrogen and to some extent also impurities such as heavy metals. The more impurities the sewage water contains, the more have to be removed and the more sludge will be produced.

Chemical treatment has long been used for removal of phosphorus, and if pre-precipitation is used it can be used for removal of both phosphorus and particulate organic matter. This pre-separation of organic matter means that less organic matter will go to a later biological step, that the energy demand for removal of organic matter will decrease dramatically and that the microorganisms can be kept a longer time in the system. A higher sludge age means better conditions for biological nitrification, but if most of the organic substance is removed, there may be a lack of organic substance for denitrification. However, the sludge contains organic material that, if it can be transformed to an easily biodegradable carbon source, can be used in the denitrification process. The transformation of particulate organic matter is an idea that was tested in the HYPRO project and the results that are presented here are partly the results from HYPRO but also spin-off effects from HYPRO.

The idea is to extend the water cycle (Fig. 1) to also include the recycling of precipitation metals and the reuse of phosphorus, organic matter and perhaps also heavy metals.

R. Klute / H.H. Hahn (Eds.)
Chemical Water and Wastewater Treatment III

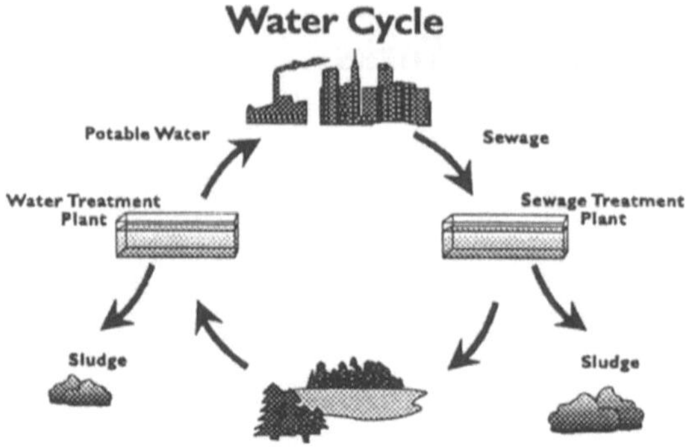

Fig. 1. The water cycle

Aluminium Recovery from Drinking Water Sludge

Water works sludge containing metal from the coagulants can be dewatered, but normally not to a concentration higher than 15%. The low percentage of dry substance means that even though it is a rather low amount of dry sludge, the volume is significant.

Table 1 shows the characteristics of a Swedish water work's sludge, using aluminium sulphate for surface waters. It has been documented that acidification of aluminium sludge dissolves aluminium and reduces the volume and dry weight of the sludge. Big savings in disposal of sludge can be achieved. The problem is that the dissolving process in sulphuric acid is nonselective. Along with the aluminium, other substances that are soluble under the same conditions will also be solubilized. A brown liquid containing most of the heavy metal will be the result. If the sludge in Table 1 is dissolved with sulphuric acid and compared to the commercial aluminium product, Kemira ALG, the heavy metals content is very high. Compared to products used in sewage treatment, the impurities are also too high. The metal concentration is too low for the coagulants. Another use of the water works sludge is to send it directly to the sewage plant. If it is discharged continuously for 24 hours, the problem in the sewage plant will most probably be negligible, perhaps even a small effect of phosphorus elimination can occur. In Sweden phosphorus in sewage is restricted to below 0.5 mg/l. If the municipality is using aluminium in surface water treatment and aluminium for phosphorus removal in sewage treatment, the ratio of consumption is about 1 : 3.

The sludge can also be used as a source of raw materials for coagulants. Figure 2 shows this production method in principle. The thought is that everything should be done at the potable water plant except the purification step. The purification step is the most investment-demanding step and it would be advantageous to collect material from more than one plant for the purification step.

The process can be divided into two parts: the dissolving of $Al(OH)_3$ and the reaction to form alunite. However, it is possible to do this in one step. The separation (dewatering) of alunite is much easier than the dewatering of the drinking water sludge. A dry substance content of between 50 and 60% can be reached.

Table 1

	Sludge	Re-calculated sludge	ALG
DS [%]	12		
VS [%]	49		
Al [%]	2.4	9.1	9.1
Fe_2O_3] [%]	0.12	0.4	0.015
Si [%]	5	1.7	–
Pb [mg/l]	60	200	2.0
Mn [mg/l]	40	140	–
Cu [mg/l]	15	50	1.0
Ni [mg/l]	7	25	1.0
Cr [mg/l]	5	20	1.0
Cd [mg/l]	0.5	2	0.05

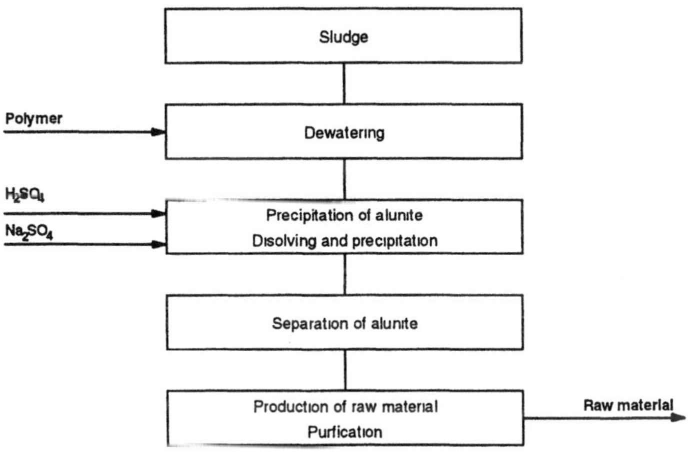

Fig. 2. Principles for aluminium recovery from drinking water sludge

The results presented here are partly from lab scale trials from Sweden (Stockholm), Germany (Gelschenwasser) and Finland (Oulu and Helsinki), but also from bath trials in 800 l reactors with sludge from Stockholm.

Alunite Precipitation

The reaction is basically

$$Me^+ + 3\,Al^{3+} + 2\,SO_4^{2-} + 6\,H_2O \Longrightarrow MeAl_3(SO_4)_2(OH)_6 + 6\,H^+$$

where Me can be, for example, Na, K or H_3O^+.

Figure 3 shows that the reaction temperature should be between 130 °C and 150 °C. The pH drops when alunite is formed.

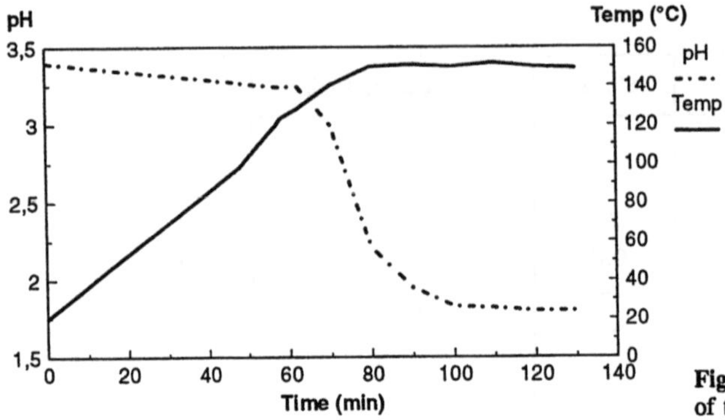

Fig. 3. Temperature dependence of the alunite reaction

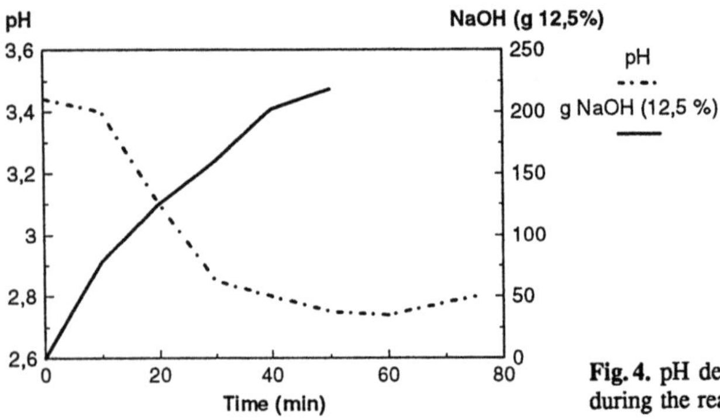

Fig. 4. pH decreases when NaOH is added during the reaction

Figure 4 shows that the reaction time is less than one hour. Figure 3 showed that the reaction works over a rather wide range and that it was not the pH that was limiting for the reaction in Figure 4.

The Quality of the Recycled Raw Material

The raw material (alunite) that was produced in the reaction still contains organic substances and some inorganic substances like iron.

The concentration of inorganic matter depends on the initial pH, and Figure 5 shows the relationship between concentration of one metal and initial pH in the raw material. The actual concentrations will, of course, also depend on the sludge. Other metals behave similar to iron.

Figures 6 and 7 show a comparison between the recycled and the virgin raw material. The diagrams show no difference in phosphorus and turbidity purification of sewage water.

The recycled raw material can also be used for PAX production. However, recycled alunite contains SO_4^{2-} and to avoid stability problems the alunite has to be mixed with other raw materials.

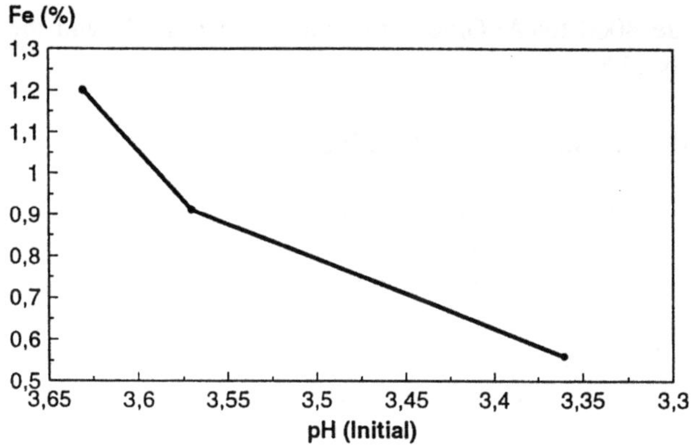

Fig. 5. The relationship between pH and iron in the raw material

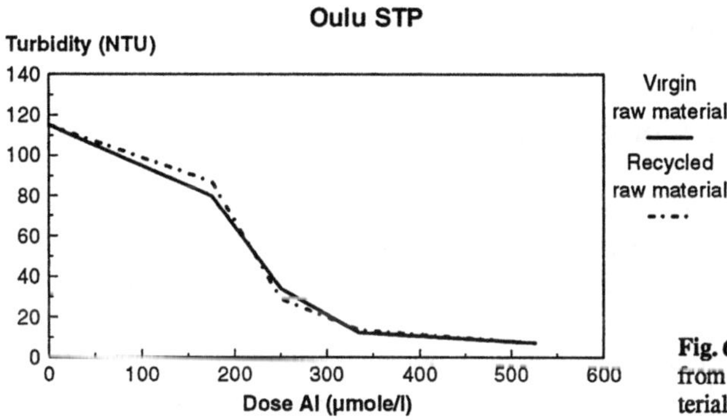

Fig. 6. Turbidity reduction of AVR from virgin and recycled raw material

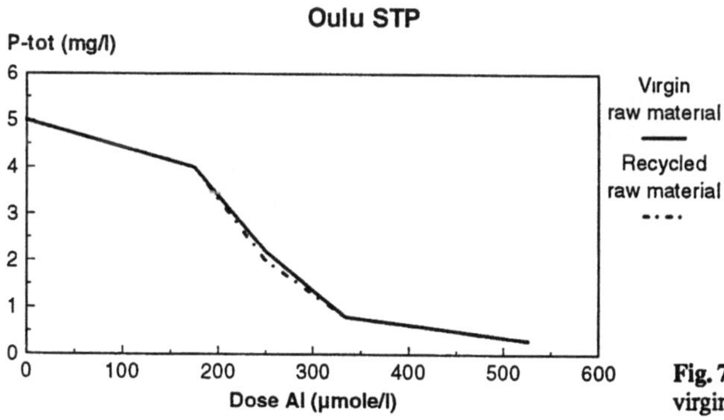

Fig. 7. P_{tot} reduction of AVR from virgin and recycled raw material

Economy of the Process

The economy of the process does not come from the production of a cheap raw material. The value of the raw material is less than the production cost, but it can still be of economic interest. To cover the investment and variable cost the disposal fee is between 200 and 250 SEK/ton. This, of course, depends on the size of the equipment and the dry substance in the sludge. For purposes of this calculation, these values are 4000 ton ALG/year, the interest rate is 8 % and the disposal of dry substance is 15 %.

Summary of Recovery of Aluminium from Drinking Water

There are several ways to recover aluminium from drinking water sludge. One very interesting way is to produce a new raw material for coagulant production. It is possible to recover over 90 % of the aluminium from sludge. Impurities like organic substances and heavy metals can be reduced to acceptable levels. Coagulants produced from the recycled raw material perform as well as coagulants produced from virgin raw material. The value of the raw material does not cover the cost of the process. However, it is still economically interesting if there is a disposal fee for the sludge. The process is very interesting but there is still optimizing work to be done before it becomes a commercial process.

Beneficial Use of the Components in Sewage Sludge

Eutrophication along the coastline and in lakes in the nordic countries has resulted in increasing demands on purification of sewage water.

Nitrogen removal demands a high ratio of easily biodegradable organic matter to nitrogen in the sewage water. In most of the nordic countries this is not the case. It is possible to hydrolyse the particulate organic matter or to use an external carbon source. The hydrolysis can either be done on sewage water in big biological tanks or on primary sludge in small tanks.

In the EUREKA project HYPRO, biological, chemical and thermal hydrolysis of sewage sludge was studied [1] on a laboratory scale as well as on a full scale.

When thermal/chemical hydrolyses were tested, a spin-off effect was that metals were dissolved and it was possible to separate inorganic substances from the organic substance.

This made it possible to extend the HYPRO process to make beneficial use of the components in the sewage sludge, see Figure 8.

Thermal Treatment of Sewage Sludge

Thermal treatment was tested on a full scale in Tampere, Finland and in Västerås, Sweden. The trials were part of the HYPRO-project.

The trials in Tampere and Västerås produced a carbon source and separated organic and inorganic sludge in three different streams, see Figure 9.

The thermal treatment (wet oxidation) of sludge has been used for a long time to increase dewaterability. Normally thermal treatment is done at temperatures between 200 and 250 °C without the addition of chemicals. One of the problems has been the production of dissolved organic substance. However, if the dissolved organic substance can be used as an easily biodegradable carbon source for nitrogen removal, it can be beneficial to the sewage plant.

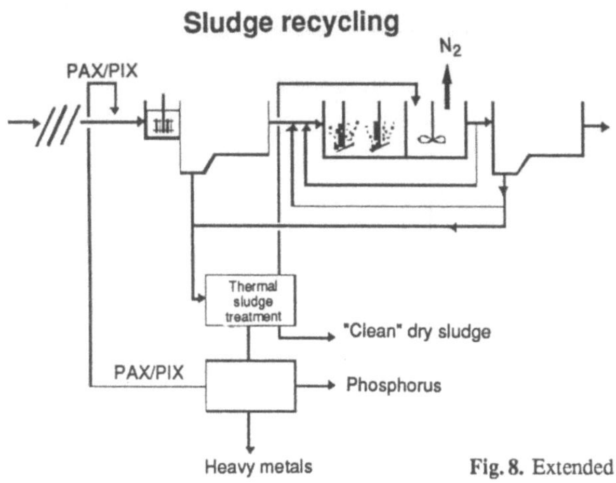

Fig. 8. Extended HYPRO-processes

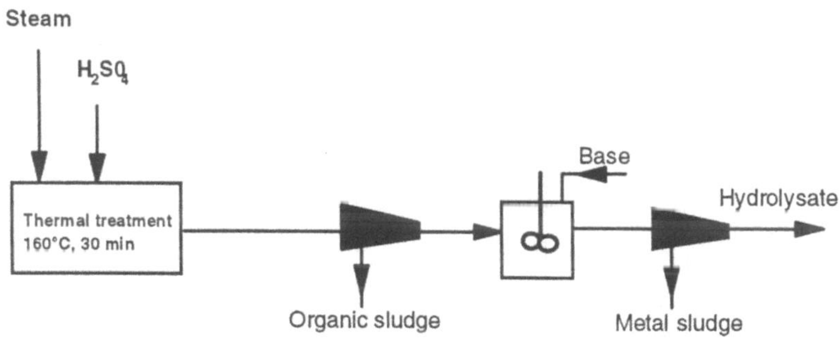

Fig. 9. Principles for the thermal treatment of sludge

Carbon Source for Denitrification

The thermal treatment can, of course, be done at different temperatures and different pHs and this will result in different yields [2]. The hydrolysis will also result in a dissolving of phosphorus and nitrogen. Phosphorus is not a problem since it can be recovered, but the nitrogen might be. Approximately 80 % of the nitrogen will be dissolved. It should be compared to the normal nitrogen release in sludge handling. It is also interesting to note that nitrogen is usually dissolved under rather mild conditions. COD dissolution increases at decreasing pH, see Figure 10.

The value will change depending on sludge type; and this is a pre-precipitated sludge.

The nitrogen removal trials in Tampere were divided into three periods: two periods where hydrolysate produced at different pHs was used as a carbon source and one period where acetate was used as a carbon source. The consumption figures in Figure 11 are rather high, but the COD consumption figures are approximately the same for thermal hydrolysate and acetate. The total residence time in

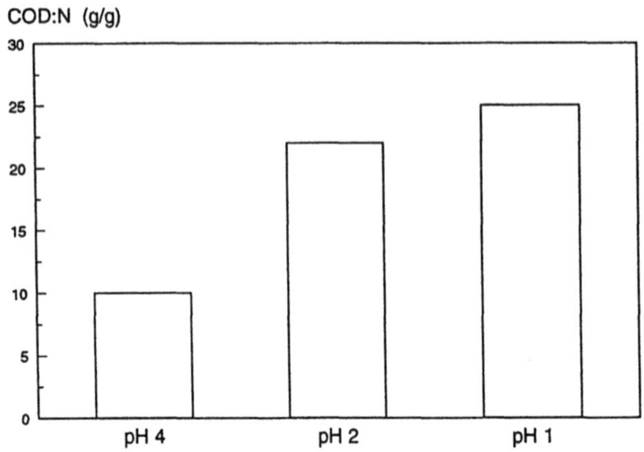

Fig. 10. COD/N ratio in the dissolved organic phase in Tampere

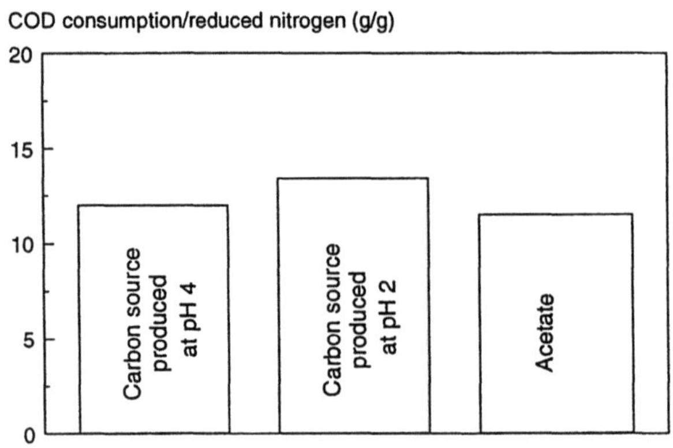

Fig. 11. COD consumption for different carbon sources in Tampere

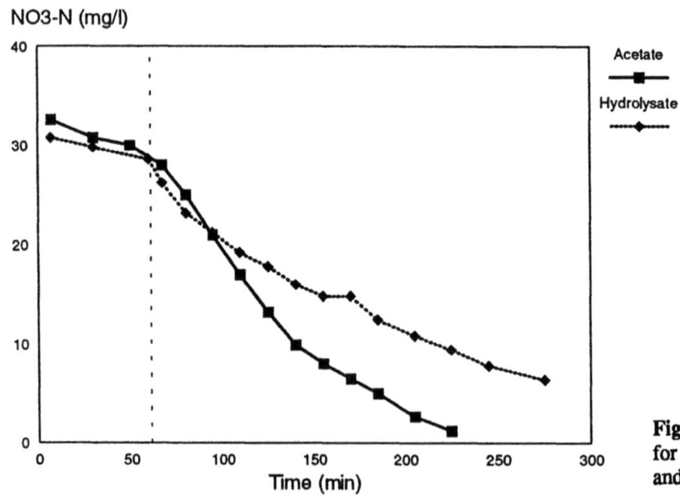

Fig. 12. Denitrification rate for thermal hydrolysate and acetate

the biological zone was 10 hours divided into approximately 50 % aerobic and 50 % anoxic zone.

The volume was not limiting in the trials so the denitrification rate was not the maximum rate. The denitrification rate was measured in the laboratory. The denitrification rate (Fig. 12) was slightly lower than that for acetate but comparable to that for a commercial carbon source like ethanol or sugar.

The effluent COD was not affected in Tampere (Fig. 13). That means the hydrolysate (oracetate) is not contributing slowly biodegradable organic matter in the effluent.

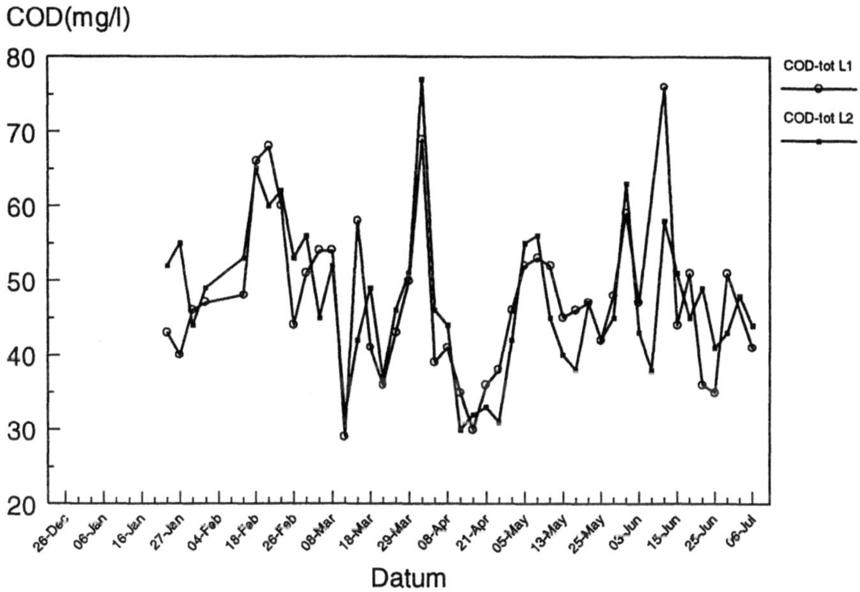

Fig. 13. COD value in reference line and the line where a carbon source was dosed

Sludge Quality

The thermal treatment at low pH does not only dissolve organic matter but inorganic matter like iron, aluminium, heavy metals and phosphorus as well. Heavy metals can be dissolved to avery high degree at pH 1 [3]. At pH above one the dissolving grade decreases down to almost zero at pH 7, see Figure 14.

The coagulants also dissolve at a low pH [2] to almost 100 %. It is important to separate the liquid phase from the particle phase to end up with an organic sludge with a low heavy metal content. The dewaterability also improves dramatically after the thermal treatment and it is easy toget a good separation. Figure 15 shows the CST value before and after the thermal treatment.

Full scale dewatering in Tampere also had very good results. Centrifugation of primary sludge (and polymer) gave a dry substance content of 18–20 %. The thermally treated sludge had a dry substance content of over 40 % without a polymer.

Table 2. Dewaterability of thermally treated sludge

	Organic sludge DS [%]	Metal sludge DS [%]
Centrifuge	40–45	35–40
Camber filter press	50	

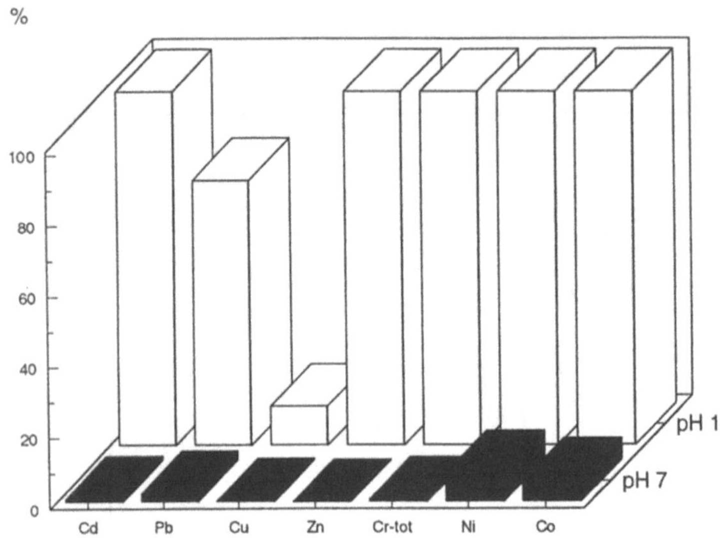

Fig. 14. Dissolving grade of some heavy metals at pH 0.9 and pH 7

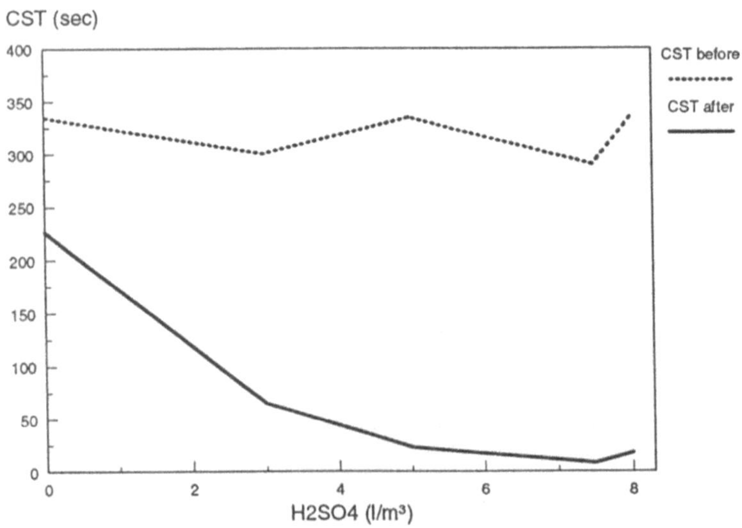

Fig. 15. CST values before and after thermal treatment

It should be mentioned that the trials in Tampere did not intend to produce a dry sludge. The purpose in Tampere was to hydrolyse particulate matter and to use the dissolved organic matter as a carbon source.

Metal Sludge

The metal sludge can be separated after pH adjustment (Fig. 9). pH adjustment can be done with different bases. Lime gives the highest DS. NaOH or $Mg(OH)_2$ gave lower dry substance values, but still they produced less sludge volume at approximately the same removal of heavy metals from the carbon source. The separation is almost complete at pH values around 8.5.

The separation of coagulants, heavy metals and phosphorus can be done in separate steps.

Separation of coagulants and heavy metals can normally be done with existing production methods. Normal production methods are not designed to separate phosphorus and iron/aluminium. Phosphorus and coagulants can be separated with extraction. Extraction is a well-known method to separate different elements from each other, but is usually used to produce more expensive elements than iron or phosphorus.

Coagulant Recovery

Before the coagulants can be extracted, the metal sludge (Fig. 9) has to be dissolved in acid. To have a successful extraction with negligible solvent loses (crud formation), the small amount of organic substance in the solution has to be removed. The iron yield from the metal sludge is over 90 %. It is difficult to end up with a highly concentrated product, and concentrations between 6 and 8 % iron are more likely in the product. Normal product contains between 11 % and 14 % iron.

Separation of Heavy Metals and Phosphorus

Extraction of coagulants will produce a refined product containing heavy metals and phosphorus. The heavy metals can be precipitated as sulphides.

Phosphorus can be precipitated as calcium phosphate or magnesium ammonium phosphate. The yield of phosphate depends mostly on the losses to the organic sludge, and a total yield of 70–80 % can be expected.

Economy of the Process

Sewage sludge is a complex raw material. It is possible to recycle coagulants and phosphorus as new products. However, the value of this product does not cover the whole process cost. The value of the carbon source and decreased disposal costs for the sludge have to be considered to make the process interesting. Another benefit is that polymer is not needed for dewatering the sludge. The value of the process is perhaps not the economy. The interesting thing is that it turns the sludge from a disposal problem to a raw material at approximately no extra cost.

Summary of Recycling of Sewage Sludge

It is possible to recycle coagulants, phosphorus and organic substances from the sewage sludge. Heavy metals can also be separated from the sludge. The value of the process covers its cost. The use of the dissolved organic substance was tested on a full scale and in sewage treatmentworks. The separation of organic and inorganic substances has also been successfully tested on a full scale. The dewaterability of the sludge is very good and it is easy to reach a high dry substance content. The recovery of coagulants and phosphorus was tested on a lab scale and works satisfactorily. There is still optimizing and full scale work to be done. It is our intention to test all parts together on a full scale starting late 1994. The full scale project will be a cooperative effort between Helsingborg's STP, Alfa Laval and Kemira Kemi AB.

The cost of the process is covered by the value of the process. The most important thing is that it creates a "value" for the sludge.

References

[1] Karlsson, I., Göransson, J., Rindell, K.: Use of Internal Carbon from Sludge in Biological Wastewater Treatment. In: Chemical Water and Wastewater Treatment II, R. Klute and H.H. Hahn, (Eds.). Springer, Berlin Heidelberg New York 1992, pp. 329–339

[2] Smith, G., Göransson, J.: Generation of an Effective Internal Carbon Source for Denitrification Through Thermal Hydrolysis of Preprecipitated Sludge. The 6th IAWPRC Conference on Design and Operation, 1991, pp. 216–223

[3] Ericsson, B., Held, H.: Inverkan av slamhydrolys vid förhöjt tryck och temperatur på slammets egenskaper, 1991 (in Swedish)

Jonas Göransson and Ingemar Karlsson
Kemira Kemi AB, Kemwater
Box 902
S-251 09 Helsingborg
Sweden

Phosphate Fixation in Sludges from Enhanced Biological P-Removal During Stabilization

N. Jardin and H. J. Pöpel

Abstract

During the start-up phase of an enhanced biological phosphorus removal (EBPR) plant, the fixation mechanisms of phosphorus in the excess sludge were investigated. The major part of the eliminated phosphorus was bound as polyphosphate (poly-P) and was estimated to be some 50 to 70 % of total P. During anaerobic stabilization the poly-P was completely hydrolysed but only a part of the released phosphate remains in solution whereas as much as 60 to 70 % of total P was fixed by physico-chemical mechanisms.

Among the possible counterions for P-precipitation, aluminium, magnesium, and calcium are discussed in detail. In addition to the sludge from the pilot plant, stabilised sludge samples from large EBPR and conventional treatment plants were also investigated. It was found that a major part of the phosphate was associated with aluminium which was demonstrated by means of acidimetric titration. In the pH range of 1 to 3.5, aluminium and phosphate release were strongly correlated with a molar release ratio of 1.45 M Al/M P. The amount of phosphate which was due to aluminium phosphate fixation was calculated as some 35 to 68 % of total P. Besides a precipitation as $AlPO_4$, adsorption to SiO_2 and/or aluminosilicates and complexation with mono- or polynuclear aluminium are thought to be mainly responsible for the fixation of PO_4.

Also, phosphate is bound as magnesium ammonium phosphate. With X-ray diffraction the crystallization of a solid phase was identified as struvite. The solubility product of struvite was determined to $\log K_S = -12.53 \pm 0.05$. It seems that the sequential dilution test is an appropriate method for determining the amount of struvite precipitation. In EBPR sludges from plants with the main stream process, 25 to 37 % of P was eliminated as struvite whereas in the sludge from a plant with the side stream process, the amount of struvite precipitation accounted for only 14 % which is mainly due to the reduced amounts of magnesium as a result of the stripping process.

Although in some sludge samples exceptionally high calcium levels of up to 8.5 % were reached, no indications of calcium phosphate interactions were found.

R. Klute / H.H. Hahn (Eds.)
Chemical Water and Wastewater Treatment III

Introduction

Most of the new or expanded plants are designed for the enhanced biological phosphorus removal process (EBPR). In contrast to conventional plants, the phosphorus content of the activated sludge solids from this process reaches values of up to 7 %. During wastewater treatment, phosphorus can be bound in the activated sludge mainly by means of three mechanisms: (1) as *physiological phosphorus*, that is, P for metabolism and growth of microorganisms, e.g. nucleic acids (DNA, RNA), phospholipids and nucleotides (NAD, FAD). Usually, this amount of phosphorus accounts for 1 to 2 % based on TSS. (2) In plants with EBPR, additional phosphorus can be bound as *polyphosphate*. To stabilize the negative charge of the polyphosphates, synthesis of poly-P is accompanied by an uptake of Mg^{2+} and K^{+}. The amount of poly-P depends mainly on the wastewater characteristics. At average large treatment plants the poly-P content will not exceed 1 to 2 %, whereas under special wastewater conditions a poly-P content of up to 5 % can be reached. (3) Furthermore, a *physico-chemical fixation of phosphate* can occur during wastewater treatment mainly by calcium phosphate precipitation or adsorption. During sludge treatment, all three P-fractions will be released to a different extent and at different rates. Therefore, investigations concerning the amount of P-release and possible fixation mechanisms during sludge treatment have to consider type and amount of these different P-fractions. With the centrate or filtrate from sludge dewatering the released phosphorus could be recycled back to the plant inlet, resulting in an additional phosphorus load (P-feedback) and, possibly, deterioration of the EBPR process.

From a theoretical point of view, most of the phosphorus eliminated in the form of polyphosphate should be released during the anaerobic treatment of excess sludge [20]. In contrast, at most large plants in Germany the soluble phosphate concentration of the digested sludge supernatant is often very low [24, 3]. Therefore, conditions have to be present where only a part of the stored phosphorus is released to the bulk liquid of the sludge treatment system or some of the released phosphorus is fixed by physico-chemical mechanisms.

In order to follow the fate of the total phosphorus during wastewater and sludge treatment, an EBPR pilot plant with different sludge treatment systems has been in operation over a 2 year period. Some of the results obtained during this study will be presented in the following, focusing on the objectives and methods listed below with special reference to the physico-chemical fixation mechanisms of phosphorus in stabilised sludge.

- Determination of type and extent of phosphate fixation in excess sludge

  For that purpose P-fractionations were periodically performed.

- Estimation of the amount of P-release and the resulting P-feedback during sludge digestion

  This was done by calculating ion-balances on the digester of the pilot plant and using P-fractionations.

- Investigation of physico-chemical P-fixation mechanisms in digester systems
  Besides the investigations of the sludges from the pilot plant, different stabilised sludge samples from large EBPR and conventional treatment plants were examined with indirect and direct methods. Among the indirect methods, P-fractionations, acidimetric titration and sequential dilution experiments were used to identify the different fixation mechanisms, and these were confirmed by electron dispersive X-ray spectroscopy (EDXS) and X-ray powder diffraction analysis.

Materials and Methods

The pilot plant consists of two continuous flow activated sludge systems both operated with settled domestic sewage (Fig. 1). Plant 2 was operated with an anaerobic zone for enhanced biological phosphorus removal, whereas plant 1 serves as a control without anaerobic tank. The excess sludge of the EBPR plant is withdrawn directly from the activated sludge tank in order to prevent anaerobic conditions prior to sludge treatment. Thickening of the sludge is carried out using a centrifuge in a continuous mode of operation. Thereafter, the thickened sludge is pumped into the digester (volume: 1.5 m^3), operated with a retention time of 20 days at 35 °C. The design parameters and the operating conditions of the plants are shown in Table 1 and Table 2, respectively. Because of the very low phosphorus content of the domestic sewage used for the experiments, the inflow of the pilot plant was supplemented with phosphoric acid. In addition to the stabilised sludge from the pilot plant, sludge samples from different large EBPR and conventional plants were used for the investigation of physico-chemical P-fixation mechanisms in digester systems. The characteristics of these sludges are summarised in Table 3.

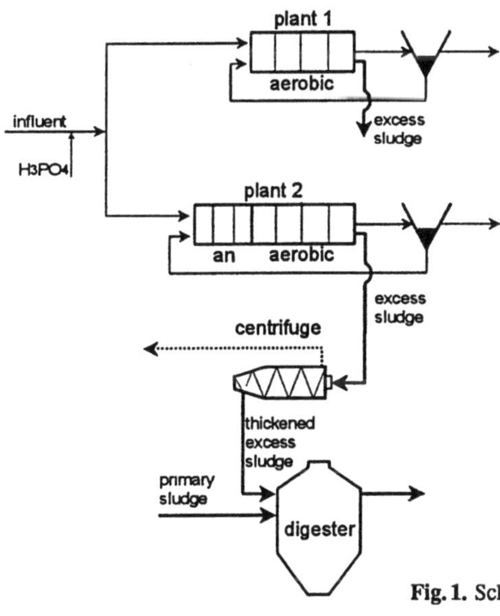

Fig. 1. Schematic diagram of the pilot plant

Table 1. Design parameters of the pilot plant (mean values during the start-up phase)

Parameter	Plant 1	Plant 2
Volume [m^3]	10	16
Influent flow [m^3/h]	2.5	4
Aerobic HRT[1] [h]	4	2.5
Anaerobic HRT[1] [h]	–	1.5
MCRT [d]	2.6	2.6

[1] Hydraulic retention time based on influent flow rate

Table 2. Average operating conditions of the pilot plants (mean values during the start-up phase)

Parameter	influent	Plant 1 effluent	Plant 2 effluent
BOD_5 [mg/l]	175	20	14
COD [mg/l]	340	82	62
$P_{tot.}$ [mg/l]	12.4	10.3	3.3
TKN [mg/l]	66	23	23
NH_4-N [mg/l]	48	19	21

Table 3. Characterization of the sludges used for the investigation of physico-chemical P-fixation mechanisms in digester systems

Sludge name	Origin	TS g/l	VS %	pH	P mg/g	Ca mg/g	Mg mg/g	K mg/g	Al mg/g	Fe mg/g
DS EBPR MS	Digester sludge from a large EBPR plant with main stream process (PHOREDOX)	38.6	51.4	7.32	29.3	51.5	9.8	7.4	17.1	15.2
stab. ES EBPR MS	Stabilised sludge from a large EBPR plant with main stream process (ISAH)	29.8	59.6	6.82	43.6	40.6	10.0	13.0	35.0	11.6
DS EBPR MS PP	Digester sludge from the EBPR pilot plant with main stream process (A/O)	28.8	58.4	7.29	46.9	45.2	11.2	16.9	23.8	10.5
DS EBPR SS	Digester sludge from a large EBPR plant with side stream process (PHOSTRIP)	25.9	55.8	7.46	29.3	57.8	7.7	6.8	26.3	12.0
DS Sim	Digester sludge from a large plant with simultaneous precipitation ($FeSO_4$)	49.3	51.5	7.8	29.6	85.4	5.8	2.0	17.0	49.7
PS	Primary sludge	40.6	75.5	5.46	10.0	31.6	3.6	5.6	17.2	10.2

Elementary analysis of P, Ca, Mg, K, Al, and Fe in the sludge samples was performed by means of atomic absorption spectrometry (AAS). Soluble Ca^{2+}, Mg^{2+}, K^+ and Na^+ were analysed by ion chromatography. Al^{3+} was determined by a colorimetric method using chromazurol S. X-ray diffraction analysis was performed using a powder diffraction system. With the aid of X-ray spectroscopy, the element distribution of the samples could be visualised for a total of 8 elements at the same time. All other analyses were performed according to DEV [9].

P-Fractionation. P-fractionations were used to differentiate between the phosphorus fixation mechanisms. For this purpose a modified method developed by Psenner [21] and Uhlmann et al. [27] was used. The fractionation consists of sequential extractions of the sludge samples with different extracting chemicals followed by incubation, centrifugation and analyses of the supernatant. In the supernatant PO_4-P (after 0.45 μm filtration) and P_{tot} were determined. The PO_4-P concentration represents the so-called dissolved reactive phosphate (DRP) and the difference between P_{tot} and PO_4-P is called the non-reactive phosphate (NRP). Table 4 summarises the extraction steps and assigns the different extracts to P species on a qualitative basis.

Table 4. Summary of the P-fractionation procedure and assignment of extracts to P species

Fraction name	Extract	P species
Original and dist. water	DRP	soluble and slightly adsorbed P
15 min, $T = 20\,°C$	NRP	P-containing suspensa
BD	DRP	part of $FePO_4$
$NaHCO_3$, $Na_2S_2O_4$		small part of magenesium ammonium phosphate (MAP)
30 min, $T = 20\,°C$		small part of P adsorbed to silicates or hydroxides
		hydrolysed poly-P in the course of extraction
	NRP	P-containing suspensa
NaOH	DRP	major part of MAP
1 M NaOH		released Mg will be precipitated as $MgCO_3$ or $Mg(OH)_2$
24 h, $T = 20\,°C$		rest of $FePO_4$
		$Fe_3(PO_4)_2$,
		released Fe will be precipitated as $FeCO_3$, FeS or $Fe(OH)_3$
		$AlPO_4$
		part of P adsorbed to silicates or hydroxides
	NRP	major part of the organically-bound P,
		poly-P
HCl	DRP	$CA_x(PO_4)_y$,
0.5 M HCl,		rest of poly-P
24 h, $T = 20\,°C$		rest of P adsorbed to silicates or hydroxides
	NRP	rest of the organically-bound

Acidimetric Titration. To investigate the pH-dependent release of phosphate, an acidimetric titration was performed. For this purpose, a 1 l sludge sample was titrated with concentrated HCl. Once a desired pH was achieved, 40 ml of sludge was withdrawn and incubated for 24 h to establish equilibrium. Usually, the acidimetric titration covered a pH range from original pH to pH = 1. After the incubation period, the pH was measured again and the samples were centrifuged at 30000 g for 10 min followed by 0.45 μm filtration. In the filtrate PO_4-P and the major cations were analysed as described above.

Sequential Dilution. All sludges were examined further with a sequential dilution test. Due to the successive reduction of total solids concentration in the course of the dilution procedure, the solubility behaviour of precipitated solids can be selectively modified without changing pH. For this test, 50 ml of sludge was filled into a 100 ml volumetric flask. Distilled water was added to make up 100 ml and subsequently, the dilution was vigorously shaken for 2 minutes. From this dilution step 50 ml were transferred in the next 100 ml flask and the dilution procedure was repeated as described until a final dilution of 1 to 1024 was reached. After an equilibration period of 24 h the samples were centrifuged, filtered and analysed as described.

Determination of Struvite Solubility Product. In order to determine the solubility product of struvite, stabilised sludge was withdrawn from the pilot plant digester and supplemented with Mg ($Mg(NO_3)_2$) and P (KH_2PO_4) to achieve oversaturation with respect to struvite. Subsequently, the sludge samples were titrated with concentrated HCl and NaOH to obtain the desired pH, followed by an incubation of 24 h at 35 °C in capped centrifuge tubes. Analyses were performed as described for the acidimetric titration.

Results and Discussion

Determination of Type and Extent of Phosphate Fixation in Excess Sludge

During EBPR operation, the excess sludge was periodically analysed by means of P-fractionation. The results of 8 such fractionations are summarised in Figure 2. As seen from this figure, the major part of total phosphorus is recovered as NaOH-NRP. In the excess sludge from plants with EBPR this fraction usually consists of organic phosphorus and polyphosphate, whereas at plants with iron or aluminium precipitation the major part of precipitated phosphorus is also found in this fraction.

A differentiation of the P-species is facilitated when the counterions are considered. Especially potassium gives valuable indications towards P-binding in the NaOH-fraction. Due to the uptake of potassium in the course of polyphosphate synthesis, potassium is expected to be released simultaneously with poly-P during the alkaline extraction. Because potassium usually participates only to a small degree in precipitation or adsorption reactions in wastewater and sludge treatment, it can be assumed that high potassium levels in the extracts are mainly the result of poly-P release and/or poly-P hydrolysis. Therefore, we looked for a relationship between potassium and NRP/DRP concentrations in the different extracts. For the NaOH-NRP fraction this relationship is also depicted in Figure 2. From this graph it can be seen that NaOH-NRP and potassium are well correlated on a high significance level. This clearly demonstrates that for the excess sludge from the pilot plant the major part of phosphorus in the NaOH-NRP fraction can be assigned to poly-P. Furthermore, for the BD- and HCl-fraction a similar correlation between DRP and potassium was found (data not shown). In all, a poly-P content of 50 to 70 % of total P was found for the excess sludge of the EBPR pilot plant.

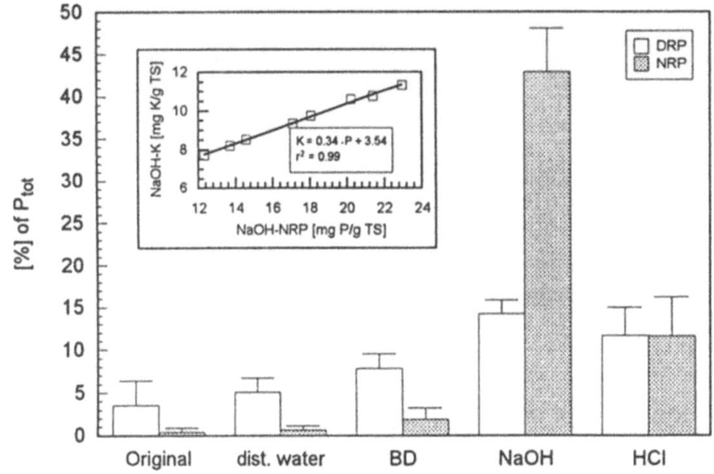

Fig. 2. Results of P-fractionations of excess sludge from the EBPR pilot plant and correlation between NaOH-NRP and potassium (bars indicate standard error)

Estimation of the Amount of P-Release and the Resulting P-Feedback During Sludge Digestion

It could be expected that at least a part of the stored poly-P is released during the different sludge treatment operations. The most probable treatment step for P-release is sludge stabilization, especially the most often used sludge digestion. To determine the fate of poly-P, P-fractionations were periodically applied to the digester sludge of the pilot plant.

Figure 3 summarises a total of 11 fractionations. Clearly, the NaOH-NRP fraction which comprises the major part of poly-P was reduced to below 5 % of total P. Considering that organic phosphorus will also be found in this fraction, the poly-P content of the sludge samples tends toward zero. From these fractionations it is obvious that a shift from the NaOH- and HCl-NRP fraction in excess sludge towards the NaOH-, HCl- and, to a lesser extent, to the original-DRP fraction has occurred. According to Table 4, the phosphorus found in the former two fractions should mainly result from dissolution or desorption of physico-chemical fixed phosphorus which accounts for as much as 60 to 70 % of total phosphorus.

The possible P-feedback can be calculated easily by performing ion balances around the sludge digester, a method described in detail elsewhere [15]. From these balances we found that under average operating conditions of an EBPR plant characterised by a phosphorus content in the excess sludge of some 3 to 4 %, the P-feedback is in the range of 10 to 20 %, whereas under excellent EBPR conditions with a phosphorus content in excess sludge of 6 to 7 % a P-feedback up to 40 % has been observed [15].

Summarising the results attained so far we conclude that:

- the enhanced *phosphorus elimination* observed in our EBPR pilot plant was mainly *due to storage of poly*-P *and*

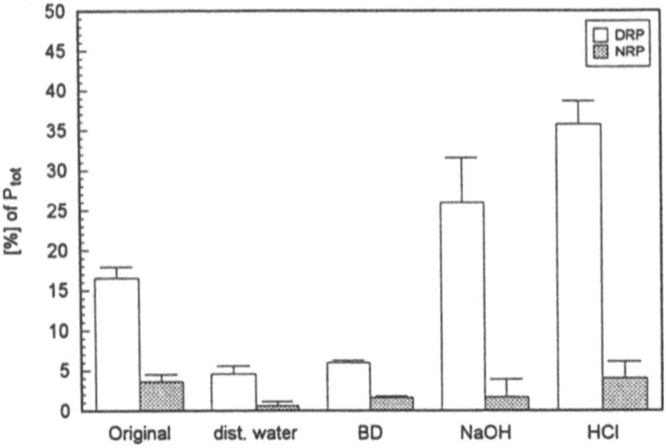

Fig. 3. Results of P-fractionations of digested sludge from the EBPR pilot plant (bars indicate standard error)

- during digestion a shift from the NRP-fractions in excess sludge towards the DRP-fractions in digested sludge occurred, indicating *physico-chemical fixation of 60 to 70% of total phosphorus.*

Investigation of Physico-Chemical P-Fixation Mechanisms in Digester Systems

Due to the high amount of physico-chemical fixation of phosphorus in digester sludge, additional investigations with regard to possible reaction mechanisms were conducted. In the following, aluminium, magnesium and calcium which are – in view of their high amounts in digester sludge – most likely responsible for the observed precipitation or adsorption of phosphorus will be discussed in detail. In addition to the sludges from the EBPR pilot plant, digested sludge samples from EBPR and conventional large wastewater treatment plants (see Table 3) are also examined.

P-Fixation by Aluminium. With an average aluminium concentration of 20 to 25 mg/g TS in digester sludge of the pilot plant, an amount of 23 to 29 mg PO_4-P/g TS could be eliminated by aluminium precipitation assuming an average composition of the solid phase of 1 M Al/1 M P. This clearly demonstrates the significance of aluminium with respect to possible PO_4-elimination reactions in digesters. Therefore, the objective of the following discussion is, (1) to determine the origin and fate of aluminium in the excess, primary and digester sludge, (2) to formulate possible mechanisms of aluminium interactions with phosphate, and (3) to estimate the amount of phosphate that can be attributed to Al-PO_4 fixation.

Origin and Fate of Aluminium in Digester Sludge. Besides industrial effluents which contribute in some cases to a considerable extent to the total aluminium in raw wastewater of municipal treatment plants, the main source of aluminium is zeolites used as a builder in laundry detergents. The most frequently used zeolite in Germany is zeolite A ($Na_{12}(AlO_2)_{12}(SiO_2)_{12} \cdot 27\,H_2O$) with an average aluminium content of 13 %. During the washing cycle the alkaline earth cations Ca^{2+} and to a lesser extent Mg^{2+} are exchanged with Na^+, resulting in mainly Ca-zeolite in the raw wastewater. To demonstrate the significance of zeolite A as one of the major sources of aluminium in primary and excess sludge, one can estimate the aluminium concentration in digester sludge if it is assumed that laundry detergents contain 25 to 30 % zeolite, a specific consumption of some 30 g detergents/(cap·d), a specific amount of digester sludge of 50 g TS/(cap·d) and furthermore, that 75 % of the zeolite is removed by primary and secondary sludge [30]. With these assumptions, an aluminium content in digester sludge of 15 to 18 mg/g TS can be estimated, which accounts for some 70 % of the average aluminium in digester sludge from the pilot plant. During wastewater and sludge treatment it could be expected that zeolite A dissolves to a high degree, particularly in the acidic primary sludge [1, 10, 7].

Evidence for hydrolysis of zeolite A during sludge treatment was provided in our experiments using X-ray diffraction and EDXS. In none of the sludges a diffraction pattern characteristic of zeolite A was found. On the other hand, in all sludges diffraction patterns were identified which can be assigned to SiO_2 (see Fig. 8). With respect to the fate of zeolite in the different sludges it seems reasonable to assume, therefore, that the zeolite decomposes rapidly within primary and excess sludge accompanied by a release of H_4SiO_4 and the formation of SiO_2 and an X-ray amorphous aluminosilicate. Alternatively, a complete decomposition of zeolite A with a release of aqueous aluminium is possible followed by a sequential reaction scheme for further aluminium reactions with complexation and polymerization, and resulting in the formation of a solid phase like $Al(OH)_3$, an aluminosilicate or $Al(PO)_4$. To what extent a complex formation of free aluminium with humic substances in digester sludges occurs is not known, but in spite of the narrow pH range of pH 4 to pH 7 for coagulation of humic substances with aluminium [8], this reaction seems not very likely to occur at slightly alkaline pH in digesters.

Considering the fate of aluminium in digester sludges, soluble phosphate is affected in more than one respect: (1) PO_4 could be precipitated as $AlPO_4$ or an amorphous analog like $Al(OH)_2H_2PO_4$ [29]. (2) PO_4 could be complexed with mono or polynuclear aluminium. Due to the high activation energy of the strong bonds of phosphate with two aluminol (AlOH) groups, these complexes do not dissociate until high pH values are obtained [33, 6]. (3) PO_4 could be adsorbed to the functional groups of aluminosilicates or aluminohydroxides by ligand exchange and (4) PO_4 could be transformed further to a surface precipitate [6, 28].

Mechanisms and Amount of Al-P-*Pixation.* To estimate the amount of PO_4 fixed by aluminium, the different sludges were investigated by means of acidimetric titration, X-ray diffraction and EDXS. Figure 4 shows the course of free Al^{3+} and of PO_4-P normalised to the total concentrations of the different sludges in

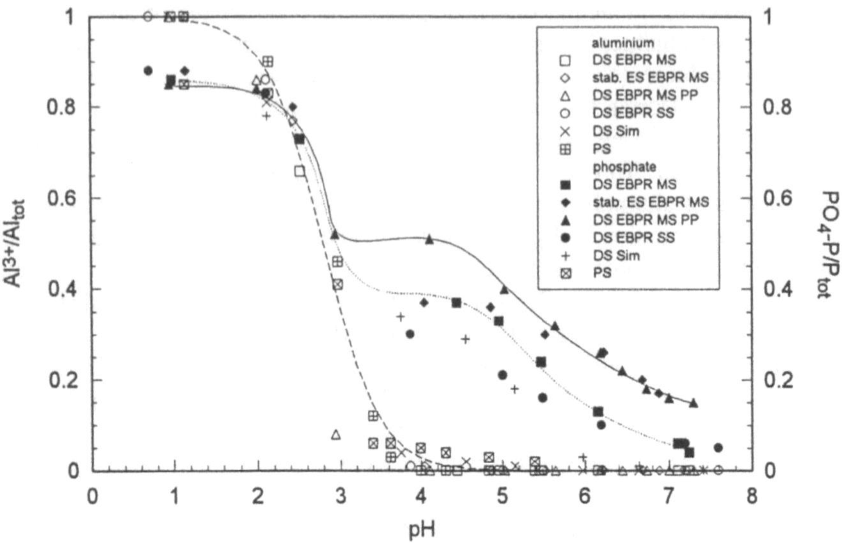

Fig. 4. Al^{3+} and PO_4-P in the course of acidimetric titration (dotted and solid line represent PO_4-P release of DS EBPR MS and DS EBPR MS PP, respectively; dashed line represents the course of Al-release of all sludges)

the acidimetric titration. For all sludges a similar behaviour was observed. Below pH 3 to 3.5 a remarkable increase of soluble Al^{3+} occurs which in all cases was accompanied by an additional strong release of PO_4. Figure 4 shows that the course of aluminium release is very similar for all sludges (dashed line) whereas at pH values above pH 3 greater differences exist for release of PO_4-P. These differences are due mainly to magnesium-phosphate interactions and will be discussed later.

For all sludges the simultaneous release of Al^{3+} and PO_4 correlates well, as shown in Figure 5 and was calculated with linear regression to a molar release ratio of 1.45 ± 0.05 M Al/M PO_4 for a pH range of 1 to 3.5.

In a further step an attempt was made to differentiate between PO_4-precipitation and -adsorption. For that purpose, the product of the actual concentrations ($C_{T,\mathrm{Al}} \cdot C_{T,\mathrm{PO_4}}$) was compared with the conditional solubility product of $AlPO_4$, calculated using the equilibrium relationships summarised in the appendix. Figure 6 shows that the EBPR sludge samples are in reasonable agreement with $AlPO_4$ equilibrium in the pH range of 2.4 to 4 and that all sludge samples are undersaturated with respect to $AlPO_4$ in the pH range of 7 to 7.6 which is a common pH range for the digestion process. Nevertheless, this cannot be regarded as verification of $AlPO_4$ precipitation because it is not clear yet, whether these results are due to dissolution of $AlPO_4$ present in the original sample or are mainly an analytical artefact which may result from a transformation of adsorbed phosphate to an $AlPO_4$ precipitate in the course of the titration because $AlPO_4$ is the more stable solid phase at lower pH. Furthermore, using X-ray diffraction and EDXS we found no indication of crystallised $AlPO_4$ solid phases.

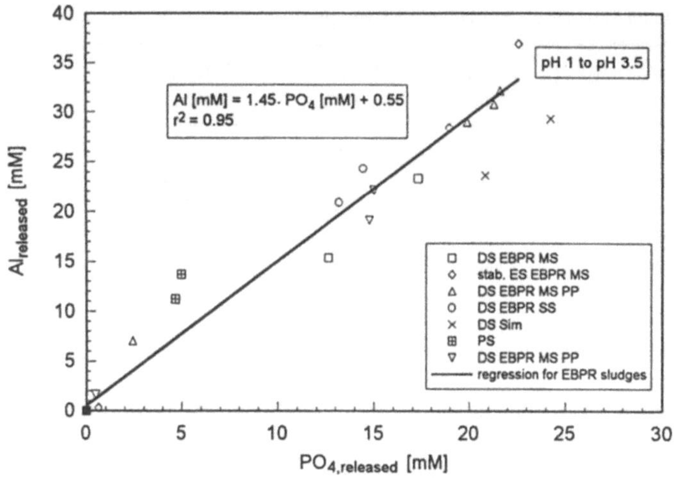

Fig. 5. Correlation of Al^{3+} and PO_4 in the pH range of 1 to 3.5

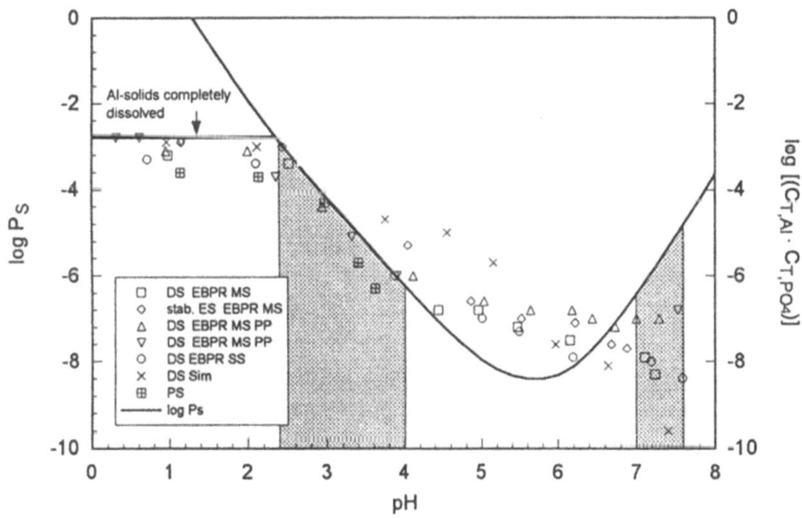

Fig. 6. Conditional solubility of $AlPO_4$ compared to the concentration product $C_{T,\text{Al}} \cdot C_{T,\text{PO}_4}$

Therefore, at present we are not able to verify the exact fixation mechanisms of Al-PO_4. But from the results found in our study it seems clear that a significant amount of phosphorus in digester sludge is bound by one of the aforementioned aluminium species. In Table 5 the amount of phosphorus which can be attributed to aluminium phosphate fixation was calculated on the basis of a molar release ratio of 1.45 M Al/M PO_4.

These data demonstrate clearly the important role of aluminium phosphate interactions in stabilised sludges which results in an amount of P-fixation by aluminium of some 35 to 68 % of total P and more than 65 % of the phosphorus

Table 5. Al and PO_4-release between pH 3.5 and pH 1

Sludge name	$Al_{released}$ mM	$PO_{4,\ released}$ mM	$PO_{4,\ released}/P_{tot}$ %
DS EBPR MS	23	16	43
stab. ES EBPR MS	37	22	52
DS EBPR MS PP	32	22	41
DS EBPR MS PP	22	15	35
DS EBPR SS	24	17	68
DS Sim	31	24	51
PS	24	9	86

bound by physico-chemical mechanisms. As far as we know, the predominance of aluminium for the physico-chemical fixation of phosphorus during the digestion process has not been reported by other investigators so far. Therefore, additional investigations should provide a better insight into aluminium phosphate interactions.

P-Fixation by Magnesium. Crystalline build-up of magnesium phosphate seems to be a common problem in operating anaerobic sludge systems. Most often these crystalline formations in pipes, pumps or digesters were identified as struvite ($MgNH_4PO_4 \cdot 6\,H_2O$) [31, 32, 4, 18, 5].

In comparison to conventional plants, digested sludges from EBPR plants often show a much higher magnesium content. As mentioned before, poly-P hydrolysis is accompanied by magnesium release. Therefore, it could be expected that struvite precipitation participates to some extent in phosphate fixation on EBPR plants. In order to investigate type and extent of struvite precipitation we present results concerning (1) the conditions and solubility product of struvite formation in digested sludge and (2) the amount of struvite precipitation. For this purpose, acidimetric titrations and sequential dilution procedures were applied to the sludge samples of Table 3, complemented by X-ray diffraction and EDXS.

From a thermodynamic point of view it can be shown that under average digester conditions, which are characterised by a slightly alkaline pH, high alkalinity and ammonium concentrations, struvite is the most stable solid magnesium phase (this was proved by equilibrium calculations on $MgCO_3$, $Mg(OH)_2$, $MgNH_4PO_4 \cdot 6\,H_2O$ and $MgHPO_4 \cdot 3\,H_2O$). To investigate possible struvite formation in the sludges the solubility product of struvite formation under digester conditions needs to be ascertained, and this was achieved by the procedure described in Materials and Methods. Figure 7 shows the result of these efforts. With the aid of non-linear regression, using the equilibrium relationships summarised in the appendix, a solubility product for struvite of $\log K_S = -12.53 \pm 0.052$ was calculated which agrees well with values reported by other authors [31, 11, 26].

Furthermore struvite crystallization was confirmed with X-ray diffraction. In Figure 8 the diffraction patterns of a stabilised sludge sample from the EBPR pilot

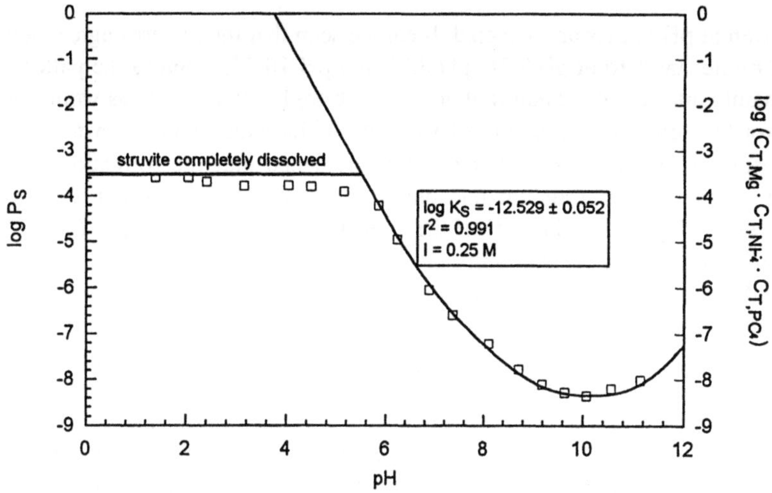

Fig. 7. Solubility product of struvite calculated from a solubility test with digested sludge

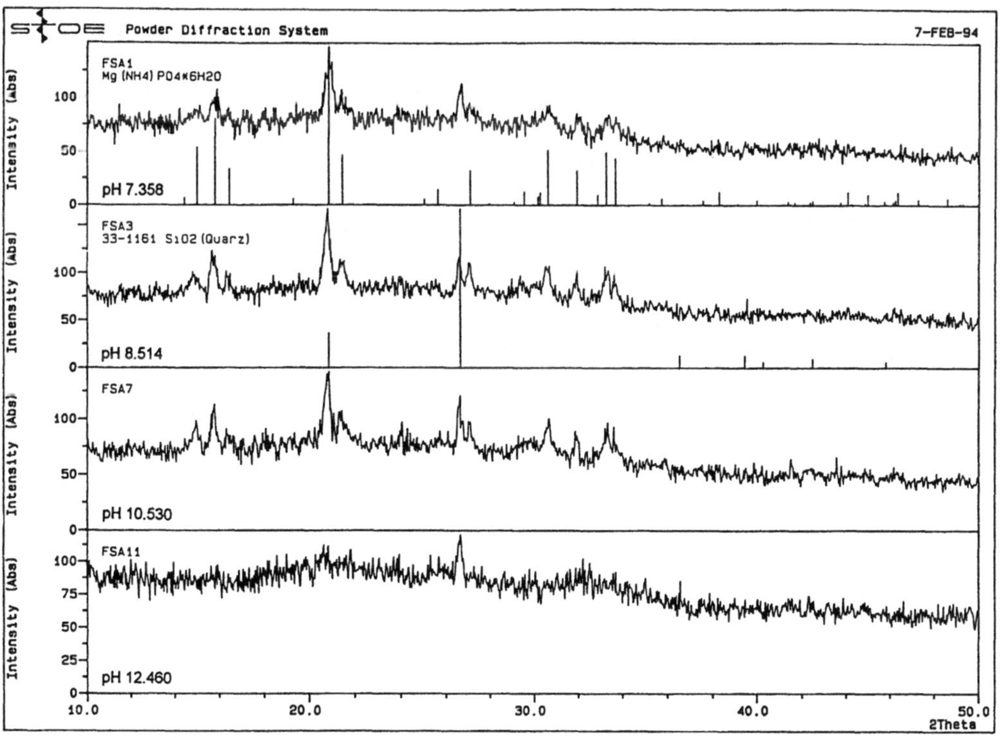

Fig. 8. Diffraction patterns of digested sludge supplemented with magnesium and phosphate at different pH

plant at different pH values are depicted. It can be seen that the patterns agree well with the struvite standard at pH 7.36, pH 8.51 and pH 10.53, whereas the patterns indicate mainly amorphous residues at acidic or strongly alkaline pH as would be expected for the solubility behaviour of struvite (acidic patterns not shown).

The results attained so far provide further evidence that the formation of a crystalline magnesium ammonium phosphate phase in the digestion process is possible even at slightly alkaline pH, if the concentration product $C_{T,\mathrm{Mg}} \cdot C_{T,\mathrm{NH_4}} \cdot C_{T,\mathrm{PO_4}}$ attains or exceeds the solubility product of struvite.

In a further experimental step the sludge samples were examined by the sequential dilution procedure, avoiding the drastic pH changes of the acidimetric titrations and possibly leading to a release behaviour of phosphorus restricted to the dissolution of struvite solid phases. The results of these tests are summarised in Figure 9 where the amount of released magnesium was correlated with the released phosphate in the course of the sequential dilution. For the EBPR sludges a surprisingly high correlation between phosphate and magnesium release was found. To verify that the observed release behaviour is due mainly to the dissolution of struvite, digested sludge from the pilot plant, supplemented with phosphate- and magnesium to induce struvite precipitation (struvite formation was proved by X-ray diffraction), was also investigated with the sequential dilution test. A comparison with the EBPR sludges revealed a nearly identical release behaviour which provides further evidence that the released amounts of magnesium and phosphate in the EBPR sludge samples are due to a dissolution of struvite solids. In Table 6 the regression coefficients for all sludges are summarised. With statistical tests it can be shown that all regression curves are identical at a significance level of 95 % and for the data from all EBPR sludges an overall slope for the regression line of 0.95 was calculated which corresponds well with the expected stoichiometry of struvite dissolution.

On the basis of the released amounts of phosphate and magnesium, the phosphorus content of the sludge samples which can be assigned to struvite precipitation was calculated. The amount of struvite-P of some 37 % in digested sludge from the pilot plant is highest, which seems reasonable in spite of the excellent EBPR conditions of the pilot plant, and was lowest for the EBPR plant with the side stream process, where a part of the phosphate and magnesium is released in the stripper and, therefore, is not available for struvite precipitation in the digester. Significant deviations from the struvite stoichiometry were found for the digested sludge from the plant with simultaneous precipitation. Obviously, concurrent release reactions for this sludge lead to a higher release of phosphate compared to magnesium, probably as a result of iron phosphate dissolution.

P-Fixation by Calcium. Due to the high amounts of calcium in the different sludge samples with an exceptionally high value of 85.4 mg/g TS for the digested sludge from the plant with simultaneous precipitation, calcium is examined further as a possible counterion for phosphate precipitation. Because the major calcium phosphate solid phases are soluble in acid, it was expected that possible calcium phosphate precipitates will dissolve in the course of acidimetric titrations. There-

Table 6. Summary of the sequential dilution tests

Mg [mM] = $a \cdot PO_4$ [mM] + b	DS EBPR MS	stab. ES EBPR MS	DS EBPR MS PP	DS EBPR SS
a	0.883	1.05	0.944	0.833
b	0.413	−0.41	−0.11	−0.18
r^2	0.995	0.993	0.992	0.913
σ_a	0.022	0.031	0.03	0.091
σ_b	0.137	0.161	0.209	0.207
Amount of released Mg [mM]	13.1	10.5	16.1	3.3
Amount of released P [mM]	14.6	9.4	17.2	4.5
Amount of struvite[1] [mg/g TS]	83	87	137	31
$P_{struvite}/P_{tot}$ [%]	36	25	37	13.5

[1] Amount of struvite was calculated using the amount of released Mg and the molecular weight of struvite (245 g/M)

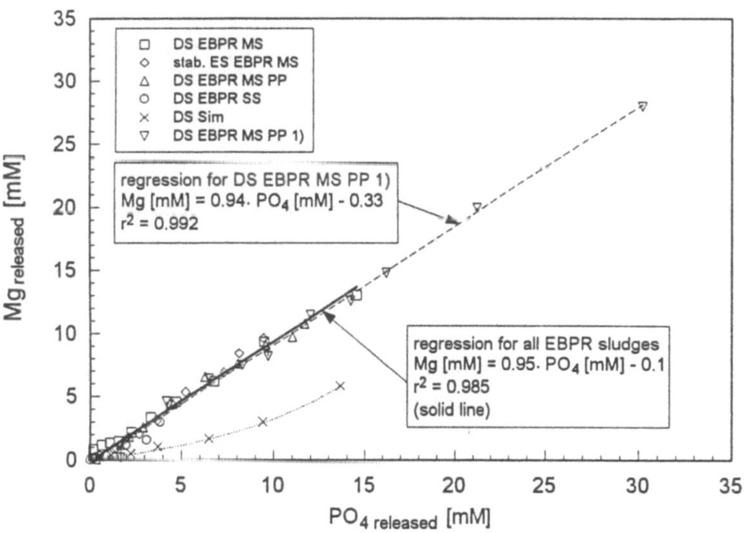

Fig. 9. Phosphate and magnesium release during sequential dilution of the digester sludge samples. 1) P- and Mg-release in the sequential dilution test of digester sludge which was supplemented with P and Mg to induce struvite precipitation

fore, at first we applied the titration procedure to the different sludge samples. Subsequently, the samples were examined by X-ray diffraction and EDXS.

Figure 10 summarises the results of the acidimetric titrations. In this graph the release of calcium is normalised to the total calcium content of the sludge samples. Some interesting information concerning possible interactions between calcium and phosphate could be obtained from this test: (1) although there are significant differences in the total calcium concentrations in the sludge samples ranging from 1200 mg/l in the stabilised excess sludge from an EBPR plant to

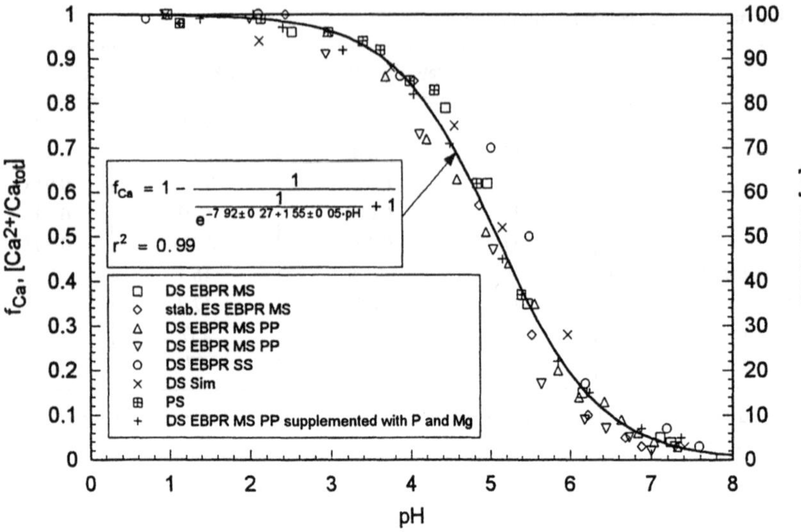

Fig. 10. Course of calcium release during the acidimetric titration. The pH-dependent release was modelled by a mathematical expression called the Kurbatov plot [16] which is often used to describe metal cation adsorption

4200 mg/l in the digested sludge from the plant with simultaneous precipitation, only minor differences are shown in the release behaviour normalised to the total concentration. Furthermore, the behaviour of the different sludges (e.g. primary sludge, EBPR sludge and digested sludge supplemented with phosphate and magnesium) is nearly the same. (2) In none of the titrations was a correlation between the release of phosphate and calcium found. Therefore, it seems likely that other calcium fixation reactions are responsible for the behaviour observed during lowering pH. Possible interactions of calcium with solid or dissolved matter in digester sludge could be:

- Adsorption to hydroxyl surface groups

  In environments with a high amount of aluminosilicates, silicon oxide or other metal oxides or hydroxides, calcium can be adsorbed by displacement of the H^+ of the hydroxyl surface groups [25, 22].

- Reaction with humic substances

  Calcium reacts easily with humic substances to form complexes with the functional phenolic-OH groups or by coagulation with humic substances [23, 25].

- Other calcium precipitation reactions

  Due to the high partial pressure of CO_2 in digester systems, at least part of the calcium could be precipitated as $CaCO_3$. Nevertheless, $CaCO_3$ precipitation could not solely explain the observed release behaviour because at a pH not lower than pH 6 nearly all of the precipitated $CaCO_3$ should be dissolved.

In addition to the acidimetric titration, the samples were examined by X-ray diffraction analysis and EDXS. In none of the samples were crystalline calcium phases identified nor did the elemental distribution of the samples show a correlation between calcium and phosphate. From these results we conclude, that although there is high fixation potential in some sludges, calcium phosphate precipitation seems only of minor importance compared to the interactions of phosphate with aluminium or magnesium.

The results obtained so far concerning the physico-chemical fixation mechanisms of phosphorus in digester sludges can be summarised as shown in Table 7. Here, the previously discussed fractions of aluminium phosphate (Al-P) and magnesium ammonium phosphate (struvite-P) are complemented by the amount of soluble P and organic phosphorus (calculated as 10 mg P/g VS) and normalised to the total phosphorus of the sludge samples. For the EBPR sludges these four fractions account for nearly all of the phosphorus present in the samples, within experimental error, whereas for the sludge from the plant with simultaneous precipitation additional phosphorus has to be fixed by other mechanisms; in this case most probably as iron-phosphate.

Table 7. Distribution of phosphorus in the different sludge samples

Sludge name	PO_4-P/P_{tot} %	Al-P/P_{tot} %	struvite-P/P_{tot} %	org-P/P_{tot} %	$\sum P/P_{tot}$ %
DS EBPR MS	3.8	43	36	17.6	100.4
stab. ES EBPR MS	17.4	52	25	13.7	108.1
DS EBPR MS PP	15.6	41	37	12.5	106.1
DS EBPR SS	4.9	68	13.5	19	105.4
DS Sim	0.4	51	12.5	18.9	82.8

Conclusions

Under the conditions used in this investigation, the enhanced phosphorus elimination observed in the EBPR plant was due mainly to a biological uptake and storage as poly-P which was estimated with the aid of potassium balances to account for some 50 to 70 % of the total phosphorus in excess sludge. During digestion of excess sludge all of the phosphorus stored as polyphosphate will be released but only a part of the released phosphorus remains in soluble form. The difference between released and remaining phosphorus can be attributed mainly to physico-chemical fixation mechanisms in the digester which account for 60 to 70 % of total P in stabilised sludge.

The major part of phosphorus is fixed by interactions with aluminium solids. This was demonstrated by acidimetric titrations. Below a pH of 3 to 3.5 a remarkable increase in soluble aluminium occurs which in all cases was accompanied by an additional significant release of phosphate. Phosphate and aluminium release in the EBPR sludge samples was well correlated with a molar release ratio of 1.45

M Al/M PO_4. Based on the results attained so far we believe that besides a precipitation of phosphate as $AlPO_4$ surface reactions with hydroxyl surface groups of silicates or metal oxides and complexation with mono or polynuclear aluminium are mainly responsible for the observed P-fixation. The amount of P-binding by aluminium was calculated as 35 to 52 % of total phosphorus for the EBPR sludges from plants with main stream processes and as 68 % for an EBPR plant with the side stream process.

In the course of poly-P hydrolysis magnesium ions are released into solution thus supporting the formation of magnesium-phosphate precipitation. Due to the usually high ammonium concentrations in digester sludge, struvite should be formed preferentially, and this was proved by X-ray diffraction analysis. The solubility product of struvite in digester sludge was determined as $\log K_S = -12.53$. The amount of phosphorus bound as struvite can easily be estimated by performing sequential dilution tests. From these tests we have calculated the amount of P-fixation as struvite from 36 % for plants with excellent EBPR conditions down to 13.5 % for EBPR plants using the side stream process.

Although in some sludge samples the calcium concentration reaches values of up to 8.5 %, we found no indications of calcium phosphate interactions. Moreover, all sludges despite their origin show an identical calcium release behaviour in the acidimetric titration. Neither in the titration procedure, nor in the X-ray diffraction investigation and EDXS analysis was there any indication that calcium phosphate precipitation takes place. Therefore, we conclude that calcium is fixed mainly by other mechanisms like adsorption to hydroxyl groups, reaction with humic substances or at least to some extent precipitation as $CaCO_3$.

Appendix. Summary of equilibrium relationships used for the calculation of $AlPO_4$ and struvite solubility (equilibrium constants were corrected for $I = 0.25$ with the Davies approximation [25] and for a temperature of 35 °C unless otherwise stated)

Reaction	log K'	Ref.
$H_3PO_4 = H_2PO_4^- + H^+$	−2.19	[12]
$H_2PO_4^{2-} = HPO_4^{2-} + H^+$	−7.19	[13]
$HPO_4^{2-} = PO_4^{3-} + H^+$	−11.96	[14]
$Al^{3+} + H_2O = Al(OH)^{2+} + H^+$	−5.29	[19]
$Al^{3+} + 2H_2O = Al(OH)_2^+ 2H^+$	−10.37	[19]
$Al^{3+} + 4H_2O = Al(OH)_4^- + 4H^+$	−22.57	[19]
$Mg^{2+} + H_2O = MgOH^+ + H^+$	$-11.44^{25\,°C}$	[19]
$NH_3 + H^+ = NH_4^+$	$-9.3^{25\,°C}$	[17]
$P_S = [Mg_T] \cdot [N_T] \cdot [P_T] = \dfrac{K_S}{\alpha_{Mg^{2+}} \cdot \alpha_{NH_4^+} \cdot \alpha_{PO_4^{3-}} \cdot \gamma_{++} \cdot \gamma_{+} \cdot \gamma_{---}}$	−12.53	
$AlPO_4 \cdot 2H_2O = Al^{3+} + PO_4^{3-} + 2H_2O$	$-18.99^{25\,°C}$	[19]

Literature

[1] Allen, H.E.; Cho, S.H.; Neubecker, T.A.: Ion Exchange and Hydrolysis of Type A Zeolite in Natural Waters. Environm. Sci. Techn. *17* (1983) 1871–1879

[2] Allgemeine Rahmen-Abwasser-Verwaltungsvorschrift über Mindestanforderungen an das Einleiten von Abwasser in Gewässer (Rahmen-Abwasser-VwV) vom 27. August 1991 (1991). GMBl, 686

[3] Baumann, P., Krauth, Kh.: Untersuchung der biologischen Phosphatelimination bei gleichzeitiger Stickstoffelimination auf der Kläranlage Waiblingen. Korrespondenz Abwasser *38* (1991) 191–198

[4] Booram, C.V.; Smith, R.J.; Hazen, T.E.: Crystalline Phosphate Precipitation from Anaerobic Animal Waste Treatment Lagoon Liquors. Transactions of the ASAE 2 (1975) 340–343

[5] Borgerding, J.: Phosphate Deposits in Digestion Systems. J. Wat. Poll. Contr. Fed. *44* (1972) 813–819

[6] Chen, Y.S.R.; Butler, J.N.; Stumm, W.: Kinetic Study of Phosphate Reaction with Aluminum Oxide and Kaolinite. Environm. Sci. Techn. 7 (1973) 327–332

[7] Cook, T.E.; Cilley, W.A.; Savitsky, A.C.; Wiers, B.H.: Zeolite A Hydrolysis and Degradation. Environm. Sci. Techn. *16* (1982) 344–350

[8] Dempsey, B.A.; Ganho, R.M.; O'Melia, C.R.: The Coagulation of Humic Substances by Means of Aluminum Salts. J. Am. Wat. Works Assoc. *76* (1984) 141

[9] DEV: Deutsche Einheitsverfahren zur Wasser-, Abwasser- und Schlammuntersuchung. VCH Verlagsgesellschaft, Weinheim 1960

[10] Drummond, D.; De Jonge, A.; Rees, L.V.C.: Ion Exchange Kinetics in Zeolite A. NATO ASI Ser., Ser. E98 (1985) 345–369

[11] Elliot, J.S.; Sharp, R.F.; Lewis, L.: The Solubility of Struvite in Urine. J. Urol. *84* (1959) 366–368

[12] Ghosh, A.K.; Ghosh, J.C.: First Dissociation Constant of Phosphoric Acid at a Number of Temperatures. J. Indian Chem. Soc. *59* (1982) 745–747

[13] Ghosh, A.K.; Ghosh, J.C.; Prasad, B.: Second Dissociation Constant of Phosphoric Acid: A Recalculation from Alkaline Solutions Using Modified Davies Equation. J. Indian Chem. Soc. *56* (1982) 489–490

[14] Ghosh, A.K.; Ghosh, J.C.; Prasad, B.: Third Dissociation Constant of Phosphoric Acid from 283.15 K to 323.15 K. J. Indian Chem. Soc. *57* (1980) 1194–1199

[15] Jardin, N.; Pöpel, H.J.: Phosphate Release of Sludges from Enhanced Biological P-Removal During Digestion. Accepted paper for the biennial IAWQ conference held in Budapest 24.–30.7.1994 (1994)

[16] Kurbatov, M.H.; Wood, G.B.; Kurbatov, J.D.: Isothermal Adsorption of Cobalt from Dilute Solutions. J. Phys. Chem. *55* (1951) 1170–1182

[17] Lindsay, W.L.: Chemical Equilibria in Soils. Wiley Interscience, New York 1979

[18] Maqueda, C.; Perez Rodriguez, J.L.; Lebrato, J.: Study of Struvite Precipitation in Anaerobic Digesters. Wat. Research *28* (1994) 411–416

[19] Nordstrom, D.K.; Plummer, L.N.; Langmuir, D.; Busenberg, E.; May, H.M.; Jones, B.F.; Parkhurst, D.L.: Revised Chemical Equilibrium Data for Major Water-Mineral Reactions and Their Limitations. Chemical Modeling of Aqueous Systems (ACS) *416* (1990) 398–413

[20] Pöpel, H.J.; Jardin, N.: Influence of Enhanced Biological Phosphorus Removal on Sludge Treatment. Wat. Sci. Techn. *28* (1993) 263–272

[21] Psenner, R.; Pucsko, R.; Sager, M.: Die Fraktionierung organischer und anorganischer Phosphorverbindungen von Sedimenten. Arch. Hydrobiol./Suppl. *70* (1984) 111–155

[22] Schindler, P.W.; Fürst, B.; Dick, R.; Wolf, P.U.: Ligand Properties of Surface Silanol Groups I. Surface Complex Formation with Fe^{3+}, Cu^{2+}, Cd^{2+}, and Pb^{2+}. J. Colloid Interface Sci. *55* (1981) 469–475

[23] Schnitzer, M.; Khan, S.U.: Humic Substances in the Environment. Dekker, New York 1972

[24] Seyfried, C.F., Hartwig, P.: Großtechnische Betriebserfahrungen mit der biologischen Phosphorelimination in den Klärwerken Hildesheim und Husum. Korrespondenz Abwasser *38* (1991) 185–191

[25] Stumm, W.; Morgan, J.J.: Aquatic Chemistry, 2ed. Wiley, New York 1981

[26] Taylor, A.W.; Frazier, A.W.; Gurney, E.L.: Solubility Products of Magnesium Ammonium and Magnesium Potassium Phosphate. Trans. Faraday Soc. *59* (1963) 1580–1584

[27] Uhlmann, D.; Röske, I.; Hupfer, M.; Ohms, G.: A Simple Method to Distinguish Between Polyphosphate and Other Phosphate Fractions in Activated Sludge. Wat. Research *24* (1990) 1355–1360
[28] van Riemsdijk, W.H.; Lyklema, J.: The Reaction of Phosphate with Aluminum Hydroxide in Relation with Phosphate Bonding in Soils. Colloids. Surf. *1* (1980) 33
[29] Veith; J.A.; Sposito, G.: Reactions of Aluminosilicates, Alumino Hydrous Oxides, and Aluminum Oxide with o-Phosphate: The Formation of X-Ray Amorphous Analogs of Variscite and Montebrasite. Soil Sci. Soc. Am. J. *41* (1977) 870–876
[30] Wagner, R.: Ergebnisse des Großversuches – Einsatz eines Zeolith-A-haltigen Waschmittels in Stuttgart Büsnau. In: Die Prüfung des Umweltverhaltens von Natrium-Aluminium-Silikat Zeolith A als Phosphatersatzstoff in Wasch- und Reinigungsmitteln; Hrsg.: Umweltbundesamt, Materialien 4/79 (1979) 58–71
[31] Webb, K.M.; Ho, G.E.: Struvite ($MgNH_4PO_4 \cdot 6\,H_2O$) Solubility and Its Application to a Piggery Problem. Wat. Sci. Techn. *26* (1992) 2229–2232
[32] Westermann, P.W.; Safley, L.M.; Barker, J.C.: Crystalline Build-up in Swine and Poultry Recycle Flush Systems. ASAE publication 13-85 (Agricultural Waste Utilization and Management), 1985, pp. 613–623
[33] Zelazny, L.W.; Jardine, P.M.: Surface Reactions of Aqueous Aluminum Species. In: The Environmental Chemistry of Aluminum, G. Sposito (ed.). CRC Press, Boca Raton 1989, pp. 147–184

Norbert Jardin and H. Johannes Pöpel
Institute for Water Supply
Wastewater Technology and Regional Planning
Darmstadt University of Technology
Petersenstr. 13
D-64287 Darmstadt
Germany

Use of Streaming Current and Other Parameters for Polymer Dose Control in Sludge Conditioning

S. K. Dentel, K. M. Wehnes, and M. M. Abu-Orf

Abstract

Studies of sludge conditioning were conducted at five different treatment facilities employing both belt presses and centrifuges. Polymer dose was varied while comparing charge measurements to full-scale dewatering efficiency. Near-zero streaming current correlated with optimum or near-optimum dewaterability under a wide range of conditions. Other assessments of sludge conditioning such as cake and liquid-stream solids, capillary suction time, liquid-stream viscosity, and liquid-stream turbidity led to a more comprehensive hypothesis concerning polymer functioning. Charge neutralization appears to be necessary, but not sufficient, for adequate conditioning: shear effects subsequent to polymer addition determine the polymer dose leading to charge neutralization when phase separation is actually effected. Moreover, such shear itself actually determines floc strength, and the polymer must be capable of maintaining floc structure under these conditions. Since mixing conditions are reasonably constant in installed sludge conditioning facilities, use of the SCD for automatic control of polymer dose appears feasible in principle. Some practical aspects of SCD design and use are also presented that enable polymer dose control to be put into practice.

1. Introduction

Since 1987, researchers at the University of Delaware have been studying use of the streaming current detector as a method of characterizing and controlling processes involving particle destabilization. At that time, the device was considered as potentially applicable in water treatment processes, and this has been borne out in subsequent years [1, 2]. However, few would have expected the SCD to be useful in sludge conditioning. A strong correlation between charge neutralization and optimum polymer dosage was not anticipated at the time since the commonly accepted conditioning mechanism – polymer bridging – seemed to exclude such a possibility. However, initial laboratory tests suggested some agreement between near-neutral streaming current and optimal sludge flocculation, and continued research provided increasing evidence to this effect. The first attempts to quantify this correlation, and to document it in full-scale applications, however,

R. Klute / H.H. Hahn (Eds.)
Chemical Water and Wastewater Treatment III

were fraught with difficulty. The streaming current value for conditioned sludge seemed to change with time – particularly while mixing continued – and conditioned sludge tended to clog sampling lines or even jam the SCD sensor (which, on early models, burned out the motor). The signal was quite noisy and vulnerable to electrical interference from nearby motors. Thus, several years of work were required to attend to these problems.

We have now varied polymer dose while comparing charge measurements to full-scale dewatering efficiency at five full-scale treatment facilities employing both belt filter presses and centrifuges. Results in all cases show a remarkable correlation of near-neutral streaming current to optimum or near-optimum dewaterability under a wide range of conditions. However, it was necessary to identify mechanisms and sources of interference in arriving at this substantiation. These are presented in this paper.

2. The Streaming Current Detector

The SCD is a means of determining the magnitude of electrical charge on particles in a suspension. The actual current is caused by movement of counter-ions beyond the shear plane of surfaces within the sensor; the fluid motion carrying these ions results from piston reciprocation within a close-ended cylinder in contact with a sampled flow [3]. The output from the instrument is indicative of the zeta potential or electrophoretic mobility of the colloidal material. A theoretical relationship has been established between the streaming current reading and the zeta potential, and this has been experimentally confirmed in carefully controlled experiments [4]. However, the reading from the SCD is often difficult to correlate precisely with these parameters due to several sources of variability.

First, it is generally assumed that the SCD's signal results from momentary attachment of colloidal material to surfaces of the piston and cylinder walls. If these materials fail to adhere to sensor surfaces sufficiently to allow complete coverage, the signal will also represent charge characteristics of the plastic sensor surfaces and thus be misleading. Such a difficulty is more likely in more dilute suspensions as encountered in water treatment. It is also likely that surface active and adsorptive materials will influence the reading due to their attachment to sensor surfaces, although this effect may be analogous to the same materials' influence on colloidal surfaces within the system, thus providing an accurate representation of surface properties. However, should the extent of deposition on sensor surfaces differ from that on colloidal surfaces, the reading would not be meaningful.

Another source of variability in streaming current readings is an observed phase shift of the sinusoidal signal response of the instrument with respect to the piston motion. The amount of phase shift varies, apparently as a function of solution conductivity and colloidal charge, and may be sufficient to nullify instrument response or even reverse the apparent sign of the streaming current. And finally, variation may also accrue from error in determination of the particle electrophoretic mobility rather than the streaming current.

3. Conditioning studies

3.1 Procedures and SCD Modifications

Table 1 shows the treatment facilities at which studies were performed. SCD determinations were either performed in the laboratory using batch samples taken from the plant, or performed on site using an on-line device. In all cases, the polymer dose was varied systematically in the full-scale process and, following equilibration at the set dose, dewatering performance was assessed using solids measurements. Laboratory jar tests were also conducted in many cases using sludges taken from the plant, conditioning them in the laboratory with the same polymer used in the plant, measuring streaming current, capillary suction time, and other indices of dewaterability, and then comparing these results to those obtained with full-scale conditioning.

Table 1. Employed WWTP characteristics including type of experiments and SCD used

Location of WWTP	Wilmington, Delaware[a]	Hatfield, Pennsylvania	Elkton, Maryland[b]	Warminster, Pennsylvania[c]	UMHJSA,[d] Pennsylvania
Size [$L s^{-1}$]	3940	285	305	360	330
Year of tests	1987–1990	1988	1988	1991–1992	1992
Sludge type(s)	Anaerobically digested	Aerobically digested	Primary and biolog. (RBC[e])	Anaerobically digested	Primary and biolog. (WAS[e])
Polymer used	Percol 728	Pollu-Treat. C316	Praestol 664-BC	Pollu-Treat. C426	Calgon WT-2530
Dewatering process	Belt filter press	Contrifuges	Belt filter press	Centrifuges	Belt filter press
Type of experiments	Batch and Full-scale	Batch and Full-scale	Batch	Batch and Full-scale	Batch and Full-scale
SC measurements	Lab & Plant	Lab	Lab	Lab & Plant	Plant
SCD used	GEN II	GEN II	GEN II	GEN II SC-4200	GEN III SC-5200

[a] Results from this plant published in Dentel and Wehmes, 1988
[b] Results from this plant published in Elicker et al., 1992
[c] Results from this plant published in Dentel and Abu-Orf, 1993
[d] Upper Moreland-Hatboro Joint Sewer Authority
[e] RBC = Rotating biological contactors; WAS = Waste activated sludge

The streaming current detectors used in these experiments were all products of the Milton Roy Company (Ivyland, PA USA), but different models were used in some cases. Earlier experiments (1987–1989) used GEN-II units which, although satisfactory in water treatment applications, proved problematic in some sludge applications. For example, Figure 1 shows early results at the Wilmington treatment facility: an increase in polymer dose clearly led to a change in the SCD response, but the change was accompanied by substantial signal noise. A second SCD employing filtrate rinsewater as the sample flow provided only a slight change in the reading with the dose change.

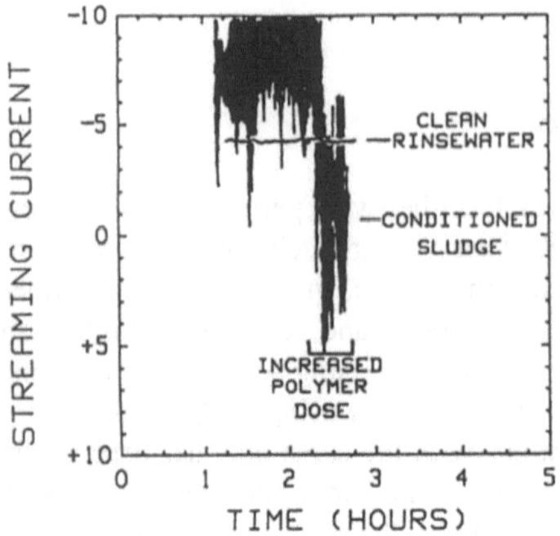

Fig. 1. Strip chart record from Wilmington facility, showing response to dose increase of cationic polymer. SCD output is shown for flows from both sludge flocculation tank and filtrate (including belt rinsewater)

Thus, later tests utilized a third-generation (GEN III) model developed as a result of these studies, incorporating modifications to circuitry, material, and sample collection systems. The clearance between the piston and cylinder was increased, as was the piston speed, in order to reduce the likelihood of clogging. The altered clearance decreased sensitivity of the instrument, although the greater piston velocity compensated to some extent. Modifications in the electronic circuitry included greatly improved filtering to exclude noise at other frequencies than that of the piston motion, and internal compensation for phase-shift phenomena in the signal rectification stage [5]. The net effect of these changes was an enhanced instrument sensitivity with decreased susceptibility to clogging, electrical interference, and phase-shift variability. Figure 2 illustrates these improvements, also with the Wilmington sludge. An additional improvement found useful in some applications is an automatic timed backwash of the SCD probe and feed lines with tap water. If the device is used for automatic polymer feed control, the dosage must be held constant at the previous level during this washing process.

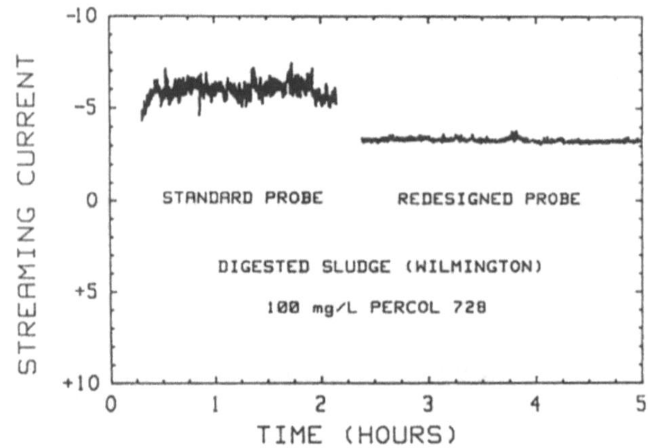

Fig. 2. Strip chart record of SCD measurements of conditioned sludge, comparing standard and redesigned probes

3.2 Streaming Current Results

Figure 3 indicates the correlation of SCD reading with dewaterability found using the GEN-II SCD in laboratory jar tests, where electrical interferences were minimal. In this case, a clear correlation of optimal dose with charge neutralization is seen. However, difficulties in maintaining flow-through sampling without clogging suggested that sampling of a centrate or filtrate would be advantageous if the same – or at least a similar – agreement of streaming current with dewaterability would be observed. Figure 4 shows results from Hatfield, PA compared to plant solids measurements, indicating that use of an SCD reading from the centrate at this plant is indicative of dewaterability trends. This is also shown in Figure 5, which shows laboratory SCD measurements for Warminster, PA conditioned sludges and centrates.

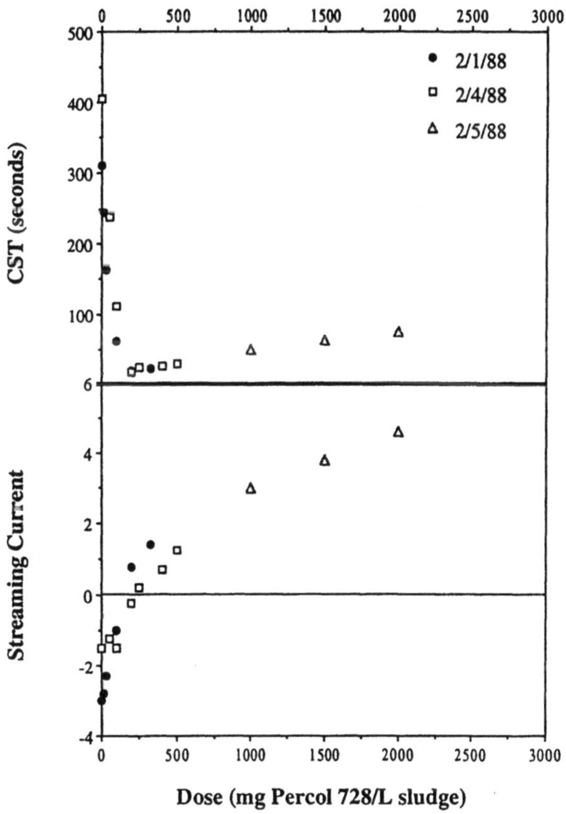

Fig. 3. Capillary suction time (CST) and SCD readings vs. dose of cationic polymer Percol 728. Laboratory batch experiments using Wilmington sludge

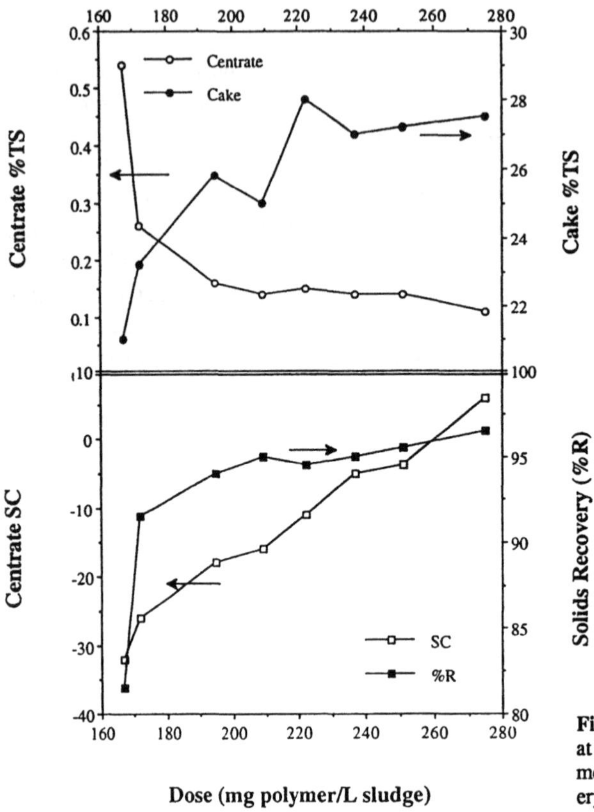

Fig. 4. SCD readings from centrate at Hatfield, with corresponding solids measurements and percent recovery as polymer dose was varied

3.3 Viscosity Measurements

Figure 5 also shows viscosity measurements taken on the centrate, generated using a laboratory centrifuge after conditioning of the sludge. The decrease in viscosity is caused by polymer-induced agglomeration of solids, leading to more effective solid-liquid separation. At the optimum polymer dose, the centrate viscosity has decreased to nearly the viscosity of water. At this polymer dose, charge neutralization has also been accomplished, and thus the adsorption of polymer onto solids will be more difficult at higher doses due to repulsion between positive charges. As a consequence, increasing amounts of polymer remain in solution, and the liquid-phase viscosity increases. It is likely that the slight deterioration of sludge dewaterability at these increased doses stems from viscosity effects rather than failure to agglomerate, since the magnitude of charge reversal is not reflected in a commensurate loss of dewaterability. These observations are supported by additional data reported elsewhere including full-scale tests. Unlike coagulation processes in the more dilute context of water treatment, polymer overdoses leading to charge reversal.

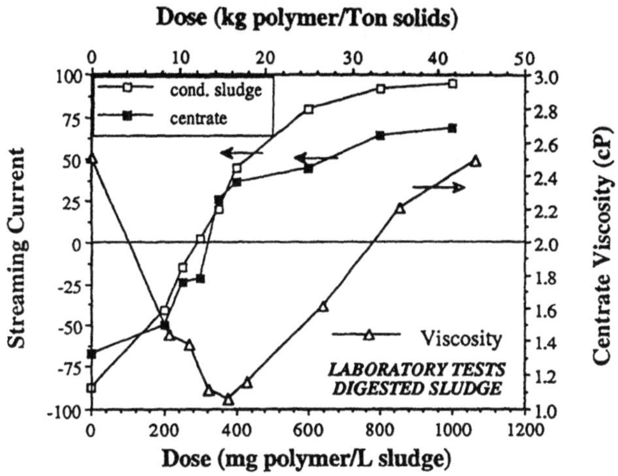

Fig. 5. Centrate and conditioned sludge SCD readings (left scale) and centrate viscosity (right scale) vs. polymer dose for Warminster sludge. Centrates obtained by 5-minute centrifugation in laboratory; viscosity measured with UL adapter on Brookfield LVT viscometer

4. Mechanisms and Control of Sludge Conditioning

4.1 Sludge Conditioning Mechanisms

These results demonstrate the importance of charge neutralization in sludge conditioning to an extent previously unappreciated, and seemingly at odds with the concept of polymer bridging often invoked as a conditioning mechanism. How can these models by reconciled? For example, the importance of physical effects such as hydrodynamic shear has been clearly documented in past research (e.g. [7, 8]), apparently consistent with a bridging phenomenon that requires some mixing for polymer dispersion and attachment, but leads to detachment or breakage with extended or intensive mixing.

Figure 6 illustrates the interaction of shear and charge effects with sludge jar tests conducted using the Warminster sludge at two different mixing speeds. Although both sets of data in this figure show correspondence of charge neutralization with satisfactory conditioning, increased mixing requires a higher polymer dose to attain this dose range. Similar data have also been reported for other sludges and polymers, and for varying mixing times as well as intensities [9]. We have shown as well [10] that the susceptibility to shear can be decreased by suitable modification of polymer structure. These results suggest that increased mixing time or intensity leads to either the shearing of flocs or collapse of extended polymer chains from particle surfaces, leading in either case to an increased number of negative sites exposed in the suspension, and requiring additional polymer to achieve charge neutralization. In this respect, shear effects represent only an influence on the governing charge neutralization mechanism. However, the increased shear that leads to higher polymer requirements may also produce a stronger floc structure and improved dewaterability under certain conditions.

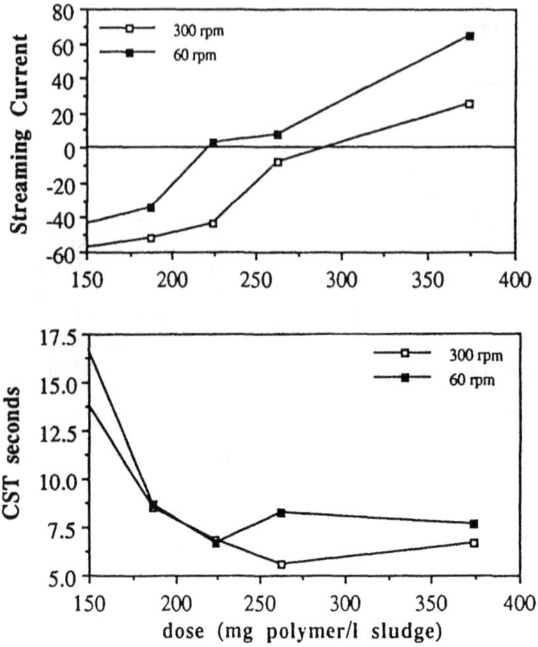

Fig. 6. Conditioned sludge SCD readings and CST results for laboratory conditioning of Warminster sludge. A 0.5 liter sample was mixed for 1 minute after polymer addition at the indicated rotational speed in a 1-liter beaker with 1″ × 3″ paddle

4.2 Control of Sludge Conditioning

In terms of process control for existing installations, mixing effects are not important in polymer dose control, since in most cases they are not adjustable. In cases where in-line polymer addition is practiced, some variation in mixing time and intensity result from changes in flow rates, but these have been found to be negligible. Thus, measured streaming current should be sufficient as a basis for polymer dose adjustment, either manually or by automatic control. We have demonstrated the on-line operability of SCD-based control, with the modifications in SCD sensor, electronics, and sampling indicated above [11]. Control based on liquid-stream viscosity is also possible in theory, based on findings exemplified in Figure 5.

5. Conclusions

1. Polymer doses leading to streaming current detector readings in the neutral range correspond to optimum dewaterability under a variety of conditions. The role of charge neutralization in sludge conditioning appears to be important.
2. Polymer interaction with the sludge can be related to changes in the liquid-stream viscosity, due at low doses to improved phase separation, and at higher doses to increased residual polymer concentrations in solution.
3. Mixing is an important factor in proper sludge conditioning. An adequate degree of mixing is needed to ensure proper dispersion of polymer, but high

intensities lead to higher polymer demand. With appropriate types of polymer, the use of energetic mixing may produce floc more resilient to high-intensity dewatering processes.

4. Though variations in mixing time and intensity affect the streaming current and optimal dose, mixing conditions are reasonably constant in installed sludge conditioning facilities. For this reason, use of the SCD for automatic control of polymer dose appears feasible in principle.
5. Several significant obstacles to actual use of the SCD in full-scale dewatering operations may be overcome by appropriate modifications. These include: use of a centrate or filtrate sample flow rather than conditioned sludge, and at an increased sample flow rate; an increased gap width and increased piston speed in the SCD sensor; greater bandwidth selectivity and zero-crossing based rectification in the circuitry; and provision for automatic backwash of the sensor and lines.

Acknowledgements. The authors wish to thank Mark Elicker, Bob Bean, and George Halfinger for assistance in conducting the research summarized in this paper. Financial support was provided by the Milton Roy Company, the Delaware Research Partnership, and the Water Environment Research Foundation.

References

[1] Dentel, S.K., Kingery, K.M.: Using Streaming Current Detectors in Water Treatment. J. Am. Wat. Works Assoc. *81* (1989) 85
[2] Dentel, S.K.: Use of the Streaming Current Detector in Coagulation Monitoring and Control. Proceedings, Workshop on Optimal Dosing of Coagulants and Flocculants, Mülheim, Germany, 1994
[3] Gerdes, W.F.: A New Instrument – The Streaming Current Detector. Proceeding, 12th Natl. Analysis Instrument Sym., Houston, TX, 1966
[4] Elicker, M.L., Resta, J.R., Hunt, J.M., Dentel, S.K.: Fundamental Consideration in Use of the Streaming Current Detector for Chemical Dose Control. In: Chemical Water and Wastewater Treatment II, R. Klute and H.H. Hahn (Eds.). Springer, Berlin Heidelberg New York 1992, pp. 165–179
[5] Schwab, G.M.: Phase Correction for Streaming Current Signal. U.S. Patent 5,121,061 (1992)
[6] Dentel, S.K., Abu-Orf, M.: Use of Liquid-Stream Viscosity and Streaming Current in Dewaterability Assessment: Laboratory and Full-Scale Studies. Submitted, Aqua (1994)
[7] Werle, C.P., Novak, J.T., Knocke, A.M., Sherrard, J.H.: Mixing Intensity and Polymer Conditioning. J. Environ. Eng. (ASCE) *110* (1984) 919
[8] Novak, J.T.: The Effect of Mixing on Sludge Conditioning. Wat. Supply *8* (1990) 53
[9] Wehnes, K.M.: Development of New Capabilities for Streaming Current Detectors. Masters Thesis, University of Delaware, Newark, DE, 1991
[10] Dentel, S.K., Abu-Orf, M.M., Griskowitz, N.: Unpublished results (1993)
[11] Abu-Orf, M.M., Dentel, S.K.: Automatic Control of Polymer Dose in Sludge Conditioning Using the Streaming Current Detector. Manuscript in preparation (1994)

Steven K. Dentel
Department of Civil Engineering
University of Delaware
Newark, DE 19716
USA

Kathleen M. Wehnes
Gannett Fleming Engineers
Cross Keys – Quadrangle
Baltimore, MD 21210
USA

Mohammad M. Abu-Orf
Department of Civil Engineering
University of Delaware
Newark, DE 19716
USA

Springer-Verlag and the Environment

We at Springer-Verlag firmly believe that an international science publisher has a special obligation to the environment, and our corporate policies consistently reflect this conviction.

We also expect our business partners – paper mills, printers, packaging manufacturers, etc. – to commit themselves to using environmentally friendly materials and production processes.

The paper in this book is made from low- or no-chlorine pulp and is acid free, in conformance with international standards for paper permanency.

MIX
Papier aus verantwortungsvollen Quellen
Paper from responsible sources
FSC® C105338

If you have any concerns about our products,
you can contact us on
ProductSafety@springernature.com

In case Publisher is established outside the EU,
the EU authorized representative is:
Springer Nature Customer Service Center GmbH
Europaplatz 3, 69115 Heidelberg, Germany

Printed by Libri Plureos GmbH
in Hamburg, Germany